14.95

KV-638-362

Scripta Series in Geography

Series Editors:

Richard E. Lonsdale, University of Nebraska
Antony R. Orme, University of California
Theodore Shabad, Columbia University
James O. Wheeler, University of Georgia

Other titles in the series:

Dando, W. A.: The Geography of Famine
Klee, G. A.: World Systems of Traditional Resource Management
Yeates, M.: North American Urban Patterns
Brunn, S. D./Wheeler, J. O.: The American Metropolitan System: Present and Future
Bourne, L. S.: The Geography of Housing
Oliver, J. E.: Climatology: Selected Applications
Jackson, R. H.: Land Use in America

Petroleum and Hard Minerals from the Sea

Fillmore C.F. Earney
Northern Michigan University

 V. H. Winston & Sons

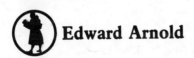

Copyright © V. H. Winston & Sons 1980

First published 1980 by
Edward Arnold (Publishers) Ltd.
41 Bedford Square, London WC1B 3DQ
and published simultaneously in the United States of America by Halsted Press, a division of John Wiley & Sons Inc.

British Library Cataloguing in Publication Data

Earney, Fillmore C F
　Petroleum and hard minerals from the sea. – (Scripta series geography).
　1. Mineral industries
　2. Marine mineral resources
　3. Offshore oil industry
　I. Title
　333.8'09162　　HD9506.A2

ISBN 0-7131-6298-8

Library of Congress Cataloging in Publication Data

Earney, Filmore C F
　Petroleum and hard minerals from the sea
　(Scripta series in geography)
　"A Halsted Press book."
　Includes bibliographical references and index.
　1. Petroleum in submerged lands.
　2. Marine mineral resources.
　I. Title.
　II. Series.
　TN871.3.E37　1980　533'.09162　80-17653

ISBN 0 470-27009-8

All rights reserved. No part of this publication may be reproduced, stored in a retrieval system, or transmitted, in any form or by any means, electronic, mechanical, photocopying, recording, or otherwise without the prior permission of Edward Arnold (Publishers) Ltd.

Typeset in the United States of America by
Marie Maddalena of V. H. Winston & Sons
Printed in Great Britain by R. Clay (The Chaucer Press), Ltd., Bungay, Suffolk.

Contents

Preface

Chapter 1 Introduction   1

PART I. THE CONTINENTAL MARGIN   7

Chapter 2 Hard Minerals of the Continental Margin   9
Chapter 3 Offshore Petroleum   33
Chapter 4 The North Sea   63
Chapter 5 The United States Offshore   91
Chapter 6 Petroleum Exploration and Production Technologies   113
Chapter 7 Transport-Storage-Transfer   133
Chapter 8 Petroleum and the Environment   143

PART II. THE DEEP SEABED   163

Chapter 9   Plate Tectonics and Mineral Formation   165
Chapter 10 Manganese Nodules   175
Chapter 11 Manganese Nodule Investment-Production Economics
            and Technology   191
Chapter 12 Manganese Nodules and the International Marketplace   227
Chapter 13 Seabed Mining and the Conference on the Law of the Sea   243

References   261

Index   283

Preface

Each day, planet Earth adds approximately 200,000 people to its population. These newborn additions comprise a small army of new demanding consumers of our planet's space and resources. With the advent of mass technology and communications, most inhabitants of the Earth have grown accustomed to or aspire for higher living standards, seeking as a minimum a dry shelter and a modicum of food to alleviate hunger.

How are we to meet these expectations? Ultimately, human skills win from earth's three kingdoms—vegetable, animal, and mineral—the raw materials that society transforms into useful products. This seemingly simplistic life-support/ environment relationship depends upon complex social, economic, and political systems. Of all mankind's life-support efforts, perhaps the most vital is the search for mineral fuels, metals, and other minerals, for without them the struggle for human survival would be brutally more difficult and disastrously less efficient. As nonrenewable resources, minerals must be obtained by finding new deposits or by creating "new" economically, exploitable mineral reserves through the use of technology. The need to augment or maintain our mineral base has long encouraged geologists, mining engineers, and entrepreneurs to seek ores in the continents' far corners. They are now looking to the oceans—earth's last mineral frontier. It is the search for and production of offshore petroleum and other minerals in this last resource frontier that is the focus of this book. The quest for continental margin and deep seabed minerals may be one of the most significant developments to occur during the remaining decades of this century, affecting just about all people in every country of the world.

Many books focus upon the technical/physical processes of ocean mining[32, 52, 202, 276, 422] or consider a limited area such as the Georges Bank of the United States' eastern seaboard, the continental shelf of the People's Republic of China, or the North Sea and its petroleum (gas and liquid hydrocarbon) resources.[5, 30, 224, 578] But only a few seemingly incomplete or out of date efforts have been made to collectively analyze and examine, on a world scale, the engineering, geological, social, political, and economic factors of continental margin and deep seabed mining.[147, 342] The purpose of this book is to present such a collective and updated view—one that can provide the student of marine resources with the main threads of contemporary efforts to exploit the oceans' mineral wealth.

The volume is organized into two parts: The first, *Continental Margin*, includes the continental shelf, slope, and rise. The second, *Deep Seabed*, encompasses areas seaward of the continental rise. Mining activities covered in this book include all extractive operations on or below the ocean floor, such as pumping, dredging, or direct access by offshore entries or by land-based tunnels to recover naturally occurring inorganic substances, though not necessarily of inorganic origin. Minerals contained in beach sands lying above the normal high-tide level are excluded. Also not examined is the industrial extraction of minerals from seawater, an important source for various salts and the metal magnesium as well as a potential source for such other metals as uranium.[a]

With the deepest gratitude, I acknowledge the assistance of scores of individuals in industry and government in the United States and other countries who provided data, maps, and photographs. I also appreciate the many permissions granted to use copyrighted material; Robert G. Burke, Editor of the journal *Offshore*, and Ben Russak, publisher of the journal *Marine Mining*, have been especially helpful in this regard. Many redrafted and reproduced illustrations included in this book are the work of Becky Heinzelman, Russell Ault, Pat Taj-Terhani, Frank Martin, and William Lensch, all of Northern Michigan University's Learning Resources Center. Their efforts were skillfully coordinated by Scott Seaman, Assistant Director of Learning Resources. Numerous photographic reproductions are the handiwork of Donald Pavloski, Tim Calloway, and Kim Marsh in the University News Bureau's Photographic Department. Jessie Luttenton of the Geography Department's Cartographic Laboratory helped design and draft several maps, and I owe a major debt to my colleague, Professor John P. Farrell, for his advice and direction in the design and preparation of several original maps. Without the frequent and diligent efforts of Dena Bovee, Roberta Henderson, Jon Drabenstott, and Marsha Larmour, all of the University's Library, I would not have had access to a multitude of reference items. I am especially grateful to Niki Calloway, Deborah Skehen, Laura Campbell, Myra Vaughan, Connie Del Bello, Jane Potvin, Pat Ogle, and Ruth Bishop for their skill and good humor in deciphering my rough draft manuscript and for typing the final copy.

Special thanks go to those who critiqued early drafts of the various chapters. Their expert counsel significantly improved the content and corrected several errors. Reviewers of this book at the manuscript stage included D. Horne and A. A. Archer of the Institute of Geological Sciences, London, and H. D. Hess,

United States Geological Survey (USGS), Menlo Park, Calif. (Chap. 2); Henry L. Berryhill, Jr., USGS, Corpus Christi, Texas (Chaps. 3 and 5); Bjorn Bratbak and Tarjei Moen, Norwegian Petroleum Directorate, Stavanger (Chaps. 4 and 8); Alex Hogg, Aberdeen College of Education, Aberdeen, Scotland (Chap. 4); Bud Danenberger and John Duletsky, USGS, Reston, Va. (Chaps. 5 and 8); Ardley R. Henemann, Jr., Ray McDermott & Co., Inc., New Orleans, La. (Chaps. 6 and 7); Peter A. Rona, National Oceanic and Atmospheric Administration, Miami, Fla., John P. Albers, USGS, Menlo Park, Calif., and John E. Frey, Northern Michigan University, Marquette (Chap. 9); Jane Z. Frazer, Scripps Institution of Oceanography, La Jolla, Calif. (Chaps. 10 and 11); C. Richard Tinsley, Continental Bank, Chicago, Ill., and John N. Nisbet, Profunda Ltd., Hong Kong (Chaps. 11 and 12); Marne A. Dubs, Kennecott Copper Corporation, New York, N.Y., and David W. Pasho, Canadian Department of Energy, Mines and Resources, Ottawa (Chaps. 11, 12, and 13); and Martin I. Glassner, Southern Connecticut State College, New Haven (Chap. 13).

The university has been supportive of my work in providing a 1978 summer research grant for preliminary bibliographic tasks, in granting a sabbatical leave during the 1979 fall term for doing field work in the North Sea area and in North America, and in making funds available for cartographic and photographic work. I especially appreciate the encouragement of Robert B. Glenn, Provost and Vice President for Academic Affairs; Roy E. Heath, Dean, School of Graduate Studies and Director of Research and Development; and Jarl Roine, Head, Department of Geography, Earth Science and Conservation, all of Northern Michigan University. Finally, my thanks go to Dr. Theodore Shabad, editor, and Victor H. Winston, publisher, of the Scripta Series in Geography, in bringing this volume to publication.

Fillmore C. F. Earney, Ph.D.
Professor of Geography

---

[a]For those interested in mineral extraction from sea water, see, for example, [91] and [234].

Chapter 1

Introduction

Minerals regulate our life in a multitude of ways. In 1974, each United States citizen required an annual average of 40,000 lbs. of mineral materials, plus enough petroleum, coal, natural gas, and uranium to generate an energy equivalent of 300 persons working around the clock. This dependence, particularly on energy, continued to increase in the mid- and late 1970s even though the Arab oil embargo prompted a massive awareness of our vulnerability. A similar condition exists in other industrialized states, and as the less developed states industrialize and improve their people's living standards, further demands will be placed upon the world's mineral supplies. As additional pressures are exerted on earth's mineral resources, their onland availability will diminish and their price will rise. An ominous demand and supply problem already exists for crude oil and natural gas. During the 1950s and 1960s, oil was so cheap and plentiful that the threat of overproduction caused most exporting countries to complain that their customers were not buying enough. That period ended in the early 1970s, and the industrial countries had lost, by that time, the capacity to lessen their increasing reliance on imports. OPEC and skyrocketing spot market prices are ever present in the news as the decade of the 1970s is ending, and most forecasts for the early 1980s are boding gloom. In addition to conservation, synfuels, nuclear power, and various nonconventional energy sources, the increasing tempo of petroleum production on the continental shelves may offer a measure of relief. It is a harbinger of the deep seabed's future as a source of energy as well as of important hard minerals. In the late 1970s, for example, the world market price for nickel has risen until deep seabed sources are nearly economically competitive with onshore mines. Prospects that are similar

2 *Introduction*

to the near breakthrough with nickel should contribute to intensified development of ocean mining.

ADVANTAGE AND DISADVANTAGE OF OCEAN MINING

Numerous assets and liabilities enter into an assessment of continental margin and deep seabed mining. Thus, governments or entrepreneurs entering the field face a complex problem of balancing the economic, political, social, and technical cost-benefit ledger. Prospects for balancing this ledger will be a primary theme throughout the book.

Assets

Several assets for continental margin and deep seabed mining stand out: (1) the ocean provides a transport mode that is usually cheaper than land-based systems; (2) infrastructural facilities such as ports that can receive ores do already exist;[64] (3) processing plants may be optimally located adjacent to markets and in politically stable areas;[531] (4) no agricultural or other productive land uses are disturbed; (5) pollution problems are not as severe as those associated with land-based mining, beneficiating, and smelting operations;[175] and (6) minerals available in the oceans may help reduce the United States' dependence on certain foreign imports and contribute to an improved balance of payments.

Liabilities

Major liabilities of ocean mining include: (1) an uncertainty in the outcome of current international negotiations within the United Nations' Third Law-of-the-Sea Conference for ownership and administration of the oceans, a situation that could jeopardize investment ventures; (2) a need for channeling capital into the development of technology capable of coping with difficult and hazardous working conditions, especially as miners push into outer continental shelf (OCS) and deep seabed waters; (3) the likelihood of some degradation of the marine environment; and (4) the possibility of price disturbances occurring when ocean minerals enter the international marketplace.

MINERAL DEPENDENCE

No state is self-sufficient in minerals. Most industrialized states depend on numerous imports, a situation that encourages great interest in ocean mining in Japan, West Germany, France, the United Kingdom, and the United States.

The overall dependence of the United States on mineral imports is well recognized, but its need for specific mineral purchases from other states is less known. Several vital metallic minerals required by the United States are potentially available from the deep seabed, including manganese, cobalt, nickel (all important steel alloys), zinc, and copper. Although only relatively small tonnages are consumed in the United States, it must import 99% of the manganese (chiefly from Brazil, Gabon, Australia, and South Africa) and 98% of cobalt (from Zaire, Belgium, Luxembourg, Finland, Norway, and Canada). Also, 71% of the nickel (from Canada and Norway), 64% of zinc (from Canada, Mexico, Australia, Honduras, and Peru), and 15% of the copper (chiefly from Chile, Peru, Zambia, and Canada) used annually have to be purchased from foreign suppliers.

Petroleum, a major product of the offshore, is the mineral most crucial not only to the import dependency of the United States, but also to its economic survivability. From 1900 to 1970, the energy consumption of the U.S. increased 13 times, while domestic production went up only 10 times.[366] In 1973, domestic oil production was 9.2 million bbl/d; in 1978, it was only 8.7 million bbl/d. Demand in 1973 was 17.3 million bbl/d, whereas in 1978 it had climbed to 19 million bbl/d.[27] In 1975, petroleum accounted for nearly one-third and natural gas for one-fifth of the total energy units consumed in the United States. Consumption of petroleum for the year totaled 5.9 billion bbl. with imports providing 37% of this amount.[477] In 1978, foreign supplies accounted for 42%[27] of total consumption. By 1979, petroleum imports met nearly 50% of the country's needs, and by late 1980 they may reach 55% (Fig. 1). This painful dependence on imports will

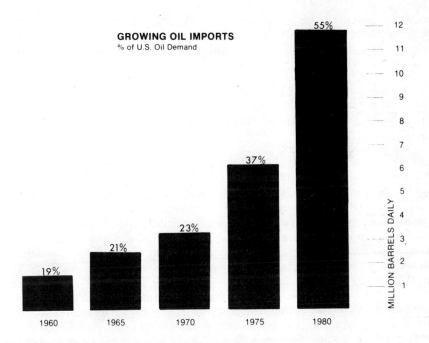

Fig. 1. United States oil imports, 1960-80.
Source: Exxon Co. With permission.

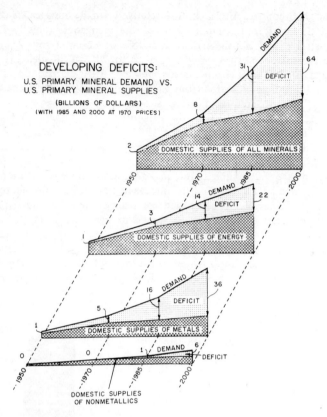

Fig. 2. Projected mineral deficits of the United States through the year 2000.
Source: [118].

persist and probably increase as demands continue to move upward while domestic production stabilizes or declines. The U.S. government seeks to reduce the national dependence on foreign oil sources, and particularly the country's increasingly destructive reliance on the OPEC cartel. United States' dependence on imports from Arab states, the dominant members of OPEC, grew by approximately 15% between 1970 and 1977, increasing from about 3% to 18%. This trend continued in 1978 and 1979.

Since the United States is now so dependent on foreign mineral sources, what does the future bode? A Department of the Interior (DOI) study projected likely trends (relative to costs) in mineral deficiencies within the United States up to the year 2000. To provide a common base and show overall trends in mineral supplies and demand, projections of cost deficits were made in 1970 prices for nonmetallics, metallics, and energy supplies. Although the DOI recognized that future prices would fluctuate, its calculations do not adequately reflect current market conditions. The cost of a barrel of OPEC oil in December 1979, rose from a prior ceiling of $23.50 to $30.00, roughly 10 times more than in 1972. During the first 6 months of 1979, petroleum prices jumped by nearly 60%. If the presence of an approximate 100% price inflation since 1970 and the problem of oil price comparability are put aside, the projected deficits for all minerals in 1985 ($31 billion) and in the year 2000 ($64 billion)[118] provide a notion of the anticipated trend of increasing mineral deficits in the coming decades (Fig. 2).

Considering the dependence of the United States and other developed states on metallic, nonmetallic, and energy imports, continental margin and deep seabed mining will play an increasingly important role in world affairs. Currently, however, efforts of several of the developed states' mineral producers to stake claims and commence work in the world's deep seabed areas continue to be frustrated. The problem stems partly from a desire of third world countries to share in this bounty of the seas. They view the oceans as a possible lifeline to greater economic development and prosperity. In the coming decades, world statesmen will wrestle with these important conflicts of interest. Their ability to reconcile such conflicts will influence the shape of economic and political affairs of the world.

Part I

The Continental Margin

Continental margins (shelf, slope, and rise) encompass a significant portion of the world's oceans. These extensions of the continental land masses are becoming increasingly important as mineral-rich zones that provide new sources of petroleum, metallics, and nonmetallics. Part I briefly exmines minerals of the continental margin that, on a minor scale, have been produced in the past, are produced presently, or could be produced in the future (Chap. 2). Subsequent chapters consider, in detail, problems and prospects for oil and gas extraction in the world's major offshore producing areas (Chap. 3); the North Sea and the United States, as two outstanding petroleum producing regions (Chaps. 4 and 5); petroleum exploration and production technologies, including an examination of platforms and seabed well completion systems (Chap. 6); pipelines, storage, and transfer facilities (Chap. 7); and environmental problems associated with offshore petroleum industries (Chap. 8).

Chapter 2

Hard Minerals of the Continental Margin

Geologically, the continental margin is an extension of the continental land mass. Although no absolute agreement exists for physically delimiting the outer margin of the continental shelf, a good compromise places it at an average depth of 425 ft. (130 m) and an average distance of 40 mi. offshore. From this point, the ocean floor's gradient increases rapidly, forming the continental slope. The beginning of the continental slope lies 9-50 mi. offshore and extends seaward with an average downward slope of 4.3°, continuing to a depth of 4,600 ft. (1,400 m) to 10,500 ft. (3,200 m)—a point where the gradient decreases significantly. This gradient change marks the start of the continental rise which gradually slopes off to meet the deep seabed at approximately the 13,100 ft. (4,000 m) isobath (Fig. 3).

Continental shelf hard mineral mining presently provides sand and gravel, calcium carbonate (algae, shells, and aragonite), coal, barite, diamonds, tin, sulphur, salt, scheelite, iron sands, and coral (Fig. 4 and Table 1). Potentially, as technology and engineering skills improve and economic conditions change, miners will likely push farther out onto the continental margin, providing an opportunity to extract several other ores such as potash, phosphorites, glauconite, and gold. The DOI, in anticipation of changes to come, was in 1979 in the process of looking carefully at the OCS as a future area of hard mineral production, because as dredging and subseabed mining become more important, concomitant impacts on the environment will develop and new management decisions will have to be made.

10   *Hard minerals*

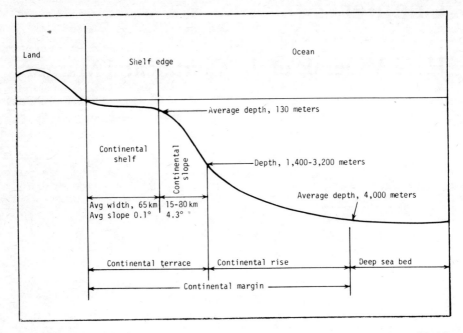

Fig. 3. Diagrammatic continental margin profile. Numbers represent worldwide averages.
Source: [324].

TUNNELING

Miners have long used land-based tunnels to extract subseabed vein and massive mineral deposits. The early Greeks mined lead and zinc in tunnels driven seaward from their Aegean shores,[238] and for centuries tin miners on England's Cornwall Peninsula have won ores from subsea excavations. Subsea mines were once a source of copper,[542] mercury, nickel, limestone,[7] and iron ore. As of the mid-1970s, more than 100 subsea mines have been or are being worked. Their depths range from 100 to 7,920 ft., and they extend up to 5 mi. from shore.[7] Sinking shafts onshore and pushing drifts seaward present formidable challenges to the engineer and potentially dangerous working conditions to the miner, and special problems of subsea tunneling extraction add to mining costs, making competition with on-land producers difficult.

Iron Ore

From 1895 until 1966, miners produced iron ore from beneath the sea at Bell Island (Wabana) in Newfoundland's Conception Bay. When the mine closed, 1.2

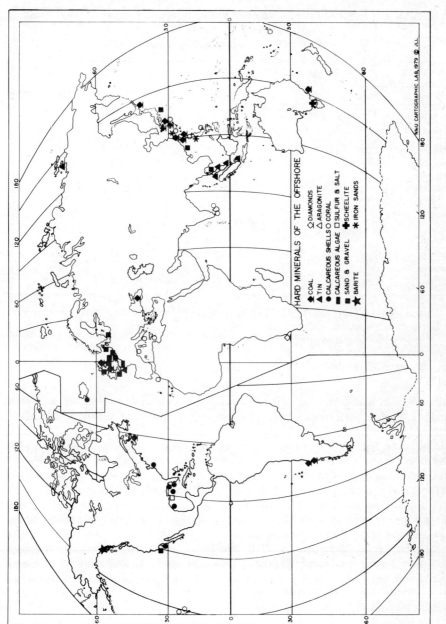

Fig. 4. Hard minerals of the offshore.

Hard minerals

Table 1. Estimated Output, Value of Production, and Percent of Total World Marine Hard Mineral Production
(thousands of metric tons and millions of $U.S.)

Year	1967-68[a]	1970[b]			1977[c]		
Mineral commodity	Value of production	Output	Value of production	% of world trade	Output	Value of production	% of world trade
Coal	35	n.a.	n.a.	n.a.	34,036[d]	n.a.	1.4[c]
Sulphur	37	1,000	26	2.4	1,500	36.0[f]	2.9
Iron ore	17	n.a.	3.8	<1.0	30	0.5	<1.0
Sand and gravel	n.a.	55,000	100	1.0	85,900	189.4	<1.0
Calcium carbonate*	n.a.	18,800	36	1.0	16,290	95.2	<1.0
Scheelite	n.a.	n.a.	n.a.	n.a.	235[g]	30.0[g]	n.a.
Tin	24	12.5	41	5.8	13.7	190.0[h]	6.1
Diamonds**	4.0	16,571	n.a.	1.0	n.a.	n.a.	n.a.
Barite	n.a.	122	1.0	3.5	378[i]	0.1	<1.0
Precious coral	2.0	n.a.	n.a.	n.a.	n.a.	n.a.	n.a.
Salt brine[j]†	.074	270	.049	n.a.	151	.054	n.a.

Sources: [a][95]; [b][16], [333]; [c][245]; [d][471]; [e][37]; [f][445]; [g][568]; [h][221]; [i][251]; [j][223].
Notes: *Calcareous shell and aragonite.
**1970 Output in carats.
†Data for 1975.
n.a. = not available.

million tons of usable iron ore were being produced annually in a subseabed drift network totalling 7.5 mi.² Efforts to develop economically competitive beneficiation techniques applicable to Wabana ore failed, and the mine closed.[159, 160] During 1961-67, magnetite was extracted in the Gulf of Finland, 50 mi. southwest of Helsinki, near Jussaro Island. The mine's reserves of several billion tons were identified by magnetometer surveys;[424] its annual production totaled about 300,000 metric tons of crude ore from which 122,000 metric tons of concentrate were extracted.[288, a] The main shaft was on Jussaro Island, with a subsidiary shaft and drifts situated on an islet, Stenlandet, immediately adjacent to the deposits. According to Mero, a major problem in the mine's operation lay "...in locating and sealing waterbearing fissures."[342]

McKelvey et al. identified, in 1970, one active land-based iron mine in the offshore of western Kyushu in Japan[325] which is no longer operating. According to an Australian government source, geologists, and mining engineers once examined undersea iron ores off Cockatoo Island in Western Australia, but no effort was made to mine the ores because they proved to be uneconomic.[580]

Coal

A 1978 report prepared for the U.S. House of Representatives places world offshore coal production at 34 million metric tons annually.[471] Mining companies extract coal from seams beneath the offshore of Australia, Chile, Turkey, Taiwan, Japan, Great Britain, and Canada. The Dominion Coal Co. of Canada operates one of the best known and largest undersea coal mines. Located on Cape Breton Island, Nova Scotia, it has been producing for more than 100 years. The mine extends 5 mi. offshore, occupies an area of 75 mi.^{2}, and employs approximately 4,000 men. Australia has one subsea coal mine, the John Darling, owned by Broken Hill Proprietary Co., Ltd. It is located 12 mi. south of Newcastle, New South Wales, extends nearly a mile offshore, and has 6.75 mi.^{2} of subseabed workings. The 60 miners employed in the establishment's undersea area produced 268,000 tons of coal in 1978. All of this coal is consumed in the production of coke at Broken Hill's Newcastle Steelworks. Coal has been worked under the sea in the Newcastle area for about a century; the John Darling began operation in 1925.[353] Japan also has undersea coal mines, providing about 30% of the country's annual domestic production.[238] In the United Kingdom's offshore (12 mines in England and 2 in Scotland,[37] 14,000 miners extract coal from tunnels thrust seaward for more than 3 mi. Ten percent of the United Kingdom's coal production is of undersea origin. The Scots have worked collieries under the Firth of Forth since at least the 1600s. Historical records show that "...in 1617, Scotland's King James VI descended a land-based shaft to inspect the mining wonder of its day."[238] Workable offshore coal reserves in the United Kingdom have been reported to be at least 550 million tons.[32]

Undersea tunneling for coal, or other minerals, is an expensive and sometimes hazardous business. Work has had to stop when overlying strata became so thin that miners heard the wash of waves. Some not so lucky, drowned when seawater has broken through, as in one episode in Japan when 237 coal miners died. A similar incident in a United Kingdom colliery took 100 lives.[238] If proper precautions are taken, however, the danger of flooding is minimal.

A more important problem is the distance miners must travel to the workface. Some mineral economists believe the limit of profitable operations may be 12 to 15 mi. offshore, although with new technology in ventilation, mining devices such as automated "mechanical moles," and speedy conveyor belt transport systems, some additional distance is perhaps possible. A better solution may be to build artificial island entries and ventilation shafts, as is done in Japan by Mitsui Mining Co., Ltd. The company started undersea coal mining in 1920, and in 1958 began construction of the first of three artificial islands in Ariake Bay, in western Kyushu; the islands provide ventilation and entry shafts to the mine. All mining is done below the seabed (Fig. 5).[b] As of January 1979, Mitsui employed 3,564 underground workers (excluding engineering and management staff), and from April 1977 to March 1978 produced 5,177,145 tons of usable coal.[360] Three other offshore coal mines in Japan are active—the Ikeshima (near Sasebo) and the Takashima (near Nagasaki), both in western Kyushu, and the Kushiro (near Kushiro) in southeastern Hokkaido. These three mines employed in 1977 a total of 4,900 workers, and produced 4,250,000 tons of coal.[365, 468] Collectively, Japan has 8,464 undersea coal miners that annually produce approximately 9-10 million tons of coal.

14 Hard minerals

Fig. 5. Artifical islands used for undersea coal mining at Ariake Bay, in western Kyushu, Japan. Courtesy Mitsui Mining Co., Ltd. A. Construction of the frame for Miike-jima, an island used for air intake ventilation and the descent of miners to the various working levels. B. The sinking of the entry shaft on Miike-jima, located approximately 4 mi. offshore. The shaft penetrates to about 2,000 ft. below sea-level. C. The first island completed, Hatsushima, is used for return air ventilation.

Scheelite

Under Bass Strait, off King Island in Tasmania, King Island Scheelite Proprietary, Ltd., mines scheelite (calcium tungstate, $CaWO_4$). Scheelite's fluorescing qualities make it useful in mineral prospecting and in process checking. The tungsten is also used in steel alloying for wear resistant cutting tools. The mine, located near the city of Grassy, employs approximately 60 underground workers and uses an onshore decline shaft to reach the undersea ores, which are produced at a rate of 255,000 tons annually.[568]

Tin

Tin mining beneath the sea has a long history. Currently, the Cornwall Peninsula of England produces tin from land-based mines that for many decades had been abandoned. The subseabed Levant Mine, which ended operations more than a century ago because of seawater seepage, was reexamined by engineers in 1960, and location of the old fissure was determined by putting a dye into the flooded mine. The dye emerged in the sea at the site of the break; in 1969 the fissure was sealed with concrete, and by the mid-1970s the mine was producing tin[238] The Geevor Mine at St. Just's on the Cornwall Peninsula also has undersea workings;

about 60% of the mine's total production (674 tons) during 1977 came from undersea operations. The mine employs 325 workers.[222]

PUMPING

Significant quantities of sulphur and salt brines, are pumped from beneath the seabed in the Gulf of Mexico along the Louisiana shore. The offshore sulphur industry recently has been experiencing severe economic problems, but should, in the long run, remain viable, because of anticipated expansions in sulphur markets.

Sulphur

A nonmetallic element, sulphur is widely distributed, occurring as native deposits, usually in association with salt domes. It is also found in molecular combinations, as in nonferrous sulfides (zinc, nickel, lead, and copper); in ferrous sulfides (pyrrhotite and pyrites); in sulfates (anhydrite and gypsum); and in crude oil and natural gas, as a contaminant.[201]

The contaminant (unwanted by-product) problem is significant to the prospects for sulphur mining. Strict sulphur emission controls have helped create a glutted market, even though world demand for sulphur has been increasing. In 1977, sulphur captured from smelter gas emissions alone accounted for the production of 5,589,000 tons. From 1965 through 1975, sulphur produced on the United States' OCS, those areas seaward of state jurisdiction, averaged annually 1.3 million tons; the value of production averaged $36 million. Yearly production values during the period fluctuated radically, falling from $53.3 million in 1968 to $24.1 million in 1970,[223] then rising again to $36 million in 1977. In constant dollars, the 1977 value does not represent a real gain over 1970. Depressed market conditions contributed to a suspension of operations at one of two of the world's only offshore sulphur mining establishments—both Frasch mines located a few miles off the Louisiana coast and owned by Freeport Sulphur Co. The other establishment, the Grand Isle Mine, continues to produce. Both mines are situated in 50 ft. of water and mining facilities consist of a series of platforms connected by bridges (Fig. 6). The decks are supported by hollow-legged base sections supported, in turn, by piles driven through them into the seabed. The platforms' design allows them to withstand the most severe hurricanes. A heliport, central control room, and living quarters for 60 workers assure uninterrupted operation. Sulphur from the Grand Isle Mine goes to shore via a 7-mi. insulated, heated, and buried pipeline. From the onshore terminal, "thermosbottle" barges (Fig. 7) transport the still-liquid sulphur to a storage and shipping site, Port Sulphur, 25 mi. to the east on the Mississippi River.[519, c]

Salt Brine

Freeport Sulphur produces brine from the salt dome associated with the sulphur deposits at Grand Isle Mine. The brine is used by Freeport in its sulphur mining

Fig. 6. Freeport Minerals Co.'s Grand Isle sulphur mine in the Gulf of Mexico in the offshore of Louisiana. This mine was the world's first and presently the only operating offshore sulphur facility. A heliport, living quarters, and hot water plant are situated to the right rear. The platform located on the far left rear is no longer in position. Courtesy Freeport Minerals Co.

process.[444] According to the DOI, production tonnage on the OCS in 1975 was 137,000 tons with an estimated market value of $54,000.[223]

DREDGING

Marine dredging is likely to become increasingly important in the coming decades, as economically exploitable onland mineral supplies become depleted. Many continental coastal areas have alluvial sediments containing a variety of metallic and nonmetallic minerals. After weathering, the continents' mineral sediments may be carried by streams to the oceans. When the sediments enter the coastal surf zone, wave action may segregate some components into concentrated placer deposits such as tin, gold, and diamonds, as well as sand and gravel. This sorting process facilitates commercial dredging. Dredgers also extract mineral precipitates (aragonite) and biogenic materials (coral, algae, and calcareous shells) that accumulate in biotically fertile offshore areas.

18 Hard minerals

Fig. 7. A "thermosbottle" barge ready to load liquid sulphur at the onshore pipeline terminal of the Grand Isle mine. When the barge is loaded, the tug will push it 25 mi., via inland waterways, to Port Sulphur on the Mississippi River where the sulphur will be transferred to a large ocean-going vessel. Courtesy Freeport Minerals Co.

Iron Sands

Dredging for iron sands in Ariake Bay of southern Kyushu and in Tokyo Bay of southern Honshu was once an important industry, but these operations are no longer active.[468] Presently, some iron sands are dredged off the shores of northern Taiwan and the Philippines. As of 1977, a total of 30 men were employed in the Taiwanese industry which produced 16,358 tons valued at $180,000.[380, 590]

Tin

Of the several minerals dredged, tin (cassiterite, SnO_2) is probably best known. In 1977, the value of offshore tin production totaled $190 million; some 13.7 million metric tons were produced, approximately 6% of the world's total output. Marine tin mining usually requires mechanical breaking of the ore before it can be brought to the surface. Bucket-line or cutter-head dredges are used rather than suction types. One of the largest operations lies 5 mi. off Bangka Island, Indonesia. The dredges work waters 30 to 50 ft. deep, removing an overburden averaging approximately 10 ft. to reach an ore assaying at 0.74 lbs./cu. yd.[552, d] Several dredges are presently active off Phuket Island, Thailand, and Burma was receiving assistance through the United Nations in 1979 to explore possibilities for offshore tin production.[114] The Soviet Union is reported to extract tin-bearing sands from deposits off the Shirokastan Peninsula in the Laptev Sea some 200 mi. east of the Lena River delta.[245]

Production of tin in the offshore can provide considerable cost advantages compared with onshore operations, though it can also prove uneconomic. For example, in the 1960s, Britain's Union Corp., Ltd. dredgers attempted to extract tin in St. Ives Bay on the Cornwall coast, but the effort failed. Tides, swells, and waves made dredging and ore transport so difficult and expensive that the firm abandoned the project.[40, 246, e] Although offshore tin dredging efforts in Cornwall have proved uneconomic, operations in Southeast Asia have provided significant savings compared to onland production costs. In 1970-71, the weighted average production cost per ton of tin concentrate dredged in Thailand's offshore was $767; onshore mining costs for Malaya and Thailand combined averaged $1,117/ton.[16]

Barite

Alaska has the world's only offshore barite ($BaSO_4$) mining establishment that is located off Castle Island, approximately 14 mi. southwest of Petersburg. The deposit lies in a synclinal fold which was first mined (about 1963) from an outcrop above high-tide. Gradually, mining extended to the low-tide mark, and by 1967 had reached into the offshore, where it is covered by detritus up to 50 ft. thick. Currently, operators use opencast extraction techniques, drilling and blasting the ore under the water at a depth of 20 ft.[542] A clamshell loads the ore into a bottom-dump barge that is towed near the shore where the ore is dumped. A dragline then pulls the ore ashore for processing. The ore was once shipped unprocessed by ocean carriers, but since 1974 a beneficiating unit (heavy media vessel) has been used to upgrade it.[102]

Exploratory drilling indicates that the Castle Island deposit contains 2.3 million tons of barium sulphate ore (deposited as a limestone replacement). The ore meets mud specifications of petroleum drilling operators which, given the current growth in the oil industry of Alaska and elsewhere, should assure a good market demand. In the mid-1970s, this establishment supplied approximately 15% of the United States' total annual barite consumption.[251] Official production data are unavailable, but output has been placed at 330,000 to 355,000 tons annually.[251, 511]

Noncommercial dredges have recovered barium sulphate concretions in many other marine areas, including deposits from the offshore of San Clemente Island, Calif.; from the Kai Islands, Indonesia; and from near Colombo, Sri Lanka. Mero contends that "...it appears unlikely...extensive deposits of these barium sulphate concretions will be found on the deep ocean floor, but those that do occur should be economically mineable.[342]

Calcareous Shells

A major portion of the world's calcareous shell production comes from marine sources. The calcium carbonate contained in the shells is a raw material for cement, agricultural lime, soda ash, paper, and poultry feed producers. Shells are also used extensively as road metal. The value of world production in 1970 totaled approximately $35.5 million;[542] in 1977; the value was $95.2 million.

Iceland has exploited shell deposits since 1953 and its dredges ply the waters of Faxaloi, northwest of Reykjavik. Because their lime content measures only about 80%, the shells must be beneficiated through grinding and flotation, a process accomplished at Akranes on the east shore of Faxafloi.[556] Dredges also extract a calcareous algae (*Lithothamnia*) off the coast of northern Brittany, in France. Producers process the "maerl" at Pontrieux, St. Brieuc, St. Malo, and LaGouesniere for use as fertilizer, cement manufacture, and water treatment. In Fiji, at Suva Harbor, a calcium carbonate, known locally as coral sand, is in exploitation.[245]

Shell production is also important in the United States. In 1977 shell output totaled 12.3 million tons, valued at $33.5 million.[446] The shell industry is especially important along the gulf coast of Texas and Louisiana, but Alabama,[274] Florida, Maryland, Virginia, and California, also produce significant amounts. Brackish bays in Louisiana and Texas contain live and dead shell reefs. Such reefs often extend to a depth of 30 ft. and may be covered with several feet of mud. The Texas industry began in Galveston Bay in 1880,[146] but as of late 1978, only one operator in Texas was active. Production in Texas during 1974-78 averaged 3.6 million yds^3, declining from approximately 7 million yds^3 in 1974 to 1.6 million yds^3 in 1978. Production value for the years 1974-78 averaged $901,000, dropping from $1.8 million in 1974 to $400,000 in 1978.[177] In coastal Texas and Louisiana, shells provide a large portion of the surfacing material and base for roads, driveways, and parking lots. In 1965, Dolan estimated that about 30% of Texas' coastal shell production went into road construction. The shells can bear transportation costs for approximately 50-70 mi. inland, and they are cheaper and much more readily available than gravel, as much of the gulf coastal zone is composed of very fine Quaternary sediments.[146, f]

Aragonite

The island state of the Bahamas has the world's only offshore aragonite mine. Aragonite, composed of 95% calcium carbonate ($CaCo_3$) forms as a precipitate from ocean waters, in this case, in association with "... the interaction of the Gulf Stream and the warm Bahamas Bank."[15] Bahamian reserves are estimated at 70 to

90 billion tons. Lying in shallow waters to the southeast of the Bimini Islands, the presently exploited deposits cover an area 22 mi. long by 2 mi. wide and have a thickness of 10 to 15 ft.

The minesite is 23 mi. southeast of Bimini, at 25°23′N. -70°12′W. To facilitate mining, Marcona Ocean Industries, operator of the establishment, used two existing cays as a construction base. Dredges filled in between the cays, creating an artificial island named "Ocean Cay" that comprises 85 acres standing about 12 ft. above sea level. The island was constructed during 1969-70 and dredging for aragonite began a year thereafter (Fig. 8).[505]

Most of the 85 workers at Ocean Cay come from Bimini and Nassau. A dredge fitted with suction pumps and a 24-in. pipe mines aragonite at approximately 1,500 metric tons per hour. A floating screening plant towed behind the main dredge first receives the ore, ridding it of most extraneous material. After the ore passes through a densifying tank, operators send it by a floating slurry pipeline to Ocean Cay to be dewatered in cyclone separators and passed across vibrating

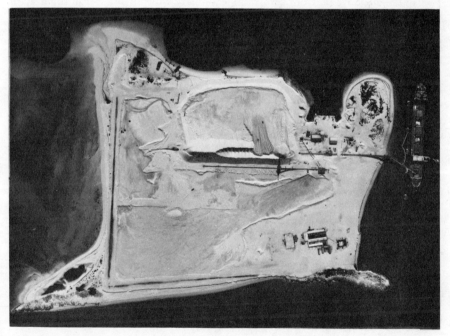

Fig. 8. Ocean Cay, a man-made island used to extract aragonite in the Bahamas. The average water depth in the immediate area surrounding the island is 6 to 10 ft., except for mined out areas where the depth is about 18 to 20 ft. The ship shown in the lower right is dredging aragonite. With an attached dewatering barge to which a floating pipeline is affixed, it moves as equipment advances in the mining process. Courtesy Marcona Sales, Inc.

screens. A conveyor belt moves the ore to a stockpile to await loading on the ship *Marcona Conveyor*, a 70,000 dwt self-unloader used to transport the aragonite to markets in the Caribbean and the United States.[15]

Aragonite production in 1977 totaled 2.2 million tons and had a value of $15 million. The cement industry is the most important market for aragonite. The small grain size minimizes grinding for the kilns, providing a saving in energy. Additional advantages are its relatively dust free composition, chemical consistency, and freedom from most iron impurities. These qualities make it an important component in agricultural liming and glass manufacturing. Many farmers and glass-makers along the United States' eastern seaboard use Bahamian aragonite.[15]

Coral

Coral is produced in many areas of the world as a decorative stone (precious coral). Production of black precious coral is locally an important industry off Maui in the Hawaiian Islands,[503] and a pink precious coral is obtained near Makapuu, Oahu. Hawaii's precious coral production was valued at $7.5 million in 1974. Taiwan has an active precious coral mining industry which uses a 2-man submersible in its extraction operations.[32] Taiwanese artists make coral figurines and decorative pieces. Divers also extract red and pink precious coral off the coasts of Algeria, Morocco, Tunisia, southern France, Corsica, Sardinia, and Naples. Craftsmen in Naples carve the coral into necklaces, bracelets, and rosary beads.

Sri Lanka has a coral mining industry that supplies lime for agriculture and gives employment to many workers in coastal villages. On the east coast, production centers around the village of Batticola, extending 50 mi. north and south. The west coast's main production area lies between a point 50 mi. south of Colombo (at Akurala) to south of Matara.[371]

During low tide, Ceylonese coral miners dig out fossil reefs in beach zones, and although illegal, they sometimes exploit numerous fringing reefs for the hard coral, *Scteractinia*.[g] The gatherers use stone ballasts to carry them quickly to the ocean floor adjacent to the reefs where large chunks of coral are pried free with knives, poles, or crowbars. A face mask, crude rafts, and baskets are their only other mining equipment. Divers load pieces of dead and living coral (the latter is richer in lime) onto a coconut-wood raft, until it begins to sink; they then tow the raft ashore, and pile the coral to dry. Next, the workers put the raw coral into small, ancient kilns 6 ft. in diameter and 9 ft. high, many of which are 100 years old. These kilns hold up to 5 tons of lime. After 2 days of firing, workers put the lime into 100-lb. bags that in 1976 sold for 10 to 12 rupees (approximately $1.00).[371]

Diamonds

In the offshore of South-West Africa (Namibia), gravel placers once supplied a rich output of diamonds. Streams carry weathered material from volcanic formations of the continental interior to the coast. Exploratory work and some mining have been under way for nearly 30 years along the shore between Hottentot Bay and Orangemund and near a colorfully named group of islands, Roastbeef, Guano, and Plumpudding.[591]

From 1961 to 1971, the Marine Diamond Corp., a subsidiary of the world renowned South African firm, DeBeers, explored and mined diamond-rich marine terraces and raised beaches. In the mid-1960s, the firm had 3 mining vessels, 11 support ships, and 2 aircraft operating in the Hottentott Bay area, earning annually $1.7 million from the diamonds acquired in 322,000 yd^3 of placer gravels.[277] Nearly all of the dredged diamonds were gemstones. Miners used a huge dredge fitted with hydraulic water jets that focused a surge of water against a conglomeratic gravel deposit varying in thickness up to 30 ft. and overlain by silts of 6 to 9 ft. As the conglomeratic gravels broke up, they were sucked into twin steel pipes. The dredge, winched laterally along the shore, cut a swath approximately 60 ft. deep, 230 ft. long, and 63 ft. wide, retaining for treatment materials less than 0.5 in. and pumping larger materials back into the trench, while taking care not to bury the unmined conglomerate.[591]

Currently, diamonds are extracted only in the surf zone. Consolidated Diamond Mines, another DeBeers subsidiary, mines diamonds by constructing a seawall, immediately outside of and parallel to the surf zone. Sands lying between the seawall and beach are then processed for the diamonds. If waves top the seawall, pumps remove the water from the work area.[504]

Although offshore mining ended in May 1971, marine exploration continues.[148] During the 1960s, DeBeers annually spent $1 million for geophysical mapping programs. Geologists seek to identify the extent of the placer gravels and to locate offshore dikes that may have acted as traps when the Benguela Current rolled the diamonds along shore. Annular drills sample the gravels every 30 ft. by sucking up a core inside the drill. Divers also explore these placers, but sediments and currents combine to make their work difficult, as visibility sometimes reaches zero. According to one report, DeBeers' exploration work has not paid for itself in the short run, but long-term prospects have encouraged a continuing prospecting program. To be competitive, offshore diamond extraction must be 4 to 5 times more productive than onshore mining.[246] If the offshore is exploited again, some engineers have suggested that a large concrete diving bell could be used to mine diamonds lying in deep waters. The bell would consist of an 80-ft. tower, containing its own internal processing plant and access to the placer's mining surface.[504] The press in 1978 reported that one firm, the Charles Anthony Diamond Investment Co. of Salt Lake City, Utah, had purchased an offshore mine lease from the South African government. The lease site, situated about 220 mi. north of Cape Town, covers a 10-mi. long, 400-ft. wide stretch of coastal surf zone. Perhaps, the offshore diamond mining industry will make a comeback along Africa's southwestern shore.

Sand and Gravel

Approximately 90% of all sand and gravel produced is used in aggregates, concrete, or as a filler; the remainder is consumed in glassmaking, in the production of abrasives, in ballasting ships, and in numerous other minor applications.[144] Aggregate supplies are becoming increasingly scarce in nearly all densely populated and highly urbanized areas. Burgeoning demands, stricter zoning ordinances, and

24   Hard minerals

"sterilization" of reserves (occurring as a result of constructing roads and buildings over deposits) are contributing to increasing haulage distances and costs. Because many major conurbations lie adjacent to ocean coastal zones, marine sands and gravels are becoming an increasingly significant aggregate source.[158,h] Some areas, where offshore waters are relatively shallow, where little or no overburden is present, and where delivery distance is not too great, have already experienced this shift. Indeed, marine aggregates are now successfully competing with onland sources, accounting for the largest volume of any solid mineral produced in the oceans. The value of world marine sand and gravel output in 1977 was $189.4 million. The main producers were Japan (41.3 million metric tons), which obtains approximately 19% of its sand and gravel from the seabed, and the United Kingdom (14.8 million tons).[115] Other producers include Denmark (9.5 million tons), the Netherlands (7.5 million tons), the United States (4.0 million tons), Sweden (950,000 tons), Thailand and Hong Kong.[7,245]

United Kingdom. Mero notes that the North Sea offshore aggregates industry first began more than 100 years ago. Miners towed barges offshore to shallow banks. When the tide went out, the barges settled on the seabed; the barge was then loaded and, with high tide, made for port.[343] From these modest beginnings, the United Kingdom today has the world's most highly developed marine sand and gravel industry. In the early 1970s, nearly 13% of the aggregates consumed annually in the United Kingdom was of marine origin, with production totaling approximately 12.5 million tons,[330] in 1977, the estimated figure was 14.8 million tons. The primary offshore production areas in the United Kingdom are the east and southeast of England. During 1977, the east coast (mainly off the Humber and Lincolnshire) and the southeast's Thames Estuary (especially the Shipwash, Sunk, and Babbard Banks) accounted for approximately 60% of the United Kingdom's total marine output. The south coast area produced 17.6%; Bristol Channel area accounted for 14.2%; the Solent dredged 5.4%; and Liverpool Bay produced 2.0%. Twenty-two companies were dredging aggregates, using 15 separate wharves in the London area and about 48 landing sites in Shoreham, Newhaven, Newcastle, Teeside, Portsmouth, Bristol, Cardiff, Liverpool, Newport and Swansea.[222] Many dredgers cannot afford to use regular dock facilities and are restricted to tidal wharves, a situation which requires work schedules to be synchronized with tides on a 12- or 24-hour cycle, and results in inefficient use of labor and capital equipment. The producers' dependence on tidal wharves also precludes the use of really large dredges.[18] Not all aggregates produced in the United Kingdom's waters are consumed domestically. Some producers deliver sand and gravel to ports in France (Dunkirk), Belgium (Zeebrugge and Ostend), and the Netherlands (Rotterdam and Flushing).[222] Dredgers working in the North Sea must contend with rough weather which contributes to considerable down time, although they continue to operate when waves are 10 to 12 ft. high. Dredges must also work with depths as great as 165 ft. The round trip for dredges averaged approximately 43 mi. in 1970, with a mean operating cost of 44¢/ton and a capital investment per dredging firm of $510,000.[330]

United States. In the United States, sand and gravel may soon become an important offshore commodity, both in value and need. In 1974, 7,000 onshore plants produced 904 million tons of sand and gravel valued at approximately $1,460 million. Of this total, urban areas in 21 coastal states consumed 25%. Only 2% of the total national production came from coastal dredging operations, including producers in tidal rivers, lagoons, and estuaries. No aggregates are currently produced in the OCS, and many coastal states are not issuing marine dredging permits. Some aggregates have been dredged in the New York Harbor area in association with navigation improvement, mainly along the east bank of the Ambrose Channel, the primary access route into the Harbor. Dredges working the area in 1974 produced aggregates primarily for highway construction in New Jersey.[552] Small amounts of sand and gravel have also been produced in coastal waters of southern California. But what of the future?

Marine sand and gravel reserves of the United States are placed at a possible 1.3 trillion tons. The ratio is about 15 to 1 in favor of sand, and when the Federal Hard Minerals Leasing Program goes into effect for the OCS, annual output is expected to reach 12 million tons or more within 5 years. The largest gravel deposits occur along Massachusetts' and Canada's Nova Scotia shores in the Atlantic and off the state of Washington in the Pacific from Cape Flattery to the Hoh River.[115] In the mid-1970s, the USGS and Woods Hole Oceanographic Institution did exploratory work on sand and gravel resources on the continental shelf of Long Island and New Jersey.[436] To assure that prime deposits are not wasted, similar mapping work is needed in other areas; this goal can be achieved only if improved exploration techniques are applied, such as better profiling devices. One proposed project, designed to investigate sand and gravel deposits in Massachusetts Bay and to evaluate prototype extraction techniques, never got under way. This failure is unfortunate because a knowledge of offshore aggregates is needed in the region, and the federal and state governments need base-line data to establish reasonable procedural regulations for mining that may become necessary in the future.

Phosphorites

Although no marine phosphorites are produced commercially, they have long been known to scientists, and could become a valuable future resource for the world's fertilizer market. Oceanographic expeditions as early as in the 1870s reported finding phosphorite concretions off the coast of southern California and South Africa. Recent exploration programs have investigated these and other large deposits along the shores of South America, Asia, North Africa, and Australia. The deposits are composed mostly of phosphates and occur as crusts, nodules, muds, and sands. Mero, a long-time investigator of phosphorites, has placed the oceans' economically recoverable reserves at 30 billion tons, enough to supply the world for 1,000 years at today's consumption rates.[342]

Geologists do not agree on the origin of marine phosphorites. Some attribute their formation to the penetration of volcanic materials into the reducing atmosphere of the sea.[193] Others hold that in oceanic regions where currents of significantly different temperatures meet, many organisms are killed by sudden thermal changes, resulting in a large accumulation of phosphatic detritus on the seabed.

Still others suggest that chemical precipitation occurs when dissolved freshwater phosphorites come into contact with seawater, or where, owing to the influence of trade winds and the Coriolis effect, "...cold waters move onto shelves abetted by movement of phosphate-rich surface water on shore."[252, 1] Deposits also occur in the deep oceans, but they are not presently economically exploitable. Mero notes that phosphorite nodules have been found at depths of only 190 ft., of as much as 8,400 ft. off Southern California and 11,400 ft. near the base of the continental slope off the Cape of Good Hope. He suggests that these exceptionally deep deposits may not have formed there, but were carried downslope from shallower waters by turbidity currents or by sediment slumps.[342]

Future phosphorite exploitation may be limited to relatively shallow waters (less than 600 ft.), and will probably lie within 40 mi. of shore. Offshore phosphorite nodules usually are not as rich as onshore deposits, and thus will incur a processing expense as the P_2O_5 (phosphorus pentoxide) is raised by calcination, grinding, and flotation. Phosphate muds and sands, on the other hand, may attract commercial producers because normally they can be used as fertilizers without processing.[542] Although now dated, Kaufman generated low and high phosphate nodule production cost estimates in 1970. Considering the progress of inflation since 1970, his high estimate best fits current economic conditions. He placed the high production cost estimate at $31.80/ton and the recoverable mineral value at $16.00/ton.[277] The 1970 market obviously precluded investment. Because phosphate prices are individually negotiated between sellers and buyers, no reliable world phosphate prices were available for early 1979. Those few prices quoted in 1979 ranged between $26.29 to $43.00/ton.[513] Garrand, an authority on phosphorites, feels that a marine phosphate nodule-producing establishment in 1979 would have to have an annual capacity of at least 1 million tons, and stresses that each phosphate deposit's economic viability depends on individual circumstances, including company structure (private or state ownership).[194] If production costs could be scaled down, enthusiasm for marine phosphate investments may grow during the 1980s. For example, recent phosphate prices have encouraged Lockheed Missiles and Space Co., a partner in a deep seabed consortium with Ocean Minerals Co., to examine the possibilities of mining phosphates off southern California's shores; however, it has now dropped the program.[453] Garrand contends that when marine phosphorites do become competitive with onland sources, the site most likely to be selected for the first mine will be the offshore of Baja California.[193] Given recent trends in demand for phosphates, it is likely marine sources will become increasingly important to phosphate deficient states. India, for example, currently imports large quantities of phosphates, but she could potentially produce some of her needs in adjacent nearshore areas.[16, 277]

Gold and Platinum

According to data presented in a United Nations seminar held during the early 1970s, gold could then be profitably dredged from marine placers, if the metal obtained totaled $1.05/yd^3 of sediment. In spite of the inflation of the 1970s, contemporary market prices for gold, which reached $800/troy oz. in 1980, may soon encourage marine placer mining. Geologists expect to see future gold production from marine sediments along the western coast of North America, in the Philippines.[64] and, perhaps, in the Soviet Far East.

Recently, geologists have carefully explored potential gold placers off the western coast of the conterminous United States and off Alaska. Especially encouraging prospects occur along coastal Northern California and adjacent Oregon, contiguous to the Klamath Mountains.[296, 354] Potentially mineable placers, 1 to 7 mi. offshore between Coos Bay and Cape Ferralo, Oregon, contain various heavy metal concentrations (including gold) that measure as high as 43%.[64] Based on beach sand sampling programs and littoral-drift and surf-action studies, geologists believe sections of the Gulf of Alaska coast may contain significant amounts of gold.[447] Indeed, in the past, scuba divers have extracted small amounts of gold from Alaska's offshore.[101] Five mi. west of Nome, geologists recently explored 22,000 acres of seabed placers, using winter ice as a drilling platform.[436] The Soviets, too, have done considerable exploratory work with gold placers. Many placers along the Sea of Okhotsk coast show good potential, but no production seems to have begun.[131]

Alaska also has been the site of past production and recent exploration for platinum placers. Until the late 1970s, the Goodnews Bay Mining Co. mined offshore platinum deposits near the towns of Platinum and Goodnews Bay. In the 1970s, Owen and Moore made intensive studies of dispersal patterns and sources of platinum placers in Chagvan Bay, 15 mi. south of Goodnews Bay.[415] Field investigations and past production activities seem to indicate that offshore southwestern Alaska may yet have considerable potential for platinum.

Environmental Impacts of Dredging

Some mineral producers contend that long-term ecological benefits result from dredging, including an increase in nutrient mixing, along with an improvement of phytoplankton development and a stimulation of animal growth. Many ecologists, however, question this contention.

Dredging for minerals can have significant impacts upon marine and coastal environments. Stationary sand dredges, for example, may make an excavation over 60 ft. deep and 250 ft. wide, depending upon the deposit's type, quantity, and depth. Trailing suction sand dredges lower the seabed by about 17 ft., varying with the material being worked, seabed stability, and the number of traverses made. As offshore contours are altered, sediment transport and deposition budgets and coastlines may be changed, creating navigation hazards, altering navigational channels and disturbing markers.[115] In addition, currents and local wave action may be altered and seabed pipelines and cables disturbed.

Biological-chemical processes affected by mining on the continental shelf are less understood than are the physical-geological processes. Severe pollution may be created during dredging, allowing the release of excessive amounts of nutrients or toxic materials—pesticides, sludges, and heavy metals[330]—that collect near shore when deposited by rivers. Some organisms might respond positively to phosphorus and nitrogen nutrients released from the sediments, whereas other biota may be harmed. The potential for toxic metals release has been important enough to concern state officials in California, Connecticut, and New York, and ultimately prohibit dredging in San Francisco Bay and Long Island Sound.[32]

Dredging may also cause an increase in turbidity and a reduction in plant photosynthesis, disruption in food chains, and destruction of biotic reproduction habitats such as stem plants to which various organisms attach their eggs. The problem of light reduction becomes critical as water depths increase, because these areas already receive only minimal light.[157]

When sediments settle out, they may affect the spawning of fish by covering gravels or by changing the particle size. Filtering mechanisms of oysters may be damaged and benthic dwellers' gas exchange capabilities may be reduced. Some bottom dwelling animals are disturbed or killed by being sucked into the dredge along with the sediments. Other consequences of offshore mineral production include disruption of migration routes of biota and oxygen depletion of the water.[414] Oxygen depletion can develop in enclosed bays when too much mineral nutrient is released, with subsequent rapid plant growth and decay. When the plants die, the water's dissolved oxygen is used in the decomposition process. Oxygen depletion can also occur when rich organic sediment layers become exposed during mining. The water's oxygen will be used in the decomposition process. Also, if dredged materials contain large amounts of free sulphides, these may be released to and oxidized in overlying waters. This chemical process could create an oxygen demand sufficient to deplete dissolved oxygen, especially in the benthic layer.[32] Dredging operations in tidal waters have been reported to deplete dissolved oxygen as much as 16-83%.[59]

Leasing and Environmental Regulation of United States Offshore Hard Minerals Mining in the OCS

The DOI, as stipulated under the *Outer Continental Shelf Lands Act of 1974*, is responsible for administering leasing and environmental control regulations on the OCS. Although the Secretary of the Interior issued proposed regulations for hard mineral mining on the OCS in 1974, no firm program has been established. The Bureau of Land Management (BLM) implements the DOI's mandate by functioning as the leasing agency, and the USGS acts as the supervisory authority. As of 1979, no actual mining of hard minerals in the OCS has been allowed (excluding sulphur and salt which are administered along with oil and gas), although the USGS has issued permits for exploration.

In 1972, Congress passed the *Marine Protection, Research and Sanctuaries Act*, giving the Secretary of Commerce authority to designate and to establish regulations for marine sanctuaries in coastal waters out to the outer edge of the continental shelf. This act would be used in such a way as to inhibit or preclude the mining of hard mineral resources (and oil and gas) on the OCS, a situation that could potentially create friction between the DOI and the Department of Commerce (DOC). Additional opportunity for jurisdictional disputes may arise, given that the Corps of Engineers, the Coast Guard, and the Environmental Protection Agency also have administrative functions relating to hard minerals mining on the OCS.

Under the DOI's leasing regulations developed in 1974, tracts for leasing can be designated by the BLM or through suggestions made by potential lessees. Nominating parties must provide the BLM with detailed plans for development, existing physical conditions, available transport modes, an environmental impact statement, and analyses of market and supply/demand situations for each suggested tract. After the BLM consults with other federal agencies, it may also hold hearings and seek input from the littoral states, before any final decision is made on the leases to be offered for sale. Leases are to be classified into three categories: (1) a 2-year exploratory lease, allowing the leasing of up to 36 tracts totaling no more than 5,760 acres each; (2) a development (sampling and refinement of mining systems) lease, for no more than 9 tracts of 5,760 acres each, representing an initial bid or a conversion of an exploratory lease; and (3) a production lease, obtained during initial bidding or through the conversion of an exploratory or development lease. The production lease is to be issued for a total of 3 tracks of not more than 5,760 acres each. Leasing sales would be awarded to the highest bidders. Annual per acre rental rates for successful lessees would be 10¢ for exploratory leases, 25¢ for development leases, and $1.00 for production leases. In addition, there would be a lease-designated royalty fee, equalling at least 2% of the annual gross production value.[32, j]

Hard Minerals Dredging Regulations in State Territorial Waters

The 30 coastal states of the United States control hard mineral mining policies relating to waters within their territorial jurisdiction. But their regulatory systems, in many instances, are reciprocally tied either directly or indirectly to federal programs. The states, for example, control water quality (as affected by dredging) through the application of the *Federal Water Pollution Control Act*. The Corps of Engineers must issue mining firms permits to discharge any dredging material or processing waste into coastal waters, but the Corps issues permits only when a given state certifies that the prospective lessee will meet "...water quality standards and effluent limitations established by the state and approved by the federal Environmental Protection Agency." Shore-based processing facilities are regulated both through local and state zoning ordinances and through air, noise, and water laws, especially in association with the *Coastal Zone Management Act of 1972* (CZMA).

The DOC, under the CZMA, financially assists states in establishing and maintaining programs for controlling industrial activities (including mining) in coastal areas. State management programs must provide for "...a definition of permissible land and water uses which have a significant impact on coastal waters, broad guidelines on priority of uses in particular areas, and an inventory of areas of particular concern." In order to qualify for funds under the CZMA, most of the 30 coastal states are developing a coastal zone management plan. However, only a few have so far actually set up specific laws dealing with the administration of their plan.[32] Two states, Alaska and California, well illustrate the divergence of the states in their approach to offshore hard mineral leasing programs.

Alaska. Given the commonly held pro-development attitudes of many Alaskans and the presence of several important marine placers, as gold and platinum,[132] it is likely that Alaska will experience considerable future growth in its offshore hard mineral mining industries. The state has established detailed regulations for development of gold and platinum in its offshore areas, although the laws are designed to provide maximum flexibility in mining firms' activities. Prospecting permits (which can be converted to active mining leases) are issued for 10-year periods for areas totalling as much as 2,560 acres. If a prospecting lease is converted to an active mining lease, it is valid for up to 55 years and renewable. There is no provision for public input relating to specific mining proposals and leases made, and as of 1978 no state agency was required to prepare environmental impact statements dealing with its decisions on prospecting or leasing permits. A mining establishment is exempt from taxation during the first 3½ years of production. Although the state requires detailed reports on proposed effluent discharges and the reporting of any likely disturbance of fish habitat, no major bureaucratic hurdles exist to acquiring various permits needed to proceed with mining.[32]

California. Under its *Coastal Act of 1976*, California has set down specific and stringent stipulations of what can and cannot be done with the coastal zone. Leases of submerged mineral lands may be obtained in two ways. One method involves competitive bidding that takes place after the State Lands Commission has determined that the state owns the lands in question, conditions for leasing have been established, and a public announcement in a widely-distributed regional newspaper. Prospective bidders must submit an environmental impact statement on each leasing unit, which is not to exceed 5,760 acres. A second method of lease acquisition involves acquiring a permit for prospecting that can be "...converted by the State Lands Commission into a preferential mineral lease." The prospecting permit is issued after statements have been submitted to the Commission that stipulate the proposed land use and the mineral to be sought. If the mineral is discovered, the permittee can apply for a preferential mineral lease which, if granted, allows mining to proceed. The prospecting permittee must pay a royalty rate of 20% on all minerals sold until a preferential lease is granted, at which time a predetermined lease rate takes effect. Permittees and lessees must maintain detailed records of all their activities, and are required to compensate the state for any damage resulting from their use of the land. Any violation of the firm's leasing contract can prompt cancellation of the lease after only 30 day's notice. Before any prospecting permits or leases can be issued, all state agencies involved in issuing permissions must submit environmental impact reports to the State Lands Commission; this report is subject to review for 60 days by all agencies and the general public and another 21 days for a last review by the Commission before a final decision is made to grant a permit or lease.[32, k]

Baram, Rice, and Lee have pointed out that the great diversity of state marine hard mineral regulations, as exemplified by Alaska and California, and the need to go through multiple state and federal agencies in obtaining permits and leases will discourage large-scale interstate or national offshore mineral investments. The delays and extra costs related to these conditions will discourage large operators that have enough capital and technical expertise to invest in and to utilize the most efficient and nonpolluting dredging equipment. Because of the many hurdles faced by industry in acquiring permits and leases and the multiplicity of both state and federal agency involvement in the regulation of offshore hard mineral mining, efforts should be made to harmonize their functions, especially under the *National Environmental Policy Act* and the CZMA.[32, l]

SUMMARY AND CONCLUSIONS

Because of depleting onshore sources of many minerals, it is likely that offshore seabed mining will become more important in the decades ahead, especially for sand and gravel. Demands for fertilizers and cement should assure a competitive place for marine shells, aragonite, and coral, and possibly also phosphorites. The future may see an expansion in diamond production and renewed dredging of coastal placers for platinum and gold. Energy shortages will demand a continuation of seabed coal mining in acutely energy-deficient regions such as Japan and Taiwan. On the other hand, prospects for some minerals, notably sulphur, are problematic. They depend on the demand situation and the stringency of enforcement control for sulphur emissions yielding large amounts of by-product sulphur which contribute to a glutted world market.

As offshore developments increasingly impinge on shore areas, the states and federal government will be called upon to more intensively monitor coastal-zone impacts; these efforts should be coordinated as much as possible, if mining for hard minerals is not to be delayed, if unnecessary inefficiencies are to be avoided, and if environmental pollutions, especially those associated with turbidity, oxygen depletion, disturbance of benthic organisms, and aesthetic amenities, are to be minimized.[113, m]

NOTES

[a]Unless otherwise noted, the metric ton (2,205 lbs.) is the measurement unit used throughout the book.

[b]For an excellent discussion of the Ariake Bay establishment, see [297].

[c]The Frasch mining system is made up of three concentric pipes. The outer pipe carries superheated water (325°F.) which passes through perforations into the surrounding sulphur formations. When the sulphur heats to 246°F., its melting point, it seeps to the base of the well. Static pressure of the hot water forced into the formation then pushes the liquid sulphur up the central pipe for several hundred feet. The smallest pipe carries compressed air down into the central pipe with the sulphur, which aerates and lightens the liquefied sulphur, thus allowing it to rise to the surface.

[d]For a detailed discussion of offshore tin dredging operations in Indonesia, see [517].

[e]Cornish offshore tin dredging is still a possibility. According to D. Horne of the Inst. of Geol. Sci., London, Marine Mining (Cornwall), Ltd. hopes by 1980 to be able to produce high-grade concentrates from dredging at sea [245].

[f]For a detailed examination of shell dredging in the Gulf Coast, see [52].

[g]Removal of the coral, especially live materials, in the fringing reefs has two deleterious effects. The reef's coastline protection function is diminished, and ameliorative action must be taken to stop shore erosion. Another serious consequence is the disturbance of very delicate reef-based ecosystems.

32   Hard minerals

^hExcept for a higher shell content and small amounts of salt, usually less than 0.1% in drained aggregates, marine materials are like those extracted onshore.

ⁱSome phosphates have been attributed to the "...inorganic precipitation of apatite within pore waters of anoxic sediments and subsequent concentration of the apatite by physical processes." See [73].

^jFor a more detailed and excellent examination of OCS regulatory functions of federal agencies, see [32]:160-63.

^kFor helpful discussions of other states' (Conn., N.J., N.Y., Tex.) regulatory systems, see [32]:221-47.

^lFor an examination of the implications for harmonizing federal and state regulatory functions, see [32]:249-67.

^mAn inclusive view of hard minerals of the offshore is contained in [113].

Chapter 3

Offshore Petroleum

A growing demand for energy over the past 3 decades and exorbitant increases in petroleum prices since the Arab embargo of 1973-74 have encouraged national governments to lease increasingly larger offshore areas for exploratory drilling and production. As early as 1970, some 75 countries had offshore petroleum exploration programs,[277] and by late 1973, about 40 countries had issued exploration licenses for waters greater than 660 ft. with some allowing work at depths that exceeded 6,600 ft.

The offshore environment makes petroleum exploration and production very expensive, as in the North Sea or along Canada's northeastern seaboard. Operational, maintenance, and supply needs of the offshore require not only large financial expenditures but also demand new skills and technologies. Large sums of capital entering these investment arenas are affecting economic, social, and environmental conditions of adjacent coastal areas, as in Scotland or New Jersey. Concern for these impacts caused governments and planners in Australia, Brazil, India, Norway, the United Kingdom, the United States, and the People's Republic of China, among others, to carefully assess their petroleum development plans and long-term national energy goals; these assessments have led to different sets of national objectives and to debate within states, as well as conflicts between countries.

Numerous variables contribute to offshore petroleum exploration and development decisions by oil companies. These variables, not necessarily unique to the offshore industry, include: (1) geological conditions; (2) size of reserves; (3) governmental attitudes and regulations concerning leasing, production, and ownership; (4) political climate; (5) location of the deposits relative to markets, processing facilities, and competing producers; (6) availability of investment capital; and (7) potential environmental impacts and regulations that must be observed.[42, 185] National governments, both with and without national oil agencies, are also concerned with these variables, as well as with: (1) domestic energy needs and a maximum autarky, (2) domestic business participation, (3) social and economic impacts, (4) balance of payments relationships, and (5) foreign relations. Although not all of these many topics can be fully developed, most will be alluded to here and in subsequent chapters (4 to 8). This chapter provides a brief view of continental margin geological conditions necessary to petroleum formation,[a] current world and regional production and reserve situations, and present exploration and development programs as they relate to investments and political conflicts both within and between states.

PETROLEUM FORMATION

Petroleum extraction on the continental shelves has become increasingly important during recent decades, and in the late 1970s drillers began to explore the continental slopes. In a few instances, drill ships have tested the continental rises, but their primary focus remains with the continental shelves. World continental shelves and slopes combined (the terrace) contain approximately 21.4 million mi.^{2}, about half of which (10.5 million mi.^{2}) is shelf; the rise covers another 7.4 million mi.^{2}. Geologists believe that more than half the area within the 1,000 ft. oceanic isobath is composed of Tertiary Age sediments that, on the average, are more petroleum productive than older sediments.[277] Although most geologists feel that the major potential for offshore petroleum lies within 200 mi. of most of the continents (Fig. 9), some are beginning to suggest that the deeper seabed areas may hold more promise than previously thought possible.[170]

Several geological conditions contribute to the potential of the continental margin for petroleum generation. These include the presence of (1) thick permeable sediments rich in organic matter; (2) structures for trapping oil, including both tectonic and stratigraphic traps; and (3) impervious strata. If any of such conditions is absent, petroleum can neither form nor accumulate.

As continental erosion occurs, streams transport vast sediment loads to the oceans where they are deposited adjacent to the coast. Sediment thickness must be at least 300 ft. (preferably 3,000 ft.)—sometimes reaching 40,000 ft. or more—so that favorable "cooking" temperatures occur (122° to 270°F. for oil and 212° to 424°F. for natural gas).[170, 229] The continental rises could provide good hydrocarbon potential because of their relatively low heat conductivity, owing to the fine grain size of the sediments.[170] On the other hand, the fine grain size is indicative that the volume of sands available for reservoir rocks may be limited. When folds, upthrust faults, shell reefs, volcanic structures, and diapirs create barriers on the continental shelf, slope, and rise, sediments can accumulate rapidly (Fig. 10). Where these formations are accompanied by impervious strata or evaporite bodies such as salt or sulphur (which may form diapirs with associated upbending of sediments, resulting in trapping mechanisms), prime conditions exist for the

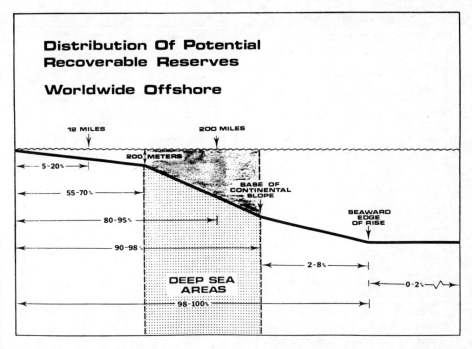

Fig. 9. Worldwide offshore distribution of potential recoverable petroleum reserves. *Source:* [84]. With permission.

development of exploitable hydrocarbon deposits. Yet, unless favorable reducing conditions are present, no petroleum can form. The basic requirement is that the organic material must exceed the free oxygen available for oxidation. Stagnant water conditions resulting from little overturn may create deficient oxygen conditions; anaerobic conditions may also result from rapid burial, as in landslides, or from turbidity currents which may isolate the organic material from bottom waters.[228]

Sedimentary basins provide the most favorable conditions for the necessary oil and gas formation processes, especially in areas where the sediments are thicker than 3,000 ft. Numerous basins occur on the continental margins, including the shelf, slope, and rise. Little is known about the geological structure and petroleum potential of major portions of the continental margins, and many geologists decry the inadequancy of our knowledge, especially about the continental slope and rise. According to one estimate, 65% of the continental shelf is poorly known and another 20% is only moderately well known.[169]

The world's orogenic belts, where plate convergence and subduction are occurring, also provide ideal sites for petroleum formation.[141] Subduction-associated "crustal subsidence allows for the accumulation of sediments, and crustal uplift provides the sediments to fill the depressions. Horizontal and vertical movements

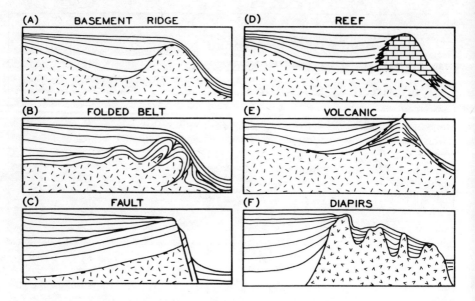

Fig. 10. Diagrammatic representation of types of geological barriers on continental margins. Vertical scale is greatly exaggerated.
Source: [228]. With permission.

within crustal plates have localized much of the deformation of these sediments, providing traps for hydrocarbon accumulation."[65] Also, subduction processes create the high temperatures needed to transform organic matter into liquefied and gaseous hydrocarbons, but if temperatures become excessively hot, the fluid petroleum is destroyed.[229]

OFFSHORE RESERVES

Some 10.8 million mi.^{2} of the world's continental shelves lie at water depths of less than 1,000 ft. About 16% of this area may have potential for petroleum. Although only a small part of this potential had been exploited by the late 1970s, approximately 21% of the annual global petroleum (crude oil and condensate) production came from offshore. Industry specialists expect offshore areas by the early 1980s to provide 33% of the world's total oil output, and by 1990 the figure may be 50%.[58, 162, 523]

A United Nations' estimate in 1973 placed world offshore petroleum reserves at 2,272 billion bbl., with over half (1,344 billion bbl.) being in continental shelf areas.[12] According to a 1974 Federal Energy Administration study, the United States has a marine and onland proved reserve[b] of only 38.2 billion bbl. of crude oil and 237 trillion ft.^{3} of natural gas. In 1975, based on existing technological capabilities and economic conditions, the USGS released estimates for the United States' inferred and undiscovered but recoverable oil and gas resources.[c] These two reserve categories totaled approximately 105 billion bbl. of crude oil and 686 trillion ft.^{3} of natural gas. The USGS data, however, are subject to wide estimating errors. It is possible (19 chances in 20) that the oil potential will be at least 73 billion bbl. and may be as high as 150 billion bbl. (1 chance in 20).[404] Most geologists feel the major promise for large oil and natural gas discoveries in the United States lies in the OCS. Since operations began in the offshore of Louisiana in 1947, fewer than 100 exploratory wells were needed, on the average, for the discovery of every 100 million or more barrels of oil or its gas equivalent of 600 billion ft.^{3}. To obtain an equal output onshore, 5,000 exploratory wells had to be drilled.[476] Such comparisons, however, can be somewhat misleading, because it is the large fields that will be used initially so that the discovery rate is likely to decline as time passes.

Giant offshore oil and gas fields occur on the margins of all the continents, excluding Antarctica. Berryhill of the USGC defined giant oil fields as those containing "...more than 500 million barrels of estimated recoverable oil; a giant gas field is one containing more than 3,500 billion ft.^{3} of estimated recoverable gas." As of 1974, a total of 70 fields (60 oil and 10 gas) were classified in these categories (Fig. 11).[42] Antarctica's continental shelf also could have major petroleum deposits. One estimate indicates that tens of billions of barrels of oil may be available on Antarctica's shelf. The Ross Shelf is thought to hold the most promise,[d] but much detailed exploratory work must be completed before the continent's potential can be adequately evaluated.

Recent exploration and development on the world's continental shelves has added substantially to oil and gas reserves and to production. In 1977, world offshore oil production contributed over 19% of the world's total crude oil production, an increase by 2.5% since 1976.[66, 369] Offshore output totaled approximaterialy 4.16 billion bbl., an amount 12% greater than that produced during the last pre-embargo year of 1973. World offshore gas production in 1977 exceeded 8 trillion ft.^{3}, with its share in the world's total of c.54.4 trillion ft.^{3} amounting to 15%.[363] The United States, and to a lesser extent Western Europe, stood out as the prime areas for offshore natural gas production since 1971. The Middle East topped all other areas for offshore crude oil output, producing 3.8 times more than the United States and accounting for 40% of the world's total 1977 output (Table 2). The pattern continued in 1978. Although accurate data for the world and for some specific regions are generally not available, the relative size and importance of the Middle East's offshore fields is indicated in Table 3). The continent of Africa had 51 operating offshore fields and the Middle East 34, but the latter produced 9 times more oil. Southeast Asia has 36 operating fields but produced only 28% as much oil as the Middle East.

EXPLORATION AND DEVELOPMENT PROGRAMS

With growing demands for petroleum, OPEC's annual and more frequent price hikes, and increasingly sophisticated exploration and production technologies,

38 *Offshore petroleum*

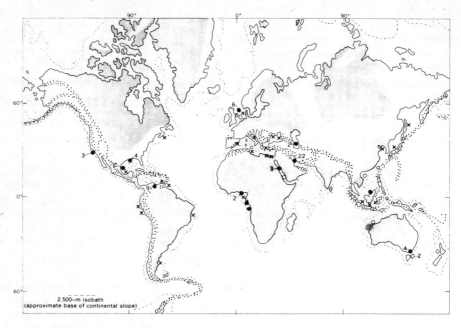

Fig. 11. World's giant oil and gas fields relative to generalized orogenic belts (stippled), belts inactive since Paleozoic and Mesozoic times (unpatterned), and regions not subjected to orogenic movements since Precambrian time (shaded). Solid circles are giant oil fields; open circles are giant gas fields; X indicates general areas of discovery either of less than giant size or fields under development. Figures indicate the number of fields in a specified area.

Source: [42].

many countries are pushing into ever more distant and deeper offshore areas. Continental shelf neighbors are beginning to vigorously compete for seabed oil and gas, and conflicts are arising between states as they seek dominion over these resources.

Investments: Drilling and Production

Oil companies and national governments worldwide have invested heavily in the offshore during the last 2½ decades, especially since 1973 when petroleum prices began to rise precipitously. Annual investments rose from $915 million in 1963 to an estimated $3.9 billion in 1974, more than a 400% increase. In Canada, for example, annual expenditures by industry for offshore drilling and other exploration work expanded from only a few million dollars in 1963 to approximately $120 million 10 years thereafter and $225 million in 1977. Although not so pronounced, a similar pattern has occurred in the United States. In 1963,

Table 2. Estimated Annual Worldwide Offshore Production of Crude Oil and Natural Gas, 1970–77[a]

Crude oil (millions of bbl)

Year	United States	Venezuela	Other Latin America	Western Hemisphere	Middle East	Africa	Asia	Western Europe	USSR	World total	Offshore oil as % of world's onshore and offshore total
1970	575.71	897.90	43.44		760.66	239.81	133.23	4.38	96.75[b]	2,751.83	16.5
1971	615.12	908.85	69.42		913.41	233.86	158.41	12.01	93.00[b]	3,004.03	17.0
1972	614.79	915.00	82.35		1,040.54	276.29	216.71	15.81	90.00[b]	3,251.49	17.5
1973	582.68	985.50	62.82		1,282.16	323.18	287.47	24.72	86.14	3,637.67	17.9
1974	532.74	756.00	73.65		1,262.73	380.34	275.08	29.87	86.25[b]	3,396.66	16.5
1975	495.28	634.04	97.40		1,070.35	364.88	323.22	116.49	83.22	3,184.88	16.3
1976	462.94	613.86	107.10		1,198.20	384.55	407.58	270.86	80.52	3,525.61	16.6
1977[c]	438.66	456.25	107.68		1,647.98	490.20	549.69	396.03	74.83	4,161.32	19.1

Natural gas (billions of ft.³)

Year	United States	Latin America		Western Hemisphere	Middle East	Africa	Asia	Western Europe	USSR	World total	Offshore gas as % of world's onshore and offshore total
1970	3,218.12	27.60		3,245.72	1,568.00	28.70	22.10	396.50	n.a.	5,261.02[d]	13.8[d]
1971	3,750.68	26.81		3,777.49	214.30	n.a.	35.80	654.90	n.a.	n.a.	n.a.
1972	3,757.42	26.30		3,783.72	252.60	n.a.	44.57	937.09	n.a.	n.a.	n.a.
1973	3,975.34	28.32		4,003.66	263.44	n.a.	102.78	1,095.00	268.36[b]	5,733.24[e]	12.4[e]
1974	4,229.75	39.91		4,269.66	259.66	n.a.	135.90	1,314.00	264.63	6,243.85[e]	13.4[e]
1975	4,257.47	82.13		4,339.60	1,396.27	114.61	132.06	1,387.91	282.51	7,652.95	16.1
1976	4,296.32	128.91		4,425.23	259.86	183.00	242.53	2,213.90[f]	328.30	7,652.82	16.5
1977[c]	4,540.01	145.75		4,685.76	333.98	79.86	464.61	2,217.00[g]	388.41[b]	8,169.62	15.0

Sources: Calculated from [369], [223], and *Offshore*.

Notes: [a]Data for prior years not available; [b]Data supplied by Dr. Theodore Shabad (pers. com., Dec. 11, 1979); [c]Preliminary data except for the U.S.; [d]Excluding USSR; [e]Excluding Africa; [f]The Amer. Pet. Inst. estimated Western Europe's 1976 production at c.5.4 trillion ft.³ and thus the world total at c.10.8 trillion ft.³. This apparently excessive estimate may have been due to figures from the United Kingdom which included a share of world output generated by British-owned oil companies. The 1976 total for Western Europe, estimated by the author on the basis of data supplied in December 1979 by *Offshore* and by the British Information Service in New York (pers. com., Dec. 7, 1979 detailing figures compiled by the U.K. Dept. of Energy), excludes the output of Denmark which did not report it. [g]Production for 1977, again without Denmark, was estimated on the same basis as for 1976. The figure exceeds that reported by the Amer. Pet. Inst. by 509 million ft.³.

Table 3. Daily Worldwide Offshore Oil Field Production during the First Six Months of 1978[a]

Region and country	Number of fields	Estimated production (bbl/d average)	Approximate % of world production
Middle East	34	4,756,278	42.8
Saudi Arabia	7	2,621,400[b]	
Abu Dhabi	5	627,000[b]	
Qatar	3	451,190	
Egypt	7	388,000	
Dubai	2	359,042	
Iran	6	278,204	
Sharjah	1	25,442	
Divided Zone*	2	n.a.	
Israel**	1	6,000	
North America	n.a.	1,286,188	11.5
United States	n.a.	1,237,800[b]	
Mexico	n.a.	48,388[b]	
Canada	n.a.	n.a.	
East and Southeast Asia	37	1,333,935	12.0
Indonesia	15	558,316	
Australia	4	397,681	
Malaysia	10	196,492	
Brunei	4	179,295	
Philippines	n.a.	n.a.	
Japan†	2	2,151	
Vietnam	n.a.	n.a.	
People's Rep. of China	2	2,000[c]	
South America	n.a.	1,536,150	13.8
Venezuela	n.a.	1,249,800	
Trinidad-Tabago	6	190,225	
Mexico	n.a.	48,388	
Peru	4	29,020	
Brazil	4	18,717	
Europe	17	1,579,166	14.2
United Kingdom	8	981,351	
Norway	4	351,028	
USSR	n.a.	205,000[b]	
Spain	2	18,461	
Italy	2	13,326	
Denmark	1	10,000	
Africa	51	528,401	4.8
Nigeria	26	205,442	
Gabon	14	125,820	
Angola	5	94,625	
Tunisia	1	45,234	
Congo	2	27,440	
Zaire	2	19,843	
Cameroon	1	10,000	
Indian Ocean	1	80,000	0.8
India	1	80,000	
World Total	—	11,100,000	100.0

Sources: [a][588]; [b][69] production data are for 1977; [c][348] for only one of the fields, the Takang in the Po Hai Basin.

Notes: *Ex-neutral zone; **Includes Israeli occupied portion of the Gulf of Suez; †Data for the Akita Prefecture field are not available; n.a. = not available.

American oil companies spent $415 million in the offshore oil and gas industry. After a decade, 1974 expenditures totaled $1,032 million. The industry's representatives estimate that between 1975 and 1980, a total of $37 billion will have been invested in the exploration for and $69 billion in the development of world offshore oil and gas deposits.[185] In 1979, nine "selected" oil companies alone expected to invest $3.7 billion in offshore oil and gas exploration and development.[71, e] This investment commitment derives, in part, from overall higher success ratios and higher yields available in the oceans than those experienced onshore. Also, once constructed, an offshore platform may provide development savings, because multiple wells (as many as 50-60) may be drilled from the same platform.[147]

In 1972, some 134 oil companies were exploring the coastal waters of 80 states.[61] By 1974, petroleum exploration was under way in offshore areas of every continent, except Antarctica. Drilling operations in the Americas extended from the Strait of Magellan in southernmost South America to Canada's arctic islands. Exploration ships were busy in Japanese, Indian, and New Zealand waters, and the Soviet Union had its drilling crews working in the Barents and Caspian Seas.[42] Efforts continue in most of these areas today. Commercial drillers in 1977 put down 2,748 wells in the world's offshore, and installed 129 platforms. The total active platforms numbered 1,038.[72, f]

Extensive worldwide investment in marine drilling rig construction is indicative of oil producers' commitment to expanding offshore oil and gas production. Sources in industry anticipate significant world growth (7-8%) in the operating drill fleet during 1979. As of September 1978, 94% of the drilling fleet of 403 floaters (semi-submersibles and ships) and bottom supported units were operating, while an additional 51 were under construction.[267]

Areas of Development and/or Conflict

Numerous world areas are experiencing rapid offshore oil industry development and related political conflicts. Several areas in North America, South America, and Southeast Asia illustrate the intense quest for the seabed's black gold. Not only are confrontations occurring between sovereign states, but conflicts are emerging within states, as subordinate political units seek to control what they feel is their rightful mineral heritage. For example, the province of Greenland is contesting Denmark's claim to petroleum resources off its shores in the Davis Strait. This issue is still unsettled even though Greenland was granted limited home rule in early May 1979.[123] And Canada's national and provincial governments are currently locked in court battles over primacy in offshore areas.

Canada and the United States. Since the early 1960s, Canada has been advancing rapidly ahead in petroleum development programs, moving from 25 million acres in 1963 to about 316 million acres in 1971. Although the annual total offshore areas granted by the Canadian government for exploration has declined since peaking in 1971, the amount is still significant, totalling some 48 million hectares, or 119 million acres (Fig. 12). Exploration leases extend from the Beaufort Sea in the northwest to the Nova Scotia coast in the southeast, with an outlier along the southern British Columbia shore.

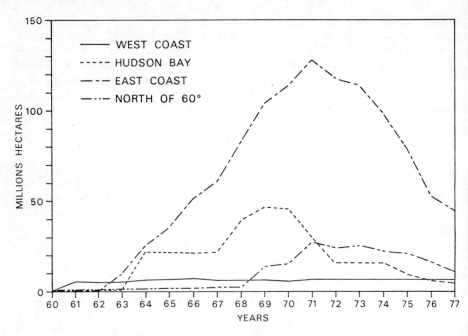

Fig. 12. Area granted for oil and gas exploration in the Canadian offshore, 1960–77. Courtesy Surveys and Mapping Branch, EMR, Canada.

With the emerging petroleum potential of the continental shelf off North America's highly urbanized and oil and gas deficient northeastern seaboard, considerable controversy has arisen over the territorial dividing line in the Northeast Channel area situated between New England and Nova Scotia.

The Georges Bank, lying within the Northeast Channel and considered to have an important petroleum potential, is the main reason for the dispute. Some oil and gas deposits have been discovered south of Nova Scotia, although their yield is too small to develop. When Canada in 1964 began issuing permits for exploration on the Bank, the United States government disputed its authority to do so. In 1970, the U.S. Department of State made its belated objection known:

> Notice is hereby given that the U.S. Government has refrained from authorizing geologic exploration or mineral exploitation in the area of the Georges Bank continental shelf. Pending agreement on the delimitation of the continental shelf in the Gulf of Maine, the U.S. Government does not acquiesce in or recognize the validity of permits or other authorizations issued by the Government of Canada to explore or exploit the natural resources of any part of the Georges Bank continental shelf, and reserves its right and those of its nationals in that area.[178]

The Canadian government argues that it informed the U.S. government in 1965 of its intention to issue exploration permits up to the median line and that:

Washington acknowledged this notification which, in effect, constituted acquiescence.[314]

Under the 1958 Geneva Convention on the Continental Shelf (GCCS), the median line will be designated as the dividing point in the absence of agreement upon a dividing line between two or more facing coastal states. The International Court of Justice (World Court), on the other hand, in considering a North Sea boundary dispute among West Germany, Denmark, and the Netherlands, ruled that a median line should neither ignore a mineral deposit's unity or the shelf's contours nor prejudice the amount of disputed area awarded to the claimants. These ambiguous stipulations are vague enough to create difficulties in adjusting many an offshore boundary dispute. But, the problem becomes more complicated because the United States and Canada use different baselines in measuring their territorial seas. The United States applies the low-tide mark and Canada uses straight base lines;[314] the position of the median line changes, depending on which system is used. Overlapping oil company leases granted by Ottawa and Washington and by Nova Scotia and Maine for exploratory work in the Georges Bank area have further complicated the problem. Oil companies have purchased leasing rights from both the United States and Canadian governments in the hope of protecting themselves.[349]

The United States has sought a settlement based not on a median line but on the center of the deepest part of the Northeast Channel. The United States argues its position on two premises: (1) the Northeast Channel is deeper than 200 m (656 ft.), a point the 1958 GCCS recommended as the outer limit of the continental shelf, and (2) the Georges Bank is a natural extension of the continental land mass, a contention supported in an earlier judgment made by the World Court for a North Sea dispute. Its opinion held that a "...natural prolongation of land territory..." is a strong argument for jurisdiction.[349] Canada contends that the 200 m isobath limitation on control of seabed resources does not apply because, technologically, drillers are capable of working in deeper waters, thereby allowing them sovereignty to the median line, a position supported by the 1958 GCCS when it defined the outer limit of the continental shelf as "...200 meters or, beyond that limit, to where the depth of the superjacent waters admits of the exploitation of the natural resources of the said areas."[100]

The state of Maine in 1968 further confused the issue when it granted permits for exploration up to 80 mi. offshore into Georges Bank, well beyond the normal 3-mi. state territorial limit. In 1969, the federal government filed a suit against Maine and all other Atlantic coastal states in an effort to settle the issue.[294] On March 17, 1975,[576] the U.S. Supreme Court dismissed the Atlantic coastal states' claims beyond 3 mi., with the exception of Connecticut whose oceanic waters were considered to be inland waters, owing to Long Island's presence offshore.[g]

Potential friction could also occur between the United States and Canada where Alaska's and the Yukon Territory's boundary abuts the Beaufort Sea, to the east of the Prudhoe Bay petroleum discoveries. Oil companies are already exploring the offshore 50 mi. east of Prudhoe Bay.[454] Canada has long used the 141°W. meridian as its western boundary, which it extends northward to the pole. If the United States should decide that the potential for oil and gas is good

44 *Offshore petroleum*

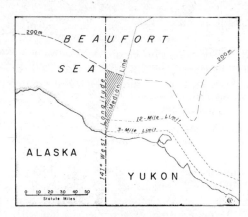

Fig. 13. Potential area of conflict in Canada's and the United States' territorial waters in the Beaufort Sea.
Source: [577]. With permission.

enough, it may argue for a median line boundary based on the coastal configuration, giving the United States some 300 mi.^{2} of ocean space lying east of 141°W. (Fig. 13). And if a national jurisdiction is assumed beyond the 200 m isobath, an even larger area would accrue to the United States, because the subsea contour bends sharply eastward.^{314}

Canada: The Provinces vs. Ottawa. Canada is not only confronted by an offshore petroleum-related dispute with the United States, but also by internal offshore jurisdictional disputes. The Canadian provinces and national government have been feuding over legal primacy in continental shelf areas since the early 1960s, when Ottawa decided the best procedure was to legally test the problem. Each province was conceded the right to present its case before the Supreme Court of Canada. Prince Edward Island and Quebec chose not to do so. Quebec contends that it no longer recognizes the Supreme Court's authority to interpret the Canadian Constitution, and consequently will not go before the Court, even though it has carried out exploratory drilling in the Gulf of St. Lawrence. In 1965, the Supreme Court decided British Columbia's case with a judgment that set the province's offshore territorial limits at the low-tide mark, a stance that went to the heart of the entire question of provincial sovereignty.$^{226,\ 262}$

Newfoundland's case is still before the Supreme Court. The Newfoundland problem is especially important because several oil and gas discoveries have already been made in waters off its northeastern shore. From 1966 to 1975, nearly 100 wells were drilled off Canada's northeastern seaboard; 6 wells brought in oil and 2 produced gas.^{33}

The Newfoundland government claims that when it joined the Confederation in 1949, it did not relinquish its sovereign rights to resources on the adjacent continental margin (shelf, slope, and rise).^{h} Many Canadians in Newfoundland

and in other provinces feel Ottawa has granted the oil companies excessively large and lenient leases, and that the national government is taking too much of the oil profits. The Newfoundlanders fear significant economic and social disruption on their island, if major petroleum developments take place in the offshore. They are especially wary of a boom and bust economic cycle, whereby vast capital inputs for various services and transport infrastructure are made that later, after the petroleum resources are depleted, would be underutilized. This fear accounts, in large part, for the Newfoundlanders' demand that offshore petroleum exploration and development be in their hands; they feel their intimate knowledge of the province's offshore will assure better management.[231, 335]

The government of Newfoundland expected the Canadian Supreme Court to settle the offshore jurisdictional dispute in 1979.[423] In anticipation of winning its case, the province's government has drafted detailed regulations for the operation of oil fields off its shores. The royalty system would be based on a distribution ratio of 75% for Newfoundland and 25% for Ottawa. A basic concern of the Newfoundland government is to control social and economic impacts upon the province. The regulations also call for:[22]

(1) preference for Newfoundland labor, goods and services;
(2) compulsory training and research development programs in the province;
(3) the landing in the province of any oil and gas produced offshore;
(4) preference for the local refining, processing and consumption of any oil and gas found.

If Newfoundland should lose its case, it may, as stated by provincial Premier Frank Moores, not abide by the decision, but will attempt to "...secede from the Confederation if necessary..." and return only on Newfoundland's terms.[376]

The People's Republic of China and East/Southeast Asia. In mid-1978, East and Southeast Asian countries (excluding India and Asiatic USSR) produced an estimated 12.0% of the world's daily output of offshore oil (see Table 3). Intensive offshore exploration and development programs should help push the region toward a more important world position, especially after the PRC expands its marine programs and numerous oceanic territorial disputes that involve areas of important petroleum potential are settled.

The PRC Moves Offshore. Estimates of the PRC's total offshore petroleum reserves differ considerably. The varied estimates occur primarily because (1) many areas of the PRC's waters have been little explored, (2) what information is available has been released by China's government only in broad generalizations, and (3) interpretations of China's offshore territorial jurisdictions vary.

Geologists in the PRC estimate that more than 40% of China's land area and 20% of its continental shelf have potential for oil and gas bearing sedimentary rocks.[564] Based on a 1978 visit to China, former U.S. Department of Energy Secretary, Arthur Schlesinger, "estimated" the PRC's offshore reserves at 50 billion bbl., but how this figure was derived is unclear. The U.S. Central Intelligence Agency earlier estimated the PRC's reserves at approximately 39 billion bbl.[303] A study done in 1977 by A. A. Meyerhoff and J-O. Willums, using a

46 *Offshore petroleum*

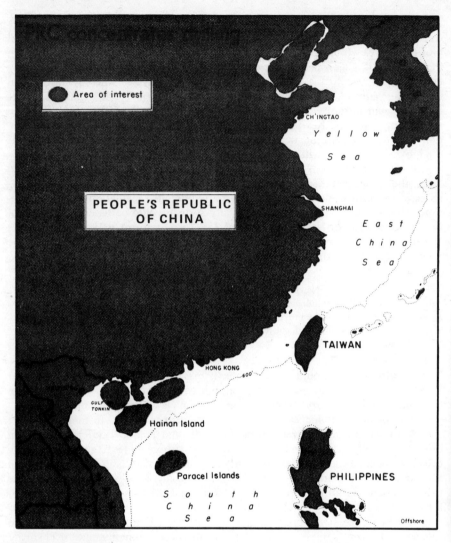

Fig. 14. Potentially oil rich areas in the offshore of the PRC. Prospects seem especially favorable in the Gulf of Tonkin, in the Liaotung Bay-Po Hai Gulf, and in the Paracel Islands area.
Source: [303]. With permission.

computer model along with data on the seismic structure and the total volume of sedimentary strata, put the PRC's offshore oil potential at 30 billion bbl.[348] Many geologists consider the Meyerhoff-Willums estimate conservative; their relatively low figure may derive, in part, from having excluded several politically sensitive areas from the data base. PRC geologists are said to put the reserves of the Po Hai Gulf alone at 37 to 75 billion bbl. (Fig. 14),[i] and a Japanese government source suggests a 112 billion bbl. reserve in the East China Sea within an undefined area centered on the Senkaku Islands (Tiao-yu T'ai).[224]

The PRC today produces 90% of its annual petroleum consumption,[406] but this favorable situation could change, considering the current Chinese drive to rapidly expand the country's industrial growth and to mechanize agriculture. The agricultural sector tripled its energy consumption from 1970 to 1979. Exploitation of offshore areas is, therefore, especially significant to China's economic development ambitions, because many onshore deposits will be depleted soon and in the past the major portion (approximately 90%) of her petroleum resources was produced in areas too near the Soviet Union and too remote from the industrial centers of the eastern seaboard.[88] Thus, the PRC is moving vigorously to develop its offshore potential.

Although eager to develop its offshore, China has attempted to follow a do-it-yourself program while selectively importing equipment and technical assistance, especially from Japan and the United States; the United States, however, has not always been cooperative. A case in point relates to the PRC's attempt to develop a domestic petroleum industry geophysical-research capability. Early in the 1970s, the Chinese tried to purchase specialized and sophisticated magnetometers (based on laser-maser technology) from American firms. The United States government disallowed the sales because of the magnetometers' potential military applications, including the measurement of missile launching pads' gravitational environment.[224] In recent years, with the shift in American policy toward the PRC, the Chinese have been doing an increasing amount of shopping in the United States. Vice Premier Teng Hsiao-p'ing, during his early February 1979 visit to the United States, noted that China hoped to rival the OPEC states in oil production, but it would need American technology to do so.[527] Teng's frank admission points out the difficulties the Chinese are having in developing their offshore petroleum fields, but they could be producing oil at a rate of 1 million bbl/d in 1985, if technical skills and equipment are adequately upgraded.

Teng's statement about technology imports officially announced what has been PRC's policy for some time. The purchases in 1977 of the Norwegian-built semisubmersible drilling platform *Borgny Dolphin*, reported to have cost $38 million, and in 1978 of an 800-ton, Japanese-built, crane, demonstrate China's commitment to expanding its offshore crude oil output, possibly by moving into deeper waters than are currently exploited. The *Borgny Dolphin* can work in 1,000-ft. waters, but to date has been used only in the relatively shallow Gulf of Tonkin.[90, 304] The PRC also recently purchased two drilling rigs from Japan and three from the United States, all contracted for delivery by mid-1980.[70] Using these units and earlier purchases as models, the Chinese have been building their own deep-water rigs.

Successful work in the relatively shallow waters of Po Hai Gulf, Liaotung Bay, the Gulf of Chihle, and Taiwan Strait[441] has encouraged the Chinese to consider plunging into deeper waters to the south adjacent to the Luichow Peninsula and Hainan Island (see Fig. 14). According to Leonard LeBlanc, "China has two reasons for concentrating activity in South China," namely: (1) to find a better quality crude oil to supplement the high-wax low gravity product found in the country's northern producing regions and (2) to assert her "...sovereignty near North Vietnam's sea boundary, which has never been negotiated, and to reinforce claims to a number of islands in the South China Sea,"[303] such as the Paracels (Hsi-sha) and the Spratlys (Nan-sha). To back up its claims in the south, the PRC recently has drilled 10 wildcat wells and 1 development well in the Gulf of Tonkin.[304]

Paracel and Spratly Islands Dispute. The South China Sea may soon become a significant petroleum source and some of its oil fields are already producing. Geologists believe many parts of the South China Sea have good petroleum potential, and several national governments seek control of the area.

Long before there was evidence of petroleum potential in the South China Sea, the PRC, North Vietnam, South Vietnam, Taiwan, and the Philippines had laid claim to several island groups and their surrounding seas.[198] But with the hint of the presence of petroleum, ownership claims became more intense.

In 1973, South Vietnam issued a decree that incorporated the Spratlys into its territory, even though the government of Taiwan, since 1946, had stationed troops on one of the islands, Itu Aba or Taiping. The PRC retaliated by issuing a decree of its own, stating that it would not tolerate any territorial infringement. On January 11, 1974, China followed up on her verbal pronouncement with an attack on a South Vietnamese military garrison stationed in the Paracel Islands, resulting in China's occupation of the entire Paracel Group, a military action dubbed the "100-minute war." During the brief but violent 2-day confrontation, both sides lost one ship, and the Chinese took at least 48 Vietnamese prisoners. The PRC moved quickly to expand military and radar installations and also constructed a building on the main island.[j]

South Vietnam's loss of the Paracels was a harbinger of events soon to come when in 1975 the Saigon government fell to North Vietnam's armies. It has been estimated that prior to its fall the Saigon government had spent $500 million exploring for petroleum in South Vietnamese waters. Work began in 1973 when 13 concessions were awarded to Shell Oil (U.S.) and to Mobil Oil. Two potentially commercial fields were discovered, but when the provisional government took over after Saigon's collapse, the wells were capped,[199] and 4 months later all previously established oil concessions were cancelled.[419] Selig S. Harrison, a well known journalist, has suggested that the possibility of major oil strikes and the intense exploration activity of the Saigon government could have helped precipitate Hanoi's final 1975 drive into South Vietnam. Hanoi may have felt that those successful strikes gave too much "credit-worthiness" to the Thieu regime.[224] After the fall of the Saigon government, the dispute over the Paracels did not subside, and Hanoi was determined to assert a reunited Vietnam's claims in the South China Sea. It thus proceeded to invade and occupy six islands in the Nanshan Archipelago, already claimed by the PRC. Consequently, Peking published broadsides castigating Hanoi's aggressions, while Hanoi was busy publishing a map showing various South China Sea islands as an integral part of Vietnam and issuing a postage stamp containing pictorial evidence of its historical claims to the Paracels. Soon after having ousted the South Vietnamese and establishing its own garrison in the Paracels, the PRC laid claim to the Nanshan Archipelago, lying 200 mi.

west of the Philippines. This claim was made even though the Philippines was in military control of five of the islands.^k In May 1976, the Philippines announced it had granted concessions in the Nanshan Archipelago (London Reefs, Tizard Bank, and Reed Bank); Vietnam, Taiwan, and the PRC vigorously protested Philippines' action.[478] But by the spring of 1977, exploration work had been done in several areas, and two strikes had been made in the South China Sea—an oil find immediately east of Palawan Island and a gas condensate discovery southeast of Taiwan.[306, 501]

Harrison has warned that the London Reefs and Tizard Bank are potentially explosive concession areas. Philippine President Ferdinand Marcos has said he will defend these and any other oil concessions up to 200 mi. off the Philippine coast. The government of the Philippines backed up this statement with the construction of a naval base on Palawan's west coast at Ulugan.[224] While the Philippinians were busy exploring in the Nanshan Archipelago, with the help of Swedish and United States companies, Hanoi was negotiating with Norwegian and United States firms to develop an expansion of the exploration program initiated by South Vietnam which, in addition to the two strikes, had provided other encouraging shows of petroleum deposits.[198] Obviously, none of these states is adhering to accepted geological and technological definitions for establishing international boundaries at sea. Because of the depth of the Palawan Trough and the China Basin (more than 9,000 ft.), a considerable portion of the South China Sea cannot be placed in the exclusive jurisdiction of the littoral states.[168]

If United States firms should become involved with the Hanoi government's exploration efforts, especially in the Gulf of Tonkin, it could create sensitive problems for Sino-American foreign relations. It is, however, unlikely the United States will allow its oil companies to jeopardize the new relationship established with Peking in December of 1978. This contention seems justified, considering that the United States government has previously discouraged companies from employing exploration vessels in the Yellow Sea in Taiwan Strait, and in the East China Sea. Recently, Petro-Vietnam, Hanoi's national oil agency, awarded three foreign-firm concessions—one each to West Germany, Canada, and Italy—for exploration between the Mekong Delta and the Spratlys. Fifteen wells are to be drilled between 1979-82. Industry specialists expect Vietnam's waters to surpass in importance those of Malaysia, but believe their productivity will be below those of Indonesia.[434]

As of 1979, the various South China Sea territorial claims were still at issue, and no settlement of the disputes was in sight. China does not want to negotiate its claims, because this would be an admission that the South China Sea islands are not an integral part of the PRC. North Vietnam and the Philippines have resisted recognizing the PRC's claims, and they continue to pursue offshore petroleum exploration programs in the South China Sea. In February 1979, two oil companies, Amoco and Cities Service, brought in four wells in Miocene reefs off the Philippines' Palawan Island; two other wells off Palawan were already producing. These new strikes cannot but add "fuel to the fire" as the three parties continue to lay claim to the area's numerous uninhabited islands, sand cays, reefs, and low-tide rock outcroppings.

Yellow Sea and East China Sea. The Yellow Sea and East China Sea have a significant potential for offshore oil-associated conflicts. Several states, including the PRC, South Korea, Japan, and Taiwan, have overlapping territorial claims. The PRC adheres exclusively to the geological principle of the continental shelf extension, whereas North Korea and Taiwan hold to a median line principle used in current Law-of-the-Sea negotiations. Japan and South Korea have further complicated matters by establishing a joint-development zone in the East China Sea, an agreement castigated by both the PRC and North Korea.[1] The PRC has frustrated South Korea's ambitions for developing the Yellow Sea's oil potential by refusing to negotiate with what it and North Korea consider a *puppet* government.

The entire issue of offshore boundaries and recognition of oil concessions in the East China Sea area are tied to recognition by the United States of the PRC as the only legal government of China and a withdrawal of recognition from Taipei.[224] When Secretary of State Henry Kissinger made his surprise visit to Peking in August 1971, the position of several oil companies with Taiwanese concessions in the East China Sea was made very uneasy. During the mid-1970s, three small oil companies with concessions in the area failed to fulfill their drilling commitments to Taiwan, primarily because of the uncertainties of United States' policies toward the PRC. Shifting policies of Washington toward Taiwan and the PRC caused Gulf Oil and Conoco to reevaluate in 1974-75 their Taiwan concessions, even though they had earlier made significant investments and had brought in substantial gas wells 60 mi. southwest of Taiwan. The United States government's overtures to the PRC and the wells' location on the mainland side of a 15-mi. undersea canyon (which may become the territorial dividing line between the PRC and Taiwan) caused the companies to decide against developing the field. Gulf has drilled four wells, all dry, in waters relatively near Taiwan which are free of dispute.[224] Gulf Oil has also been reluctant to push into areas where Japan and Taiwan have overlapping claims, because the company's financial stakes in Japan are much greater than in Taiwan.

The United States government has also overtly discouraged oil companies from doing exploratory work and drilling in waters claimed by both the PRC and Taiwan. Consequently, since 1976 Taiwan has been attempting to go it alone with its national petroleum firm, Chinese Petroleum Corp. But because of the political sensitivity of the area in which it wants to drill, near the Senkaku Islands (Tiao-yu T'ai), it has had great difficulty leasing drilling rigs. The United States government has, in fact, intervened when the Taiwan government and an American firm were about to consumate a deal for leasing a drilling rig for use within 35 mi. of the Senkaku Islands.[224]

It has been suggested that Peking and Taipei may be able to come to some kind of economic-political agreement on development of East China Sea oil. Taiwan could act as the developer of the deep water areas of the East China Sea while the PRC concentrates on inshore waters, with some mutual financial arrangement drawn up for their sharing of the oil revenues.[224] Given the recent normalization of relations between the United States and the PRC and the recognition of Peking as the representative of the Chinese people, this may be the only alternative open to Taiwan. Entrepreneurs have long pressured the United States government to help bring about such a reconciliation.

From the Gulf of Siam to the Tasman Sea: Southeast Asia's Archipelago Frontiers. Southeast Asian states are vigorously seeking to develop the potential petroleum riches of their coastal waters. In most instances, these states are dependent on western technology sources and development firms.

Cambodia, prior to its recent national devastation, had awarded as early as 1973 exploration drilling permits to a Hong Kong-based firm, Marine Associates, for work in the Gulf of Siam.[23] Oil exploration and production programs are under way 135 mi. offshore from the Malaysian port of Kuala Trengganu, on the Malay Peninsula,[173] and also farther to the southeast, immediately adjacent to Indonesian waters.[281] Malaysia also has gas fields located 80 to 172 mi. off Sarawak in 200-400 ft. of water. These fields, outlined by some 40 exploration wells, are associated with a $1-billion Petronas (the Malaysian state petroleum corporation) liquefied natural gas (LNG) plant that should be completed by 1983 at Bintulu on the central Sarawak coast. The gas will be piped ashore, processed and pumped aboard LNG tankers for shipment to Japan,[280] which has only two small offshore oil and gas fields—one discovered in 1971 at Aga, 9 mi. from the shore in Niigata Prefecture,[287] and another that began producing in 1976 in Akita Prefecture.[m]

Burma had exploration programs underway in 1975, in some waters up to 3,000 ft. deep; most operations were working in depths of no more than 600 ft.[282] Thailand has recently discovered two gas fields in the Gulf of Siam—one 250 mi. south-southeast and another 113 mi. to the southeast of Bangkok. The Thai government is making preliminary gas pipeline surveys for possible lines that would deliver the product from these fields (a distance of 383 mi.), to a landing at Sattahip and then on to Bangkok or alternatively to Songkhala in southern Thailand.[586] The World Bank is expected to lend Thailand $500 million for construction of the pipeline and shore facilities. The natural gas piped ashore should provide 25% of Thailand's energy needs.[355]

Indonesia had no offshore petroleum until 1969. In 1975, production had reached 258,000 bbl/d and by 1978 output topped 558,000 bbl/d, a 116% increase in only 3 years. As early as 1974, foreign concerns were operating 19 offshore drilling rigs in Indonesian waters, scattered from the eastern shore of Kalimantan to Java to Sumatra.[282] Yet, during the mid-1970s, her offshore petroleum industry did not develop as rapidly as it might have if investors had not been discouraged by a policy of Pertamina, the national oil administration of Indonesia, that required oil companies to reinvest in the country's oil and gas industry any profits earned from petroleum sales in Indonesia. In 1977, Pertamina set aside this reinvestment policy and instituted an intensive incentives program. Investment credits of 20% were established for all new exploration, as long as the Indonesian government receives at least "...49% of the cumulative equity in production over the life of the particular field." Previously, Pertamina allowed this cost-recovery sector of the production-sharing formula only for high-risk or high-cost exploration.[467] Between 1979 and 1983, Indonesia expects to invest $2.2 billion in its petroleum industry,[355] with a significant portion likely to go into the offshore sector.

Australia has been actively exploring its offshore areas. Foreign investors have recently been encouraged with Australia's lifting of a ban on farm-ins (to acquire an interest in a tract from the original lessee) and liberalizing of a past requirement that all petroleum developments have a minimum of 50% domestic-firm participation. In 1977-78, drillers discovered three promising oil and gas fields off Australia's southeastern coast in Bass Strait; two other fields have been operating in Bass Strait since 1970-71. Off the northwestern shelf, on the Rankin Trend in the Timor Sea, three major gas fields will be developed that, when fully operative, should produce an estimated combined output of 1.3 billion ft.^{3}/d. Australians will consume approximately 50% of this gas, with the remainder being shipped as LNG to Japan and the United States. Drilling companies in early 1979 were preparing to explore offshore Western Australia in waters 2,600 to 6,500 ft. deep on the Exmouth Plateau, lying to the west of the Rankin Trend.[355] Offshore exploratory work has also been undertaken recently in the Arafura Sea and off northern Queensland.

The New Zealand government hopes to have a $1 billion gas project, the Maui gas/condensate (liquid natural gas) field, onstream by late summer 1979. The field is located in the Taranaki Bight off North Island's southwest coast. Production from the Maui field should total 500 million ft.^{3}/d. of gas and 21,000 bbl/d of condensate, enough to provide a domestic natural gas surplus.[355]

Indian Ocean. Within the vast expanse of the continental shelves of the Indian Ocean, of Madagascar to Australia to India, geologists estimate there may be between 37 and 371 billion bbl. of oil and between 35 trillion and 1,253 trillion ft.^{3} of natural gas.[85] India hopes it may have a significant part of this potential reserve, because it is rapidly industrializing and has long had a poor petroleum production/consumption ratio. In 1973, for example, India's annual crude oil production was only about 59.5 million bbl., whereas its consumption was 164.3 million bbl. During the next decade of the 1980s, India would like to have discovered an initial recoverable oil reserve of 1 billion bbl. This amount would significantly help offset an anticipated oil consumption of 391 million bbl. annually by 1980. Thus, in the early 1970s, India reversed a long-standing policy of refusing exploration concessions to foreign firms.[256]

By 1973, India's new policy had already generated 13 participatory offshore exploration offers from foreign oil companies, including firms in Japan, Canada, France, and the United Kingdom,[256] and in 1974 a total of 40 firms had expressed an interest in India's offshore; two contracts had been signed, one for the Bengal Basin immediately south of Calcutta and the other for the Kutch Basin, bordering India's northwest coast. Collectively, the two contracts involved an investment of $23.5 million for exploration, beginning with seismic surveys.[258] Recently, however, India has been focusing most of its development efforts on the Bombay High field situated in relatively shallow water (250 ft.) in the Gulf of Cambay. Petroleum geologists expect the field to produce 80,000 bbl/d when it is fully operative.[2] Production began here in May 1976, and by the end of March 1977, some 154,000 bbl. of crude had been produced;[143] in June 1977, the World Bank lent India $150 million for further development of the Bombay High field.[150] Other capital sources funding the Bombay High include $50 million in private funds from financial interests in Hong Kong and from OPEC and the Japanese government.[355] India will be self-sufficient in petroleum within a few years, and should be able to remain so for several decades,[143] especially as it explores more of its 96,525 mi.² of continental shelf. Wildcatters have already pushed southward from the Bombay High, and were scheduled in 1979 to explore waters off the Mahanadi Basin along the coast of Kerala.

The Indians have worked out interesting contractual arrangements. India's Oil and Natural Gas Commission (ONGC) has divided the country's continental shelf into 10 administrative units (composed of about 1,100 mi.² each), delimited primarily by basins. Exclusive seismic exploration contracts are granted within these basins, and they provide for ONGC's participation in up to 10% of the venture if the firm makes a commercial discovery. The contract period is for 24-27 years. During the first 24 months, the firm must select an area of no more than 200 mi.² for drilling exploration. The area's remainder reverts to ONGC along with all data accumulated during the 2-year period. Finally, at the end of the third year, the exploration area must be reduced to 100 mi.². At the end of the 7th year, the contractor can claim only those areas that are actually producing.[258]

The contracts call for minimum expenditures at each stage of development. ONGC lays claim to all fixed assets and to all "...movable assets required permanently for operations...." Annual budgets and work programs are jointly approved and supervised by the contractor and ONGC, whereas the contractor must shoulder the entire cost of exploration and development, to be recovered from up to 40% of the oil production. The other 60% of production is shared (35:65 ratio) as profit, with ONGC taking the 65% share. The company pays no taxes on its 35% share and when its exploration and development costs have been recovered, the oil company's share is reduced to 25%. India has designed this contract so as to control "run-away profits" the companies might reap. Another important feature of the contract is that, until India is self-sufficient in oil, it will acquire (for domestic consumption), at prevailing world market prices, all of the oil produced.[258]

Sri Lanka in the late 1960s offered leases to numerous interested oil firms for exploration in waters adjacent to her oceanic boundary with India and the country has made a reciprocal agreement with India to exchange oil exploration data.[586] In 1977, the government of the Seychelles signed its first offshore oil exploration contract. The concessions cover 12,000 mi.² surrounding the islands.[587]

Africa. In the late 1970s, the greatest exploration and development activity in Africa centered along the continent's west-central coast, extending from Angola on the South to Nigeria on the north. There were 13 drilling rigs active in mid-1977 despite political unrest in several states, as in Angola, and threats of speeded-up nationalization, as in Nigeria.[107] By late 1978, a total of 50 operating fields (see Table 3) rimmed the western coastline, including Angola (5), Zaire (2), the Congo (2), Gabon (14), Cameroon (1), and Nigeria (26). Nigeria's fields are small, but collectively they make it by far the largest offshore producer in Africa. Nigeria pumps approximately 205,442 bbl/d, whereas Cameroon, with an output of only 10,000 bbl/d, is the smallest producer.[588] Although Gabon's offshore production did not begin until 1974, it produces 125,820 bbl/d, an output equalling 61% of Nigeria's production. The Loanga field 19 mi. off the coast of the Congo came onstream in 1977, after having been discovered in 1973.[587] In 1978, Angola struggled to reestablish its pre-civil war output of 130,000 bbl/d, and by mid-year, production had climbed back to 94,625 bbl/d.

In recent years, much exploration has been under way in other west coast areas. After more than 12 years of exploration, with only small shows of gas and oil, the Republic of South Africa was in 1977 still attempting to find an economically viable gas or oil field. Ghana, Benim, Ivory Coast, and Senegal also were actively seeking a paying field.[107]

In North Africa, Morocco, Algeria, Tunisia, and Libya, were attempting to explore their offshore areas but, in several cases, under a cloud of political uncertainty and international antagonism. Morocco and Spain have been negotiating the location of their Mediterranean and Atlantic boundaries, and Libya and Tunisia continue to bicker over long-standing boundary disputes, focusing especially on the Gulf of Gabes, a case both parties have agreed to put before the World Court. Libya recently rubbed salt into Tunisia's wounds by granting exploration concessions to Exxon for work in the vicinity of the disputed waters.[107]

Southern Europe. Although North Sea oil and gas discoveries hold the limelight in Europe, the Mediterranean is becoming an increasingly important area for petroleum production. In the past, the small size and known spread of encouraging geological structures on the continental shelves in the Mediterranean have tended to discourage exploration for petroleum. During the last decade, however, drillers have struck numerous, although small, deposits of natural gas and oil. Consequently, the pace of exploration is quickening as companies move additional drilling rigs into the region and into deeper waters where geological features are more encouraging.[305]

The Italian sector of the Adriatic Sea now supplies 60% of Italy's annual national oil production, and the Ionian Sea contributes some natural gas. Greece produces small amounts of gas in the northern Aegean Sea; Spain has producing fields southeast of Barcelona and in the eastern Bay of Biscay; France continues to seek hydrocarbons in its Bay of Biscay;[128, 136, 502] Portugal is also testing the Atlantic; Yugoslavia is probing its Adriatic waters; Turkey is exploring the Sea of Marmara (while delaying exploratory work in Greek-disputed Aegean waters); and Romania is drilling in the western Black Sea.[305]

Middle East. In 1977, the Middle East accounted for nearly 40% of the world's offshore oil output. Events during 1978-79 in the Middle East's Persian Gulf and Red Sea regions make it hazardous to speculate on likely offshore petroleum field developments and crude oil and gas output in the near future, but the overall recent trend has been one of rapid expansion (Table 4). Saudi Arabia leads the world in offshore crude oil production—22.9% of the total in 1977. Its output in 1977 was 2,621,400 bbl/d out of the country's offshore and onshore total for the year of 9,015,000 bbl/d,^{n} with the Safaniya field providing more than half of this total. The Saudi's nearest offshore rival in the Middle East is Abu Dhabi, whose output in 1977 was approximately 627,000 bbl/d. This figure represented a 40% increase over its 1973 output.[69] Egypt's offshore oil production has grown rapidly in recent years, increasing 205% during the 1973-77 period despite its difficulties with Israel and disruptions in the Gulf of Suez, the main area of production which centers on the July, Ramadan, and Morgan fields.[472] In 1976, Egyptian output was estimated at 231,120 bbl/d and one year later, the 1977 production had

Table 4. Growth in Offshore Crude Oil Production in Selected Middle East Countries, 1973-77 (thousands of bbl/d)

Country	1973	1974	1975	1976	1977	% change 1973-77
Saudi Arabia	1,990	2,025	1,386	1,695^a	2,621^a	32
Abu Dhabi	454	513	463	560	627	40
Iran	452	455	481	427	507	12
Egypt	131	147	165	231	399	205
Dubai	222	133	249	308	317^a	43
Qatar	n.a.	n.a.	n.a.	n.a.	244	n.a.

Source: [69, reprint].
Notes: ^{a}Estimate; n.a. = not available.

increased 72.7% for an output of 399,000 bbl/d. During the first half of 1978, output declined slightly to 388,000 bbl/d. Egypt also reported its 1978 offshore natural gas output at 180 million ft.^{3}/d. Qatar released data on its offshore production for the first time in 1977, reporting a total output of 244,340 bbl/d.[69] During the first half of 1978, Qatar was producing 451,190 bbl/d, an 84.7% increase from the previous year. Sharjah's production in early 1978 was 25,442 bbl/d (see Table 3). In late 1978, Abu Dhabi with five producing offshore fields was reported to have plans for the development of five new fields, and Dubai was expecting to develop an additional field, beginning in 1979.

The Persian Gulf region has not only great potential for future offshore petroleum production, but also considerable potential for conflicts among the various bordering Islamic states. A recent oil strike in the offshore bordering district of Sultanate of Oman and the Emirate of Ras al-Khaimah in the Horn area has created friction between these two parties.[455] Considering the fears of Saudi Arabia concerning the political turmoil in Iran under the Ayatollah Khomeini, the Persian Gulf could become the focus of bitter rivalries between Iran and Saudi Arabia. The political sensitivity of the region is reflected in oil workers' varying references to the region—those along the eastern shore speak of the Persian Gulf; those on the western shore speak of the Arabian Gulf.

Although the paramount role of the Middle East as the principal exporter of oil and dominant factor in OPEC's price manipulations is not the focus of this book, it should be noted that its offshore production, particularly that of Saudi Arabia, could experience a sharp decline in the early 1980s. To wit, Saudi's Oil Minister, Sheik Ahmad Zaki Yamani, threatened in late 1979 to reduce that country's petroleum output in an amount which may be a significant share of OPEC's possible drop of as much as 3 million bbl/d in 1980, as estimated by a U.S. interagency task force headed by DOE Deputy Secretary John C. Sawhill. The Saudis, who have held the price of their Arabian light crude at just $18.00 a barrel in 1979, while other OPEC members have broken through the cartel's ceiling price of $23.50, have in fact "subsidized" their Western clients. Their ability to earn more while producing less will probably be reflected in the Middle East's offshore oil statistics next year. Saudi Arabia and most other exporters of the region have realized that unmined black gold, offshore and onshore alike, will be worth more in the years to come than the inflation-ravaged dollars for which it is sold today.

Soviet Union. The relative importance of the Soviet Union's offshore hydrocarbon production has been declining recently. In 1977 offshore production accounted for only about 1.9% of the national petroleum output, whereas in 1970 it was 3.7%[29]—roughly 268 million metric tons less than in 1977.[o] Because the country yet has potentially large and untapped onland sources, it is anticipated that the Soviets will continue to proceed cautiously in developing offshore oil and gas resources. But, the offshore potential is there. The Soviet Union has a vast continental shelf area—more than 2.5 million mi.^{2}. Much of this shelf lies in the Arctic Basin, a likely source of major deposits. Soviet interest in the Arctic Ocean centers on the Barents and Kara Seas. The Kara Sea is especially open to optimism because of the nearby onshore Tyumen fields. It is of note, however, that some estimates view the costs of developing the offshore Arctic for exploitation to be likely 3 or 4 times greater than those of tapping the onshore fields of West Siberia.[142]

The first Soviet offshore programs were developed in the Caspian Sea, on the Neftianye Kamni (oil rocks) 25 mi. east of Azerbaijan's Apsheron Peninsula (Fig. 15). Production facilities were linked to shore by a causeway. The field's annual gas production totaled 264 billion ft.^{3} in 1975,[p] and the objective for 1985 is 424 billion ft.^{3},[466] roughly equal to the country's planned total (offshore and onland) of 432 billion ft.^{3} for 1980.[142] Whether this production goal can be achieved is debatable, given the failure of this field to meet previously announced targets. Oil production from the Neftyanye Kamni field in 1975 was 28,120,000 bbl. Two nearby and good prospects (the Neftyanye Kamni 2 and the 28th April fields) are currently being examined by petroleum geologists. Recently, drillers have shifted their primary efforts to an area extending 20 mi. offshore of Cape Sangachaly to the southwest of the Apsheron Peninsula in the vicinity of Bulla Island and Duvanny Island. In 1975 the field's oil production totaled approximately 33,300,000 bbl. A potential giant gas-condensate field (Andreyev Bank), lying to the southeast of Bulla-More, could add to the region's future output. The 1985 target for the Caspian Sea's overall annual crude oil output is approximately 160 million bbl.[142] Because of the civil strife in Iran during late 1978 and 1979, and in light of Ayatollah Khomeini's abbrogation of the Soviet-Iranian Defense Agreement of 1921, important gas supplies exported by pipeline to Azerbaijan were shut off. The mid-winter interruption in early 1979 created considerable hardship for the republic, and it is likely the Soviet government and energy planners will be doubly motivated to proceed rapidly in developing the Caspian's petroleum potential.

In the early 1970s, the Soviets began exploring the Black Sea. They have now brought three gas fields onstream—one on the Golitsnaya Uplift, about 50 mi. west of the western tip of the Crimean Peninsula, and two in the Sea of Azov, immediately east of the Crimean Peninsula. The Black Sea is a likely place for

Exploration and development programs 57

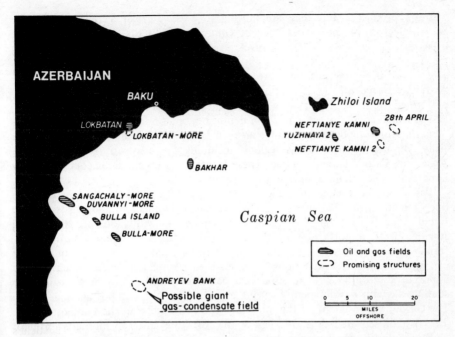

Fig. 15. Caspian Sea oil and gas production in the Apsheron Peninsula region of Azerbaijan, 1978.
Source: [29]. With permission.

the Soviets to push offshore development, because the water is relatively shallow and there is a ready market nearby in the Ukraine. On the other hand, a shortage of adequate pipeline-laying barges will probably slow development in the area.[440]

The Russians, Poles, and East Germans have recently formed a joint exploration group, Petrobalt, to examine the southeastern Baltic Sea for hydrocarbons,[440] and in the mid-1970s the Soviet government collaborated with Japan in a successful oil exploration project off northeastern Sakhalin (4 mi. east of Okha) in the Okhotsk Sea, a region some geologists feel will become a major producer. The Japanese, whose dependence on distant imports from the Persian Gulf is a growing threat to their economy,[q] invested $100 million and provided $42.5 million in export credits for the equipment used; they are slated to receive a portion of the oil output,[64] which Western sources have reported as presently having a flow of 6,000 to 7,000 bbl/d. This strike could prove especially important to the Soviets because their Far East consumes a much larger amount of petroleum than it produces and because of their tense relationship with the PRC. The Soviet government has not said much about its Sakhalin find, perhaps because of the heavy foreign capital investment[29] and the use of a Norwegian drilling rig in the discovery effort.[529]

It is apparent that with the Soviet Union's dependence on foreign equipment and technology that their offshore drilling industry is somewhat behind other offshore producers, but they are moving to close the gap. Perhaps one of the best indicators of the Soviets' recent commitment to the development of the offshore is a purchase in 1976 from Armco of their first semisubmersible drilling rig. The unit was expected to be put to work in the Caspian Sea where drillers sorely need it as they push into ever deeper waters off the Apsheron Peninsula.[440] As relatively recently as 1975, 90% of the Russians' subsea wildcatting had to be performed from fixed platforms because they had only four mobile drilling rigs, two of which could not work in water depths beyond 66 ft.[466]

Latin America. Several offshore areas in Latin America are emerging as potentially important petroliferous regions, including the coasts of Brazil, Uruguay, Argentina, Colombia, and Mexico. These frontier regions chould become especially important as petroleum sources for the Western world.

Based on the theory that the continents of Africa and South America were once joined, many Brazilian geologists believe truly major oil deposits will be found along Brazil's east coast, where it matches up with fields off Nigeria and Angola. Brazil imports 80% of the oil it uses, and its President Marques Neto, former head of the state oil monopoly Petrobas, has vigorously promoted the offshore search. His convictions seem justified. Wildcatters recently have made important discoveries along the Brazilian coast from the mouth of the Amazon River southward, with development focusing especially on the Campos district.[225, 426] The Garoupa field is one of the most important producers in the Campos district. The field, discovered in 1974, lies 50 mi. offshore and 155 mi. east northeast of Rio de Janeiro, and is expected to produce 200,000 bbl/d when fully onstream (Fig. 16).[248]

Other South American offshore areas that are already producing or with petroleum potential include Uruguay, where two dry holes had been drilled by 1977. The Peruvians recently granted leases for exploration, and Venezuela, already a large offshore producer, is making plans for exploratory work. However, its most recent new discovery of vast onland reserves, reported to be the world's largest in November 1979, may prompt a shift in Venezuela's exploratory input. Less endowed Colombia has been developing fields near the Venezuelan border, which created some conflict between the two states.[12]

Geologists consider Argentina's offshore one of the most potentially petroleum-rich areas in the world. To date, it remains one of the least explored areas, even though preliminary work was done by the Lamont Institute and the Argentine Navy more than 20 years ago. As of mid-1978, a total of 33 wells had been drilled, but only one produced oil. Argentina's coastal infrastructure is ideal for the support of a major offshore petroleum industry, but a high inflation rate coupled with endemic political unrest may preclude significant developments for the near future. The Argentine government lays claim to the entire continental shelf, and this stance has created political disputes with the United Kingdom over the Falkland Islands, lying 300 mi. offshore. In 1975, when the British brought in an exploration rig, the Argentine government protested, even though Britain has had a settled population in the Falklands since 1765.[306] A long-standing territorial controversy with Chile in the Tierra del Fuego region, considered an excellent prospect for petroleum, also poses a dilemma for entrepreneurs and the Argentina government.

Fig. 16. Oil fields off Sergipe, Alagoas, Rio de Janeiro.
Source: [225]. With permission.

Venture capital for many portions of the Argentine offshore may be in short supply until these disputes are settled.[405]

One of the world's most recent and potentially important petroleum discoveries is in the southern reaches of the Gulf of Mexico in the Bay of Campeche, lying off the coasts of Mexico's states of Veracruz and Tabasco. Geologists and officials of Petroleos Mexicanos, Mexico's national oil corporation, claim that the Campeche Trend may extend at least 200 mi., perhaps more, into the Gulf of Mexico. Proven reserves are placed at 40 billion bbl. Some estimates put the Campeche Trend's potential reserves as high as 200 billion bbl.[347] Once the United States and Mexico have reconciled recent political frictions, this petroleum-rich region will likely assume an increasingly important role in supplying energy-hungry United States markets that lie within pipeline distance.[329]

SUMMARY AND CONCLUSIONS

The world's continental margin region will become increasingly important as the supplier of crude oil and natural gas, and as relatively accessible near shore/ continental-shelf reserves deplete, drillers will push their efforts into ever deeper seabed zones, perhaps even onto the outer continental slope.

During the next several decades, as petroleum energy shortages become even more acute and aggravated by concomitant price increases, tension among states for the black gold of the continental shelves will intensify, especially in frontier areas. Many sensitive oil-related political problems will continue to exist in various corners of the world's oceans, but they will probably go unheard in the World Court, as in the East and South China Seas disputes. Interested observers should keep a good focus on waters shared by the PRC and Vietnam. Given the early 1979 fratricidal conflict between these two Communist countries, a potentially explosive situation exists in the Gulf of Tonkin where both parties are pushing ahead with exploration programs. Disagreements between the PRC and Taiwan over the East China Sea bear watching, as do those between Argentina and Great Britain, concerning South Atlantic waters lying between Argentina and the Falkland Islands.

Although frictions may continue between the United States and Canada over the delimitation of territorial waters in the Northeast Channel, separating Nova Scotia and New England, no threat exists for major antagonism. Some potential for disagreement over territorial claims in the Beaufort Sea off Alaska and the Yukon Territory, is not to be discounted, but it should pose no threat to continued congenial relations between the two neighboring countries. The same cannot be said for relations between Ottawa and its provinces. Citizens of Newfoundland have shown grave concern for what they feel is an effort by the national government to preempt its provincial resource heritage. It is possible this controversy between Newfoundland, now before the Canadian Supreme Court, and Quebec's refusal to place its case before the courts, could contribute to a further weakening of the Confederation.

As to the more immediate future, while exploration and development programs for offshore oil are in evidence in late 1979 throughout the non-Communist world, the political future of Iran and of most Arab members of OPEC appears to mortgage all reasonable forecasts for the early 1980s.

NOTES

[a] L. Weeks provided a detailed discussion of this topic [569]:953-64.

[b] Proved reserves include those discovered oil and gas deposits that can be produced under existing technological and economic conditions.

[c] Inferred reserves include deposits not yet proved but expected to be present in already known fields; undiscovered recoverable reserves are those expected to be present in yet unknown fields.

[d] Antarctica's continental shelf will be difficult to exploit because of very thick pack ice and large icebergs. Seismic surveying from ships would be nearly impossible and hazardous. Icebergs scour the bottom so that wellheads would have to be put below the seabed surface. See [592].

[e] For a detailed look at costs, investments, and projected production in the offshore, see [36].

[f] Useful exploratory work also has been done under the auspices of the U.S.' national Deep Sea Drilling Project (DSDP) program. As of March 1977, 418 holes had been drilled. Drilling has been done through sediments 1 mi. thick in water depths as great as 3.1 mi. Drillers have not, however, penetrated very far into the continental rises. They fear a major oil strike could not be contained and would pose an environmental hazard. The drill ship does not carry the

expensive equipment needed to control the oil flow that might occur. Despite the DSDP program, the gas and oil potential of both the continental rises and marginal basins must be inferred from experience with onshore oil fields and known geological processes. For a map showing the worldwide distribution of the DSDP holes, see [170]:5.

gFor details of the U.S. vs. Maine case, see [522] and [544].

hFor legal opinions favoring this view, see [317] and [524]; for an unfavorable view of Newfoundland's position, see [291].

iDetailed discussion of the Po Hai Gulf is provided in [220].

jSee [419] for a thorough examination of the teritorial controversies associated with the Paracels and other South China Sea islands.

kIf the 12-mi. territorial principle and a 200-mi. economic zone concept were applied to the PRC's claims, the South China Sea would become a *mare clausum* [478].

lThe PRC is currently selling Japan some oil. Many Japanese oil experts feel the PRC is selling oil to Japan merely to "cool" any interest in developing the country's offshore areas until the PRC can establish a 200-mi. economic zone in the East China Sea [224].

mA third discovery is potentially commercial; it lies 5 mi. off the Honshu coast, near Joban in water 500 ft. deep, making it presently uneconomic to exploit.

nCentral Intelligence Agency's statistitians reported the Saudi Arabian total (offshore and onshore) crude oil production to decline in 1978 to 8,065,000 bbl/d and rise again, though not above the 1977 level, to 8,990,000 bbl/d. during the first 6 months of 1979.

oThe most recently published crude oil production statistics establish the Soviet Union's total output at 353 million metric tons in 1970; 546 million in 1977; 572 million in 1978; and 640 million planned for 1980. See [142].

pThis compares with the Soviet Union's offshore and onshore total for 1975 of 289 billion ft.^{3} which is estimated to have increased to 372 billion ft.^{3} in 1978. See [142].

qFor example, during the third quarter of 1979, Japan paid 64% more for imported petroleum products than in the corresponding period of 1978.

Chapter 4

The North Sea

The emergence of the North Sea and adjacent Atlantic waters as a major oil and gas producing region is the most important petroleum industry development to occur during the decade of the 1970s. Without North Sea petroleum, the political and economic consequences of the Arab oil embargo initiated in 1973 and of OPEC's subsequent price increases may have been more severe. The Arab embargo and the North Sea area's political stability stimulated the sea's littoral states and world oil companies to rapidly explore and develop the region's oil prospects, even though the numerous major discoveries of the early 1970s were also responsible for the effort. One estimate indicates that, by the early 1980s, 20% of Western Europe's oil requirements will be met by North Sea producers.[488] The Organization for Economic Cooperation and Development has made a more conservative estimate, projected only for the EEC, that is probably closer to reality. It contends that only about 10% of the EEC states' needs can be met by North Sea production in 1985.[397, a] On the other hand, Peter Odell and Kenneth Rosing of Netherlands' Erasmus University have developed a simulation study of the long-term production-consumption ratio that shows the North Sea could produce 75% of Western Europe's annual petroleum consumption sometime between 1982 and 1983. Production is projected to remain above the 75% consumption threshold until between 1995 and 1996 (Fig. 17). Their work has been criticized as probably overly optimistic. Where does the truth lie? This question is hard to answer until the fields are more precisely known. But assuredly, some increases above current reserve estimates can be expected, because of the likelihood of improved recovery rates and new discoveries.[323]

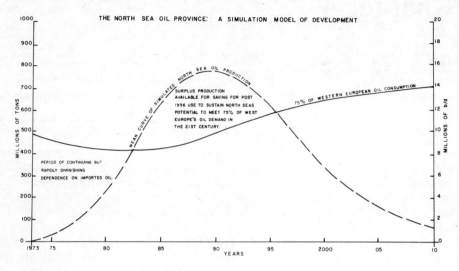

Fig. 17. Simulation model of North Sea oil province development, 1973-2010, as compared with 75% of the future demand for oil in Western Europe.
Source: [401]. With permission.

SHARING THE BOUNTY

As the rudiments of the North Sea's geological structure became known in the late 1950s and early 1960s, an intense interest developed in the area's petroleum potential, with an ensuing scramble to establish territorial sectors. A willingness to enter bilateral agreements and accept the median-line principle established by the 1958 GCCS helped Denmark, Norway, the United Kingdom, and the Netherlands to agree on mutually acceptable North Sea boundaries. West Germany, not a party to the GCCS, was determined not to be left out of the division of potentially prime North Sea waters. Consequently, West Germany submitted its case to the World Court. In 1969, the Court issued a decision granting West Germany 13,900 mi.^{2} of North Sea waters which, after small adjustments were negotiated with Denmark and the Netherlands, gave her an area 50% greater than she would have received under the median-line principle. Under the bilateral agreements, the United Kingdom acquired 46% of the total North Sea area (95,300 mi.^{2}) and Norway received 25% (51,200 mi.^{2}), with the Netherlands, Denmark, West Germany, Belgium, and France dividing the remaining 29%. If the seven states sharing the North Sea had had a more detailed knowledge of the area's geology, it is likely the bilateral agreements worked out in a relatively short period during 1965-66 would have been more difficult to negotiate.[145, 326, 578]

NORTH SEA OIL AND GAS: PRESENT AND FUTURE

The first North Sea hydrocarbon discoveries were gas deposits in the United Kingdom's and the Netherlands' near offshore. As petroleum geologists pushed their search northward, they made additional major gas and oil finds. These discoveries now extend from the Netherlands' coast to the 62°N. parallel, between the Shetland Islands and Norway (Fig. 18).

In 1973, the North Sea's proven reserves totaled an estimated 12 billion bbl. of oil and 65 trillion ft.^{3} of natural gas. By 1977, the United Kingdom alone estimated its crude oil reserves at between 16 and 30 billion bbl. In 1978, the Netherlands had 9.3 trillion ft.^{3} of gas reserves;[b] Denmark claimed gas reserves of 2.47 to 2.83 trillion ft.^{3} and an oil reserve of 300 million bbl. The Norwegian government placed its national reserves at a conservative 22.8 trillion ft.^{3} for gas and 4.9 billion bbl. for oil. Petroleum companies gave a more optimistic estimate of Norway's reserves—25.8 to 27.6 trillion ft.^{3} for gas and 5.9 to 6.3 billion bbl. for oil. But many in industry are convinced that the truly major North Sea finds have been made and that increases in output will come primarily from extending production in fields now under development. In mid-1978, North Sea crude was flowing at 1.5 million bbl/d—1.1 million came from the United Kingdom; Norway produced 400,000 bbl/d; and Denmark pumped 10,000 bbl/d.[171] The North Sea's most active production and exploration area today is the northern half, shared by Norway and the United Kingdom and extending from Ekofisk on the south to Magnus on the North. Of the 45 major oil fields operating or under appraisal north of 56° latitude, Shell/Esso, British Petroleum, and Conoco each control 5, Texaco and Phillips 4 each, and Hamilton 3. Fourteen other companies are working the remaining 19 fields. Two major oil fields also produce gas (Ekofisk and Brent); Statfjord will begin gas production in the mid-1980s. A few fields are important primarily for gas—Frigg, East Frigg, Heimdal, Sleipner, and Odin.

Exploration

Exploratory success of the various littoral states of the North Sea and the adjacent Atlantic has varied considerably. Oil companies are now looking at waters west of the Shetland Islands and Ireland and to the Celtic, Irish, and Norwegian Seas. These frontier areas may become increasingly important for oil and gas production. In anticipation of this potential. Eire and the United Kingdom have been at odds over the delimitation of the Irish Sea's territorial limits; they also disagree about the limits of their control in waters west of the Hebrides in the vicinity of Rockall.

Intensive exploration on Eire's continental shelf has been under way since the early 1970s. But despite the shelf's large area (217,000 mi.^{2}), drillers have had only limited success. Marathon, finally confirmed in August 1973, a commercial gas strike, the Kinsale Head field, 30 mi. southeast of Cork in the Celtic Sea. The field's estimated reserve was put at 1 trillion ft.^{3}. Delivery of 15 million ft.^{3}/d to the Irish shore began in late 1978, and is expected to reach 125 million ft.^{3}/d sometime in 1979.[331]

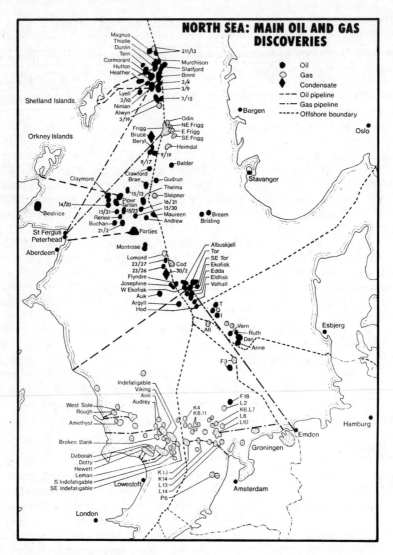

Fig. 18. North Sea: main oil and gas discoveries. Gas production predominates in much of the southern North Sea, whereas the far north is primarily concerned with oil production. Some discoveries (designated by numbers and letters) have not been named. A gas pipeline connects the Ekofish field with Emden, West Germany, on the continent, and an oil pipeline joins Ekofisk with the United Kingdom, at Teesside. Two gas and four oil pipelines connect the British sector's fields in the north with the Shetland and Orkney Islands and the main island.

Source: [35]. With permission.

The government of Eire also hopes to find oil on its continental shelf. In 1974, Esso made two small oil discoveries. These strikes raised the hopes of oilmen, and by June 1975 the Irish government had issued 23 licenses and granted 70 firms nonexclusive exploration rights.[62] In 1978, Marathon, Shell, Elf, Deminex, and Phillips were each operating a rig in Irish waters.[363] The drillers produced mostly dry holes, but one small noncommercial well drilled by Phillips, located 100 mi. west of the mouth of the Shannon River, was brought in at 730 bbl/d, once again reviving Eire's hopes for a major strike.[558]

The United Kingdom has explored portions of its Celtic Sea for gas and oil, but commercial quantities have not been found, resulting in a recent lowering of the United Kingdom's estimated offshore oil reserves. Department of Energy officials reduced the top oil reserve estimate from 32.85 billion bbl. to 30.7 billion bbl. and its low estimate from 21.9 billion bbl. to 16.8 billion bbl.[537] The United Kingdom has found commercial quantities of gas in the Irish Sea 30 mi. off the Lancashire coast.[263]

Extensive exploration is under way in waters west of the Shetland Islands, but success has been limited. Only one strike (in Block 206/8) has occurred. British Petroleum (BP) brought the well in during the summer of 1977. A much heralded second well drilled during 1978 in the same block showed no commercial quantities of crude.[481] In addition to BP's find, wells drilled in the area by other companies (Phillips, Esso, Elf, Mobil, and Texaco) have shown the presence of oil and gas.[364] Thus, many oilmen remain convinced that the area has good potential for major reserves (Fig. 19) and some feel that geological conditions in the entire territory between Great Britain and Iceland and Greenland are favorable for petroleum deposits.[58] Geologists believe reserves in the area west of the Shetlands could measure 4-13 billion bbl., although the recovery rate might reach only 20% because the oil is very heavy and lies in a thin and widespread structure.[26] The commercial prospects of this frontier area have seemed good enough to investors that the value of BP's stock increased by $540 million in the months after its strike.[481]

The Netherlands and Norway have had considerable success in finding gas off their shores. As of early 1978, the Netherlands had declared 11 of its 51 gas discoveries as commercial fields; it also had 1 small oil field. West Germany has not been so fortunate, drilling more wells than Norway between 1964 and 1973 without success of finding discoveries of commercial quality. Since then, one yet undeveloped gas and oil field has been brought in near the juncture of the sectors of the Netherlands, United Kingdom, and Denmark.

Norway, with a much smaller petroleum demand than West Germany, has had an exploration success ratio of 1:5. Perhaps as a consequence of this initial success, only 6-7% of Norway's North Sea territory has been explored. Its problem has not been in finding oil and gas, but in choosing a market for them. Surplus gas has been an especially difficult problem, because domestic demand is not there, and also because a deep trench lying off the Norwegian coast presents significant engineering problems for pipeline construction.[179] High prices and relatively good accessibility have encouraged construction of a pipeline from the Ekofisk field southward to the continent. A similar dilemma exists for the more recently discovered Statfjord field. Its gas production may eventually be piped across the

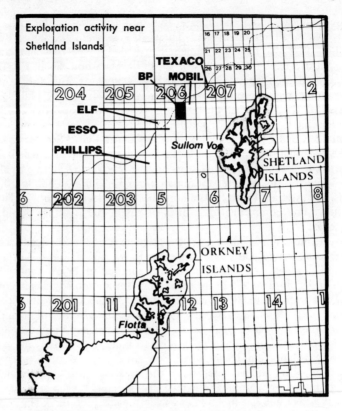

Fig. 19. Exploration activity near the Shetland Islands. Several oil firms have initiated exploratory work in a region 30–50 mi. west of the Shetlands.
Source: [364]. With permission.

Norwegian Trench to Sotra, near Bergen. A study completed in the fall of 1978 showed the technical feasibility of a Statfjord-Sotra pipeline, and final recommendations for crossing the trench were scheduled for 1979.[350] Producers have also given serious thought to tying the field in with British trunklines or, perhaps, even pumping the gas south to Denmark or West Germany.[435]

Investments and Entrepreneurship

As of late 1978, the North Sea had 100 to 120 production platforms in operation, and another 50 should be in place by 1990. The United Kingdom in 1978 had 17 oil and 7 gas fields onstream or under development; Norway had 9; the Netherlands 2; and Denmark only 1.[138, 529] These projects have required vast capital inputs from oilmen and other entrepreneurs. Some oilmen were skeptical

Table 5. Time and Costs Requirements for Finding and Developing an Oilfield in the North Sea[a]

Exploration, construction and production	Time span (years)	Direct employment (hundreds of men)	Capital investment	Operating expenditure
			(millions of 1973 U.S. $)	
Exploration[b]	2-6[c]	2-4	24-147	
Construction[d]	5-6[c]	10-20	612	
Production	16-20[e]	3-4	122-245[f]	612-735

Source: [42].

Notes: [a]Field size, 1 billion bbl. of recoverable reserves and 250,000 bbl/d capacity; [b]2,000-4,000 km of geophysical surveys, 5-30 exploration-appraisal wells; [c]Overall range for exploration and construction, 7-12 years; [d]Planning, design, production well drilling, production and transport facilities; [e]Includes build-up, 3-5 years; plateau, 5 years; and decline, 8-10 years; [f]Dependent on type of secondary recovery scheme or other necessary operations.

when in 1972 predictions were made that development of North Sea oil and gas fields up to the year 1990 would demand an investment of $100 billion. The subsequent Arab oil embargo of 1973 and inflating oil production costs soon made this figure seem reasonable. By the mid-1970s, the costs of developing an oil drilling and production complex seemed to approximate nearly $1 billion[273] (Table 5). To illustrate, the cost of developing the Forties field in 1970 was estimated to require £300-400 million ($720-960 million); in 1976 this figure had reached an equivalent of £800 million ($1.44 billion),[179] and in 1979 the cost was still climbing. In some fields, development costs have become so high that they are now marginal producers facing difficulties competing with oil fields elsewhere in the world. Rates of return on investments during 1978 in various fields dropped precipitously from the 1977 level. For example, in the British sector, Union's Heather field's rate of return plummeted from 22.1% to 6.0%; BP's Buchan's declined from 22.7% to 7.7%; and Shell/Esso's Cormorant's dropped from 12.3% to 6.2%. In the Norwegian portion of the Frigg field, the return was down from 10.9% to 6.3%.[388] In 1975, estimated oil field investment costs per daily barrel of capacity in North Sea fields were about $4,000. In offshore Africa, the cost was $480-$720, while in the Mideast it was only $240.[261] Limited geological knowledge, difficult climatic conditions, and relatively deep waters help make North Sea development efforts take longer. For example, 7 to 12 years may be required in the North Sea to bring a field into full production (the Forties field took nearly a decade), whereas the Gulf of Mexico normally requires only 3 to 4 years.[42] Drilling an exploratory well in the North Sea may require 3 to 4 months; in the offshore of the United States, the same well might take only 4 weeks. Time is critical, when one considers the costs of equipment use and labor. In the mid-1970s, daily operational costs for a drill ship ranged from £6,000

($13,320) to £8,000 ($17,760); a jack-up from £10,000 ($22,200) to £12,000 ($26,640); and a semi-submersible from £15,000 ($33,300) to £25,000 ($55,500). Forty to 60% of the charge is for use of equipment and the rest is allocated for labor, materials, the use of supply boats and helicopters, and other support functions.[323]

Ekofisk

The North Sea's first major crude oil and gas discovery lies in Norwegian waters immediately north of the juncture of the other littoral states' sectors (see Fig. 10). The field, dubbed Ekofisk by its joint operators, Statoil (the national petroleum agency for Norway), and the Phillips Petroleum Group, is the epitome of the modern oil field complex (Fig. 20). After its discovery in 1969, exploratory drilling had, by 1970, confirmed Ekofisk as a truly major field. Current estimates place recoverable reserves at 3.7 billion bbl.[54]

The Ekofisk complex sits on an upper Cretaceous limestone reef lying 10,000 ft. below sea level, and the main field covers about 36 mi.^{2} Most geologists consider the Ekofisk formation a "...conventional anticlinal trap—a large flattened dome dipping away at the edges—covered with a shale band and what oilmen call gumbo, a glutinous clay to make a perfect seal."[49]** The Ekofisk complex consists of six fields other than Ekofisk itself. West Ekofisk and Cod came onstream in 1977, Tor and Albuskjell in 1978, Eldfisk in 1979, and Edda is scheduled for production in 1980.[54] By 1977, Ekofisk already had 30 oil wells onstream with a collective crude oil output of 325,000 bbl/d (in comparison, the entire state of Oklahoma had 60,000 wells producing a total of only 450,000 bbl/d).[264] Daily production in early 1978 from the West Ekofisk field alone reached 80,000 bbl/d of oil and 400 million ft.^{3}/d of natural gas, and in its entirety Ekofisk produced 400,000 bbl/d of oil and 800 million ft.^{3}/d of gas.[586] In the early 1980s, when all seven Ekofisk-area fields are onstream, crude oil output is expected to peak at 575,000 to 625,000 bbl/d. The Phillips Petroleum Group reported in late 1978 that its total investment in Ekofisk had reached $5.5 billion.[537]

Part of this investment has gone toward creating a home away from home for oil field workers. As in several of the other major North Sea oil fields, many of Ekofisk's 1,000 workers are provided living quarters and various amenities at the worksite, including a 7-story, hotel-like building situated on its own platform that has 106 two-bed bedrooms, a dining room, coffee shop, chapel, cinema, hospital, and recreational facilities. In addition, a ship maintained in the area, the *Sedco/Phillips SS*, provides living quarters for 150 workers and facilities for diving work; the ship also serves as a fire station and hospital.[247]

NATIONAL POLICIES

The several states with North Sea petroleum industries have significantly different development policies. Denmark, for example, in 1963 gave one group (Dansk Undergrunds Consortium) headed by a shipbuilder Arnold P. Moller a complete

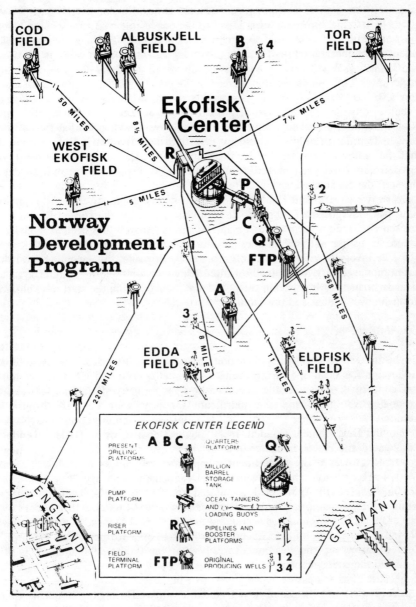

Fig. 20. Cartogram of the Ekofisk complex. The Ekofisk field was the first major oil discovery in the North Sea. Numerous platforms, a 1-million-bbl. storage tank, processing facilities, offshore loading buoys, and pipelines make up this $5.5 billion complex.
Source: [425].

monopoly of offshore petroleum concessions. The concession is valid until the year 2012. In 1972, a "protocol" was added to the concession strictures whereby Moller must attempt to "...see that Denmark benefits directly to the maximum possible extent from continental shelf development" by utilizing Danish workers, platforms, and refineries. West Germany has heavily subsidized one domestic group (Deminex) made up of 7 oil companies that merged in the late 1960s. The German government now also has a large interest in a merged corporation composed of Gelsenberg and Veba; it controls 51% of Gelsenberg and 43% of Veba, in effect making the corporation a national oil agency. Aside from the relatively poor potential for hydrocarbons in Denmark's and West Germany's North Sea sectors, David B. Keto has noted that the policy of severely limiting the number of participants in the national quest for petroleum has probably contributed to the slow pace of exploration experienced in these states' waters.[285, 578]

Norway's government has limited its concessions to a relatively small number of consortia, composed of many small companies, whereas the United Kingdom has granted concessions to numerous major oil firms. Norway has been very conservative in investing national monies in the oil industry. The government does not becomes involved with ownership until a strike is made, thus avoiding the risk of losing money in dry holes. In the United Kingdom, on the other hand, the government is financially involved in the petroleum industry from the start of exploration through development and production.[110]

United Kingdom

The history of the United Kingdom government's interest in oil resources and production has been one of increasing involvement. Prior to 1970, the government was concerned only with the manufactured gas industry. Relations between the national government and the petroleum industry were handled through the Admiralty, a government Economic Development Committee, and the National Economics Development Council. In 1970, the Department of Trade and Industry was created and placed in charge of all activities dealing with petroleum production and trade. However, the Department's inept management led to the establishment in 1974 of a new and separate agency, the Department of Energy.[403]

Use Policies. In 1956, coal accounted for 84.9% of the United Kingdom's primary fuel consumption while oil's share was only 14.6%. By 1966, coal's share had dropped to 58.7% and oil accounted for 37.5%.[28] The trend continued, and by 1974 Britian was spending £3.5 billion a year to pay for petroleum imports.[124] Recently, however, the government has been attempting to strengthen the position of coal in the energy economy.

Because of an increasing dependence on oil in the mid-1960s, the United Kingdom sought to maximize the use of its petroleum resources. During the administration of Sir Alex Douglas-Home in 1964, passed the *Continental Shelf Act*, whereby all ownership of North Sea waters was vested in the United Kingdom. The government offered exploration licenses for 960 blocks of 100 mi.^{2} each. Other rounds of licensing followed in quick succession—round 2 in 1965-66, round 3 in 1970, and round 4 in 1970-72.[567] Thus has the government sought self-sufficiency for the country through rapid petroleum resource development, a policy not accepted by many United Kingdom citizens, but one which has taken the country from a position of no crude oil production in 1974 to a spectacular growth of output from 86 million bbl. in 1976, to 280 million in 1977 and 326.5 million in 1978.^{c} The policy paid off rather handsomely when a near 60% self-sufficiency in oil was attained in 1978. A few years earlier, MacKay and Mackay argued that rapid development was probably wise because future petroleum prices might weaken. On the other hand, while supporting growth, they also pointed out that strategic considerations and the need to develop a domestic offshore technology might call for slower development.[323] Although government officials maintain that exploitation of the North Sea's resources must be designed for the maximum public good, especially in Scotland and other regions that have experienced the most difficulties from an antiquated industrial structure,[551] many Scots resent the hurried development pace and consider control from Westminster as an usurpation of Scotland's resource heritage. The Scots' discontent with the current situation was reflected, in part, in the favorable devolution referendum vote cast in early March 1979 for greater home rule. But those defending the national government's control of petroleum resources note that Scotland, under the *Act of Union* in 1707, relinquished sovereignty over offshore territories.[124]

In 1974, Peter Odell estimated that by 1984 crude oil output in the United Kingdom's sector might reach 300 to 500 million bbl. yearly.[124] A 1974 government report forecast that production for 1980 would be 750 million to 1 billion bbl. Although many in industry and government considered this estimate highly optimistic,[551] the government in early 1978 was still predicting a 1980 production total of approximately 750 million bbl. annually. Yearly production in February 1978 was rated at about 400 million bbl., with five major producing fields: Beryl, Brent, Claymore, Forties, and Piper.[56] Other small operating fields included Argyll, Auk, and Montrose. Additional fields under development at the time, have now been completed or are nearing completion—South Cormorant, Dunlin, Heather, Ninian, Statfjord (U.K.), Thistle, Buchan, and Murchison. Awaiting final approval for development in 1978 were BP's Magnus field, Shell's Fulmar field, Mesa's Beatrice field, and the Phillips Group's Maureen field. By mid-1979, platforms were under construction for these new fields, all of which are needed if the United Kingdom is to reach and maintain self-sufficiency in petroleum. The United Kingdom's Offshore Operators Association (UKOOA) contends that, although autarky may be reached by 1980, to remain so after 1995, some 60–95 wildcats/year will be needed during the 1980s and early 1990s.[357]

Although the initial North Sea petroleum development policy of the United Kingdom was one of "full speed ahead," it recently became somewhat more conservative, perhaps because the breakneck pace to gain self-sufficiency has been criticized as contributing to resource waste. Inadequate pipeline facilities, for example, have made gas flaring a common practice in many fields. The government is no longer allowing flaring on many platforms. And while Anthony Benn, Minister of Energy, is attempting to "persuade" the oil companies to comply with the new nonflaring policy, force cannot always be used because some of the original leasing agreements stated that the producers could not be made to reduce production. In

1977, however, the Piper Group sought the government's permission to increase production. The government took advantage of this request to tell the Group that production could be increased only if it consented to a workable program for conserving the field's gas. The government's ultimate objective is to tie all North Sea fields to a pipe trunkline network.[39] Some resource geographers argue that both oil and gas are too precious to burn, even onshore, and where possible, should be husbanded for use primarily by petrochemical industries.[87] Several platforms have had to shut down until gas injection systems could be installed, as Shell's Brent B platform. The Brent B field will soon pipe its gas ashore to St. Fergus on Scotland's northeast coast. Some fields—Auk, Argyll, and Montrose—have little gas. Others with much gas, such as Ninian, Thistle, and Piper, have had to plan for its use. Consequently, Piper has now been connected by a 16-in. gas pipeline to the Frigg pipeline at its midway pumping platform. The dry gas is sold to the British Gas Croporation, and two gas separation plants at Flotta terminal in the Orkneys handle Piper's liquid gas output. A major part of Thistle's gas production will be removed from the oil at a facility at Sullom Voe in the Shetlands.

In addition to the gas flaring issue, producers in 1978 faced a new worry that the national government might impose production limits, a move that could jeopardize their adequate return on investments and ability to meet loan payments.[171] As of late 1978, the United Kingdom's Department of Energy was considering a proposal to place a ceiling on North Sea oil production, not to be implemented until the country reaches self-sufficiency in 1980 when 2.25 million to 2.50 million bbl/d could be produced. The current output target for 1980 is 2 million bbl/d. This policy should bear no surprises for industry, because as early as 1974 Energy Secretary Eric Varley was considering a curb on output after 1982. He qualified his statements, however, by noting that any field found after 1975 would be allowed to accrue a 150% return on investment before cuts would be made.[124]

Fiscal and Leasing Policies. Recently, oil companies have been less than enthusiastic about exploring for new petroleum fields. Wildcatters drilled only 39 wells in 1978, the smallest number since 1974.[510] A survey of UKOOA, made by the Department of Energy to determine why there was a 35% drop in exploration and appraisal drilling in the United Kingdom during 1978 from the preceding year, found that a newly instituted "unattractive" fiscal regime was the most important factor.[181, 386] What fiscal policies could cause this drastic downturn, and why were these policies implemented?

Based on evidence of the tremendous response to an experimental tender bidding (£37 million—$89.5 million) on several tracts in the fourth round leasing sale, many government officials began to feel that oil companies reaped huge profits and probably remitted overseas more than one-half of their after-tax money. Estimates place such annual remittances at as much as £1 billion ($2.42 billion). The government, therefore, argues that it is merely attempting to establish a more equitable tax and royalty system, to secure better balance-of-payments relationships, and to assure greater public control of the industry. To achieve these goals, the government in 1975: (1) implemented a Petroleum Revenue Tax on oil firms' profits from the continental shelf; (2) closed a major tax loophole, which allowed oil companies to write off taxes paid in the Middle East; (3) moved to extend its control to both production and pipelines and to institute licensing policies allowing the national government to acquire ownership (while sharing in exploration and development costs, including those already incurred) of a certain portion of the profits or the oil produced; and (4) established a British National Oil Corporation (BNOC).

BNOC began operating in 1975, and by January 1977 the British government had acquired firm participation agreements with Gulf, Conoco, Tricentrol, and Ranger. Tentative agreements had been reached with BP, Shell, Esso, Chevron, Deminex, Texaco, Mobile, Texas Eastern, Amerada, Hess, ICI and Santa Fe, Murphy-Odeco, the Occidental Group, and the Union North Sea Group.[320] By mid-1978, nearly all companies had agreed to the government's participation program. Under the program, BNOC has the option of buying up to a 51% partnership in all leases granted during the first through the fourth round of licensing. It has automatically received 51% of all subsequent leases, which had reached the seventh round in late 1978.[363, d, e]

During October 1978, further policy changes had been or were soon to be realized when the government introduced legislation to (1) increase the Petroleum Revenue Tax (PRT) from a rate of 45% to one of 60%, an effort UKOOA feels is a breach of faith because of government assurances given in 1975; (2) reduce from 75% down to 35% the write-off allowed against the PRT for capital expenditures for production commencement or increases; and (3) lower the volume of oil allowed free of the PRT from 7,450,000 bbl. per field each year to only 3,665,000 bbl. The latter policy is designed to obtain as much tax as possible from large fields while encouraging the development and operation of small fields.[525, 538, f] Indeed, the petroleum industry has become an important component of the British economy. J. Dickson Mabon, Minister of State of the Department of Energy, estimated that in 1977 the United Kingdom's oil industry contributed over 3% of the GNP, some $5.1 billion in mid-1977 prices.[320]

When the sixth round of bidding on oil blocks was held in late 1978, the government had established a new leasing package and an overall average tax increase of 5% (rising from 70% to 75%) on new licenses, including royalties, the corporation tax, and the PRT. The government hopes these changes will bring in an added $700 million/year by the mid-1980s, but some in industry feel the effect will be to diminish overall revenues, as companies refuse to participate under the new program. Esso refused to join in the sixth round licensing, but 100 companies did participate, seeking to acquire some of the 46 blocks placed on bids, the smallest number ever offered in a British sale.[172]

On December 18, 1978, the Department of Energy established a variable royalty rate (a full waiver to 12.5%), depending on the economics of a given field. This policy was instituted to make small marginal fields economically viable, and it seems to have borne fruit: three development programs of the type the Department of Energy had in mind (Shell/Esso's North Cormorant, Phillips Group's Maureen, and British Petroleum's Magnus) are in process as of mid-1979.[560, g]

BNOC helps administer all of this legislation. Many oilmen feel that BNOC puts many roadblocks in their way while they are trying to produce oil and gas. BNOC could argue, as Mabon has pointed out, that it is merely attempting to see that oil companies operating in United Kingdom waters have (1) technical competence; (2) ability to find funds to carry out exploration programs and pursue development if a strike is made; and (3) willingness to contribute to the overall development of the British economy, a favorable balance of payments, and the employment of British trade union workers.[320] The government has, in fact, already assured that British trade union workers and the economy will benefit, by requiring that all crude produced in the United Kingdom's sector be landed in the British Isles before being exported.[255, h]

Norway

Oil producers began exploratory drilling in Norwegian waters during 1966, and a total of 30 dry holes had been drilled before oil was discovered in 1969 at Ekofisk. By mid-1975, 125 wells had been drilled in the 91 concessions granted by the Norwegian Royal Ministry of Industry. These concessions covered an area of 11,208 mi.^{2}, or 21.9% of the country's North Sea territorial waters. A total of 22 petroleum strikes had been made, including the giant fields of Frigg, Ekofisk, and Statfjord.[54, 497] Annual crude oil production rose from 2.3 million bbl. in 1971 to approximately 130.1 million bbl. in 1978, of which 92.2%[54, 240] was exported (Table 6).

The Petroleum Department of Norway's Den Norske Creditbank estimated in 1975 that total investments in the oil industry and oil-related activities between 1963-74 amounted to $4.5 billion. This sum was equal to about 60% of the Norwegian government's total national budget for 1975. Some entrepreneurs estimated that from 1975 to 1980, investments of up to $14 billion would be needed for Norway's North Sea petroleum industry, an amount equal to 62% of the country's GNP for 1975. Such large capital demands make it increasingly difficult for oil companies to maintain a large equity share in oil fields. They simply cannot generate adequate capital internally.[497] Consequently, other entrepreneurial sources are being drawn into the investment structure. Large investments are being made in crude oil and gas exploration (seismic surveys and drilling), production, and transport (pipelines and terminals). A significant volume of funds is also entering Norway's petrochemical industries, and bankers expected a 3 billion Norwegian kroner ($564 million) investment in petrochemicals during the 1975-80 period (Table 7). Norway now employs pver 300,000 workers in North Sea oil production, support, and administrative industries (exceeding employment in her fishing industry), and her investments in oil are nearly as much as in all other Norwegian industries combined.[563]

From the beginning of its involvement with oil production, Norway's government has been concerned about the negative impact the petroleum industry might have on its overall economy and populace. Its basic precept, when compared with the policy of the United Kingdom, is "caution." Governmental economists and industrial planners feel that the country's small population (4.1 million), long oriented toward agriculture, fishing, and shipping, could neither immediately consume its petroleum nor cope with the rapid influx of oil-industry capital. Not all Norwegians agree with this perspective and many in government feel the present capital flow into Norway's economy is inadequate.

Table 6. Norway's Crude Oil Production, Exports, and Wells Drilled, 1971-78

Year	Production	Exports	No. of wells drilled
	(millions of bbl.)		
1971	2.3	1.8	17
1972	12.2	12.6	16
1973	11.9	11.5	26
1974	12.8	14.9	38
1975	69.3	58.3	50
1976	103.5	102.2	31
1977	101.6[a]	101.7	54
1978	130.1[a]	119.9	69

Sources: [54], [240], [350].
Note: [a]Including natural gas sales totaling 2.5 billion m^3 in 1977 and 12.8 billion in 1978.

Cautious demanding leasing policies seem to have slowed the production growth rate in Norway's petroleum economy during 1976 and 1977, although 1978 saw a significant increase in output over 1977, as new gas deposits came onstream. The rate of increase in government petroleum revenues declined from 1976 to 1978, although net petroleum revenues grew by 103%. Within the petroleum industry,

Table 7. Expected Investments Associated with Norwegian Oil Activities, 1975-80

Area of investment	Approximate investment (billions)	
	Norwegian kroner	U.S. $[a]
Seismic surveys	0.140	0.026
Drilling of exploratory wells	3.000	0.564
Field development[b]	40.000	7.518
Drilling platforms[c]	5.000	0.940
Petrochemical plants	3.000	0.564
Miscellaneous	3.000	0.564
Total	54.140	10.177

Source: [261].
Notes: [a]Author's calculations based on an average 1975-78 Norwegian kroner value of 5.32 to 1.00 U.S. $; [b]Including pipelines and terminals; [c]Investments in excess of those already ordered and financed.

Table 8. Taxes and Levies on Petroleum Production and Fixed Capital Formation

Year	Taxes and levies on petroleum production, based on currency values for a given year (millions)		Gross fixed capital formation for oil and gas production, drilling, and pipelines, based on 1978 prices (millions)	
	Kroner	U.S. $[a]	Kroner	U.S. $[b]
1970	–	–	313	59.7
1971	–	–	704	134.4
1972	–	–	1,237	236.1
1973	69	12.0	3,108	593.1
1974	121	21.9	5,615	1,071.6
1975	209	40.1	7,615	1,459.4
1976	1,958	358.9	10,812	2,063.4
1977	3,126	587.4	14,327	2,734.2
1978	3,820*	728.9	9,625	1,836.8

Sources: [384], *[240] by pers. corresp. in mid-1979.
Notes: Author's calculations [a]based on kroner and dollar exchange rates for a given year; [b]based on a 1978 kroner value of 5.241 to 1.00 U.S. $.

gross fixed capital formation grew steadily until 1977, then declined sharply, nearly 33% in 1978 (Table 8). And annual government revenue surpluses (based on all Norwegian industry) experienced up through 1974 were replaced in 1975 by deficiencies that reached 8.2 billion kroner ($1.57 billion) in 1978.[583] In 1977, because petroleum royalties did not keep pace with expectations, Norway experienced a balance of payments deficit.[285] However, expanded exports of oil, gas, and drilling platforms in 1978 helped turn this situation around.[583] For several years, some international economists have advised Norway's government to step up its oil production so that it can benefit while petroleum is still in demand, because before the end of this century other energy resources will be "...more economical than oil."[584, i] Surely, these economists are taking the short view in policy advice, and must not have considered petroleum as a raw material in petrochemical industries.

Initial responsibility for developing a national oil policy in Norway was in the hands of civil servants and financial and legal experts. Then, as it became clear that Norway was the owner of great potential petroleum wealth, numerous "interested" parties became involved; the result, according to the U.S. Office of International Energy Affairs, has been a somewhat poorly defined national energy development plan,[403] a rather amusing judgment, considering the disarray of the United States' energy policies.

Ownership Structure. The government's goal for the petroleum industry is to meet problems associated with two basic assumptions: (1) foreign enterprises will seek to explore Norway's oil and gas resources, and (2) Norway should benefit from the exploitation of these resources while keeping control in its hands. Thus, foreign interests became deeply involved in Norway's petroleum industry, though within a governmental participation framework. The operators include (1) Phillips Petroleum Group, a United States dominated enterprise; (2) SAGA Petroleum, a multi-ownership organization composed of about 80 companies, all Norwegian except for one firm, that are joined into 4 large consortia; (3) Petronord, a group of mainly French interests, with 20% of the ownership vested in Norsk-Hydro; (4) Norsk-Hydro, an independent company, which is 51% owned by the national government; and (5) Den Norske Stats Oljeselskap (Statoil), Norway's national oil company.

The Norwegian Storting created the state-owned Statoil in late 1972 under the *Norwegian Companies' Act*. When Statoil first began operating in early 1973, it had only two employees. As of mid-1979, Statoil employed between 500 and 600[54, 358] people and the government had invested in it between $170 and $340 million.[403]

Like any other company, Statoil pays taxes and royalties to the national government. It has financial interests in a new Norwegian petrochemical complex, in a retailing chain, and in transportation facilities such as the oil pipeline connecting the Ekofisk field with Teesside[392] and the gas pipeline tying Ekofisk to Emden.[403] Immediately prior to the Phillips Group's start of construction on the Teesside line, the Norwegian government demanded that Statoil be made a 50% partner.[392] Statoil's main interest, however, is directly in the crude oil production sector. Any commercial oil strike results in the establishment of a joint venture composed of the private company and Statoil. Some Norwegians are beginning to worry about Statoil's power which, with the full development of the Statfjord field, will accrue vast amounts of money, and many feel present legislative control of Statoil is not stringent enough. In theory, the Storting sets policy for Statoil, but some observers of the Norwegian oil industry feel that the national oil "giant" will soon unduly control the government.[358]

Use Policies. As noted earlier, the Norwegian government has moved cautiously in developing its petroleum resources by opening only a limited area of its North Sea waters to exploration and development. This is nowhere better demonstrated than in the government's policy of not allowing oil drilling north of the 62°N. parallel. Several factors account for Norway's reluctance to move north of this point. Foremost are desires to (1) conserve a precious national resource; (2) restrain inflation in the national economy; (3) reduce the social effects associated with shifts in the industrial structure, as when farmers or fishermen abandon their rural locations to cluster in coastal nodes to work in oil-related industries such as platform construction; (4) maintain a relatively unpolluted environment, especially in association with the fishing industry of the western fjords and skerries; (5) allay any fears the Soviets may have about the use of platforms near their Arctic Ocean waters and shipping lanes as intelligence gathering points or as helicopter landings in moving NATO troops to Norway.[358] Recently, the Norwegian government has been under heavy pressure by industry and labor to open waters north of 62° to the drillers bit.[377] This pressure is especially strong because of delays in developing fields south of the 62nd parallel and because increasing production costs have led the government to curb domestic oil consumption and to reduce public spending, as in road construction. Oil-earning estimates for the period 1977-81 have been cut recently from 142 billion kroner ($27 billion) to 119.5 billion kroner ($22.8 billion).[176] Given this financial situation, along with the government's newly acquired seismic knowledge of the area (obtained in investigative programs since 1975) and a solution of anticipated environmental problems, the government may open areas north of the 62nd parallel in 1980, with Statoil as the primary developer. Government oilmen are already considering a stepped-up timetable for leasing and production in waters south of the 62nd parallel.[171] Many Norwegians would like to see their North Sea holdings soon generate $10 billion annually (at 1979 prices). This huge capital flow could bring the Norwegian people better roads and schools, improved welfare services, and reduced taxes, which in 1975 took 40% of the average paycheck.[316]

Leasing Policies. As in the United Kingdom, Norway uses discretionary licensing. The government does not auction licenses, but rather "...awards licenses with a specific, negotiated work program for exploration." The holder of concessionary rights has control of a given block (approximately 200 mi.^{2}) for 6 years. The oil company may explore for and produce petroleum during this period; it then must request an extension of these rights for another 30-year period, but 50% of the total block area needs to be returned to the national government at the time of the application for lease extensions. The state has thus managed to acquire blocks of prime potential,[403] without incurring initial exploration costs.

Like the United Kingdom, Norway's leasing and taxing policies recently have become more stringent, although Norway's regulations have been more rigorous from the start than those of the United Kingdom. In 1975, the government proposed a 90% tax on excess profits of the oil companies, but the excess profits definition had not yet been defined, and some oil companies threatened withdrawal from the area if the tax were instituted.[403] Industry has since continued to remind Norway's government that oil entrepreneurs need assurance that their investments are safe and that they can operate only in a stable economic situation.[387] Further pressure was put on producers in late 1978 when applicants for concessions were required to submit detailed statements outlining their equipment purchases made from Norway.[176, j] This policy implies that preferential weight is given to those applicants trading with Norwegian suppliers.

United Kingdom and Norway: Policy Implications Compared

Both proponents and opponents of the national petroleum industry policies of Norway and the United Kingdom are sometimes tempted to argue their case by pointing to the policies of the other. At least one author cogently identified significant differences in the situations of the two states that made it unwise for them to seek parallel policies in 1974. He pointed out that, comparatively, the United Kingdom's area is small (94,194 mi.^{2}) relative to its 1974 population of 55 million, is highly industrialized, and has a large balance of payments deficit—£2 billion ($4.68 billion) annually for oil alone in 1974. Norway, on the other hand, is large (125,032 mi.^{2}) in relation to its 4.1 million people, has a scattered and more rural population, employs many people in agriculture and fishing, and has a small balance of trade deficit. In addition, Britain's imports of petroleum are much greater than Norway's (more than 10 times in 1974).[270] Britain still depends heavily on domestic coal supplies and has considerable nuclear power production, whereas Norway has neither of these energy sources. Norway extracts nearly all of its hydrocarbons for export, while Britain produces mostly for a domestic market. The two economies also have different capacities to absorb the capital inflow generated by the petroleum industry; Norway must be cautious of a large inflow. Finally, Britain is afflicted by chronic unemployment in some regions, whereas Norway in certain regions has a labor force inadequate to meet the demands of a burgeoning oil industry.

ONSHORE IMPACTS OF OFFSHORE PETROLEUM

Preceding discussions touched on several regional and local impacts of offshore oil and gas production on the governments and peoples of the North Sea's littoral states. Some of these impacts (opportunities and difficulties) merit closer examination. The main problems focus on industrial space needs vs. ecological viability and aesthetics; stable community life vs. rapid social change; and locally based and financed economies dependent on indigenous labor vs. international consortia having ties to foreign capital, markets, and labor.

Scotland

Scotland exemplifies the various benefits and problems associated with a burgeoning offshore oil industry. The economy of many areas has been dramatically stimulated, but difficulties have also emerged, including social, economic, and environmental problems.

Socioeconomic Impacts. Coastal areas with established service industries (harborage, airports, hotels, freight handling and storage, engineering facilities, and technical schools), with large numbers of skilled seamen, and good proximity to offshore petroleum fields have experienced severe pressures from abrupt influxes of population and new businesses, such as heliports and the supply and servicing of drilling equipment (Fig. 21). In 1974, the Secretary of State for Scotland announced that for the first time in 40 years Scotland had a net in-migration of population.[124]

The movement of people into northeastern Scotland vividly illustrates the social and economic problems of the North Sea oil boom. Some citizens are not happy with the presence of the oil industry here and elsewhere in Scotland. For example, children and adults from the United States, England, and even from other areas of Scotland have had some difficulty being assimilated into communities, additional schools have had to be built, and some increase in prostitution has occurred. By far the most important problem, however, is housing, which has become difficult to obtain and much more expensive. In the Cromarty Firth area, pressures on housing and labor were for a period so great that many small firms

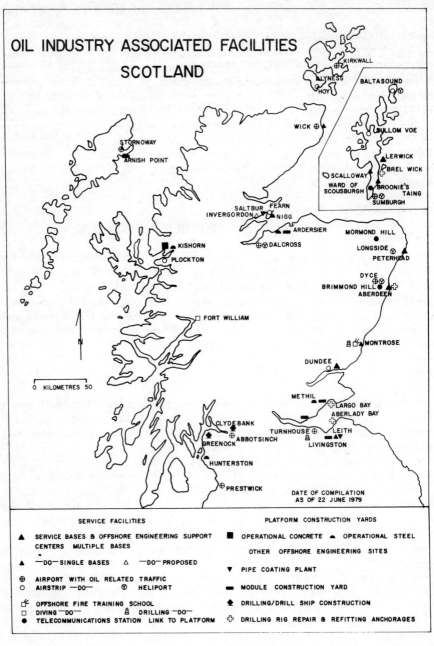

Fig. 21. Onshore engineering, service bases, and airports associated with the North Sea oil industry.

Sources: [491]; redrafted on the basis of updated information supplied in mid-1979 by [235]. With permission.

that might have located there were frightened off.[30, 255] By 1979, though, a modest housing surplus had developed in some areas as a result of the rush to supply the market. In addition, government agencies have also stepped in to build dwellings for key workers in oil-related businesses. In the past, because housing was so expensive, many needed additional "spin off" personnel, such as teachers, social workers, planners, and construction workers, were unable to come to some of the most rapidly growing oil centers, as in Aberdeen, dubbed by some with the fancied title "Europe's Offshore Capital." Once a relatively quiet university and fishing town, Aberdeen now bustles with oil-related activity—as early as 1975, some 4,000 people were employed in more than 200 oil-associated businesses.[30] During the spring of 1979, the estimate was 26,000 jobs in 400 companies which consider themselves to be 100% oil related.[390] Best situated of the major ports to service the early oil industry developments in Scottish waters, Aberdeen has emerged as the main North Sea drilling company headquarters and supply base. To illustrate, of the 29 rigs operating in the United Kingdom's sector in 1975, 21 were serviced from Aberdeen. Lerwick, in the Shetlands, also services the more distant fields to the far northeast of Scotland.[490, 491] and Lyness, in the Orkneys, supplies some of the needs of drilling rig operators active to the northwest of the Orkneys.

With the development of the petroleum industry came considerable disruptions of traditional employment patterns. Scottish industries, such as textiles, fishing, agriculture, food processing, forestry, and boatbuilding, have been most adversely affected, especially in small communities like Peterhead, Kishorn, Shetland, Orkney, and Lewis. Not only have new workers entered these small communities, but local employees in long-established industries have been drained off to work in oil-related industries. Women have left relatively low-paying jobs in food and textile industries for hotel, catering, and other expanding service businesses. Younger girls leaving school, who might otherwise have gone into the textile and food industries, tend to take secretarial jobs with the oil companies. On the other hand, labor demands in oil and oil-related industries have not always kept pace with declines in traditional Scottish industries. West-central Scotland alone experienced a net loss of 41,000 manufacturing jobs between 1970 and 1973.[491] It is possible that the oil industry's competition for labor may have helped the demise of some of these businesses.

Based on tabulation methods used by Scotland's Department of Industry, Smith, Hogg, and Hutcheson showed in 1976 that overall employment (direct and indirect) in Scotland's oil industry was between 59,300 and 67,900 workers. Of this total, 44,200 to 45,800 were workers classified as directly employed by the petroleum industry; the remaining 15,100 to 22,100 were categorized as "spin-offs," or indirectly employed.[491]

Numerous oil-related service activities have developed recently. Notable examples are a $3.5 million deep-sea diving training center established at Fort William[233] and drilling technology schools at Livingston and Montrose. A newly formed Petroleum Industry Training Board offers courses, at various locations, in rig fire fighting and rig safety. Several universities and technical institutes now provide advanced study in engineering and petroleum geology.[491] Catering services also provide employment for many workers who keep the offshore oil field personnel supplied with food and beverages. Helicopters are used extensively to ferry workers and supplies to and from platforms.[k] The diversity of oil industry services can be illustrated by a 1974 listing of almost 3,000 oil-related offshore supply companies.[l]

Spatial-environmental Impacts. Frictions have developed between the oil companies and fishermen in the Shetland Islands, especially in the shellfish industry. Preparation for pipelaying work for the Brent field aroused the fears of fishermen that associated pollution would destroy their livelihood. Consequently, the Shetland County Council refused permission to oil companies for a floating base off Shetland to service the Brent pipeline project.[255] Problems have also surfaced between the oil industry and fishermen in Aberdeen. Several new oil industry service bases were constructed in the harbor while a long-needed refurbishment of Aberdeen's fishing wharves was delayed, a situation that has helped create resentment among fishermen.[30]

A major problem of the oil industry, of small communities, and of those concerned with the environment relates to the need for supply centers and platform construction yards (steel and concrete) that require large labor supplies, plenty of level land, and good access to building materials and to the sea. Platform construction is highly labor intensive, for a single platform may need up to 2,000 men over a 2-year period. Steel platform construction demands many skilled welders. Concrete platforms, weighing as much as 250,000 tons, may require 15,000 tons of reinforcing steel, 25,000 tons of cement, and 100,000 tons of gravel and sand.[30] The need for sand and stone in the construction of concrete platforms and other facilities, such as harbors, has put stress on local areas' aggregate sources, forcing many communities to take protective measures, as at Ninian's Isle tombolo and the Lang Ayre and Laward beaches, south of Lerwick in the Shetlands.[255] The quest for sand and gravel in constructing harbor facilities at Peterhead on the coast, northeast of Aberdeen, has been pursued despite the potential for shoreline erosion and damage to ecosystems.[30]

Platform construction yards should also be located as close as possible to the platform installation sites, because their bulky mass makes towing a slow process and an expensive and dangerous task[200] (Fig. 22). While under tow, high waves and wind may put a platform under stress, causing it to fracture. Although not ideally suited for proximity to North Sea oil fields, the northwest and west of Scotland have deep and protected nearshore waters that are attractive to platform construction firms and to engineering companies testing diving equipment. Several areas in the region stand out as platform construction sites: Loch Kishorn, Loch Fyne, and the Firth of Clyde where concrete platforms have been built at Kishorn and at Ardyne Point. Another yard at Portavadie, on Loch Fyne, failed to generate orders while one at Hunterston, on the Clyde Coast, is now to build a steel structure. Major module construction yards operate at Leith and Bumtesland on the Firth of Forth, but yards at Clydebank and Dumbarton on the Firth of Clyde have been closed. As of mid 1979, a construction yard at Arnish Point on the Isle of Lewis was converting a drilling rig to function as a production unit, and Kishorn and Hunterston yards are sharing work on a gravity-type steel platform (see Fig. 21).

Fig. 22. The 13,000-ton Brent "A" platform jacket was towed from its construction site at Methil, Scotland. Supported on two barges in a catamaran arrangement, the 500-ft.-high jacket was moved about 3 mi. offshore where it was released from the barges. Then, under its own buoyancy, tugs towed the jacket to Shell/Esso's Brent field 112 mi. northeast of Shetland.

One prime site that does not have a construction yard is Drumbuie, a small village located near the mouth of Loch Carron, a few miles from Kishorn. Local community residents and aesthetically and conservation-minded individuals throughout the United Kingdom fought a vigorous legal and political battle to save the area from the noise, dirt, and socioeconomic disruption that would have occurred had the Secretary of State for Scotland not ruled in August 1974 to deny planning permission to two firms desiring to build concrete production platforms on lands and in waters near the village.[m] The village's population of 24, its crofting economy, and its traditional patterns of life would have been profoundly disturbed with the influx of the some 600 workers needed to operate the facility. The Drumbuie case was a significant victory for opponents of indiscriminate land and water use, because the permission applicants had direct and strong support from Scotland's

Highlands and Islands Development Board and indirect assistance from the Conservative Party in the House of Commons. The Conservative Party attempted to persuade the Commons to pass a bill that would have allowed the government to nationalize urgently needed prime platform construction sites. The various political parties recognized the bill as directed primarily at Drumbuie, an effort especially resented by the Scots as an usurpation of their long-held planning prerogatives.[n]

Other important needs of the offshore oil industry are landfalls for pipelines and sites for tanker terminals. England has four landfalls, including Teesside, which services the Ekofisk complex, and Easington and Theddlethorpe on the Humber Estuary and Bacton in East Anglia, all three of which service various gas fields in the southern North Sea. Scotland also has four major pipeline landfalls—St. Fergus and Cruden Bay on the mainland, Flotta in the Orkneys, and Sullom Voe in the Shetlands.

The United Kingdom hopes to have Sullom Voe handle one-half of the country's daily crude oil production. When completed in 1981, the Sullom Voe Association port facility will have a crude oil handling capacity of 1.4 million bbl/d, an oil refinery, and a LPG processing plant. Construction at Sullom Voe began in 1976, with building costs originally estimated at $500 million; workers employed were expected to number 1,200 during the anticipated 2-year construction period. Before the project is finished, construction costs will have totaled $1.6 billion, and 6,000 workers will have been employed during the 5-year construction program.[518] The facility covers 220 acres and provides harborage for support vessels and tugs, handling 200,000 dwt tankers at four separate berths.

The Sullom Voe Association, composed of the Shetland County Council and oil companies with interests in the Brent, Cormorant, Dunlin, Thistle, and Hutton fields, jointly planned the facility. The Shetland County Council insisted that all port and support facilities be located at one site in order to reduce the environmental impact. The Council's effort seems to have produced a fairly even balance of social and environmental costs and economic benefits. Based on Parliamentary legislation passed in 1974, the Council holds 50% of the shares in the Association and has the right to levy a tax on each barrel of oil passing through the port,[30] a significant source of revenue. In early 1979, however, Shetland residents were angered by a poorly handled bunker fuel spill and also by independent shippers' dumping of ballast water into the nearby sea. These derelict shippers tarnished the image of the Sullom Voe Association oil company members, although their tankers were reported to have been complying with local ballast-dumping rules. In an effort to control the situation, the Association first threatened to turn away the offenders, until a ballast water handling facility could be completed (scheduled for mid-1979), and then in April required all tankers to enter port with a 35% ballast without which they were not to be loaded.[559]

Norway

Of Norway's total anticipated gas and oil output in 1980, the national government expects domestic needs to require only one-sixth. The surplus will be available for export, with the accrued earnings ready for domestic investment in industry and social services. The government expects to use half of its annual revenue from 1980 onwards to pay off a foreign debt of 40 billion kroner (equivalent to $7.6 billion in 1978) and to provide developing states with more foreign aid, to be offered through agencies such as the World Bank.[584] Taxes of 50.8% in 1978 were distributed as follows: 26.5% to the central government; 21.3% to the municipality of the company's head office; 1.7% to a tax equalization fund; and 1.3% to the special tax for assistance to developing states. The Norwegian government's total income from the average North Sea field, excluding government participation, is estimated to be 55 to 68%.[205]

As in the United Kingdom, investors in Norway have made large capital commitments in platform construction, a tremendous boon to the shipbuilding industry, although recently platform builders have suffered from a lack of major orders. But unlike Scotland and England, Norway is fortunate to have many good sites for concrete platform construction. Protected deep-water fjords with numerous low-lying, flat areas containing sandy clays are available; they have good load-bearing capacity and are relatively watertight. These qualities help reduce soil failure when constructing very heavy concrete submersible platforms that, in the initial stages, are constructed onshore and then floated into deep water for completion. Good deposits of concrete aggregates (sand and gravel) are also usually available to construction yards.[575]

Businessmen in cities such as Stavanger, Norway's oil capital (population 85,000 with a foreign contingent of 7,000), anticipated considerable petroleum-related growth. For example, Stavanger's banking firm Rogalandsbanken places full-page advertisements in the journal *Offshore* extolling the social and service amenities available to businesses seeking to locate in the area.[209] Engineering firms have reaped a bonanza in supplying services and equipment to oil-related industries, as in the gas turbine manufacturing industry. Norwegian producers supply many of the turbines used by North Sea pipeline operators. Norway is also rapidly entering the oil refinery and petrochemical fields. Prior to its involvement in the oil industry, the country already had considerable know-how for the operation of these industries.[584] In 1978, four petrochemical plants were operating, one at Mongstad, near Bergen; one at Sola, near Stavanger; and two in the Valloy Tonsberg-Slatentangen area, approximately 100 mi. southwest of Oslo.[321, 358]

With growth in the oil industry and its support enterprises, the Norwegian government expects a labor pool deficit of 60,000 to 85,000 by 1980. In only one year, 1973-74, employment in the oil industry jumped from 7,000 to 16,000. Many people in the mid-1970s were complaining that intense oil industry labor demands had caused numerous workers to leave long-established labor-intensive industries such as agriculture, forestry, fishing, and engineering works.[584] If and when the oil industry wanes after the turn of this century, this trend could create difficulties for local areas, unless provisions are made to diversify the industrial base so that surplus labor can be absorbed.

SUMMARY AND CONCLUSIONS

During the past decade of the 1970s, petroleum resources of the North Sea have contributed to a significant restructuring of production-consumption relationships in Western Europe, especially in Norway, the United Kingdom, and the Netherlands. Some observers have suggested that if those littoral states which are also members of the EEC acted in concert to set oil industry policies, their impact on Western Europe's petroleum economics and on environmental management (as in pollution control) could be significant. On the other hand, Norway, one of the major petroleum powers, is not a member of the EEC and, in part, did not join it for fear of an outside political force dictating policy on its continental shelf and its resources. Thus, Keto notes, "the outlook from the Community perspective for a greater integration of North Sea policies and greater Community influence over offshore activities is not bright."[285] Consequently, these states will likely remain rather vulnerable to the whims of the OPEC cartel, which can do much to manipulate world petroleum prices.

Pressures associated with the quest for petroleum have helped create some friction between states, as in the case of Eire and the United Kingdom, and also within states, as between Scotland and the United Kingdom. On the other hand, petroleum has been the catalyst for various cooperative efforts. Joint cost-reduction decision making by Norway and the United Kingdom for development of fields straddling their territorial waters, as the Statfjord field, or for determining the need and feasibility of new pipeline systems,[391] and the successful establishment of the Bonn Agreement, providing for reciprocal assistance among several of the littoral states in oil spill monitoring and cleanup or other accidents, are good examples of mutually administered concerns.

National policies toward petroleum development by the various littoral states, such as the United Kingdom and Norway, have been different. The British have sought rapid expansion of the industry, aimed at national autarky, whereas the Norwegians have been more cautious, seeking to avoid adverse impacts on the national and on local economies. Scotland demonstrates the potential for social and economic impacts of offshore oil on onshore areas, especially when the oil fields depend upon service centers in markedly rural areas like the Highlands, the Orkneys, and the Shetlands which have become closely tuned to black gold. The various states with North Sea petroleum resources are intensively involved in their exploitation, concerning themselves with varying degrees of state ownership and participation and the regulation of exploration, production, safety, and pollution control. Some in industry fear British and Norwegian governmental control may stiffle future incentives for petroleum investments; but, considering the current world price situation for petroleum products, along with increases in demand, producers should be able to reap substantial profits while contributing amply to the financial coffers of the national governments.

Exploration programs are being pushed into frontier areas, and it appears that Norway will shortly open its waters north of the 62° parallel. If areas north of 62° and west of the Shetlands prove petroleum productive, the importance of the North Sea-North Atlantic region will be enhanced. But even without these new areas, the North Sea will continue to be an oil region of the first order of significance for several decades.

NOTES

[a]Because North Sea crudes are very light, Europe will continue to be dependent on heavier crudes from Venezuela and the Middle East for blending to produce the correct structure for needed industrial fuel oils [30].

[b]P. Odell, a leading authority on North Sea petroleum resources, has noted that the Dutch government releases few data for gas reserves and understates figures that are released [400].

[c]Production figures supplied by British Information Service in New York (pers. com., Dec. 7, 1979, quoting figures compiled by the UK Dept. of Energy).

[d]For a detailed discussion of the 1974 White Paper proposing the establishment of BNOC and its participation functions, see [120].

[e]D. B. Keto presents an excellent examination of both exploration and production licensing in Great Britain [285]:84-89.

[f]For a useful analysis of taxes on offshore petroleum in the U.K., see [155].

[g]Information on Shell/Esso's, the Phillips Group's, and BP's project status (as of June 1979) was provided by A. Hogg [235].

[h]During the summer of 1979, the Conservative Government of Margaret Thatcher proposed an extensive reorganization plan for BNOC; the plan was largely scrapped by late 1979. BNOC's special privileges were under fire, especially its automatic right to 51% interest in all licenses, a policy the government feels has been discouraging new wildcat ventures—only 13 were drilled in the first half of 1979 in comparison to 67 during the course of 1977. See [557].

[i]P. Odell and K. Rosing have expressed concern that Norway may be too conservative in the development of its resources [401].

[j]A useful discussion of Norway's leasing policies is available in [285]: 101-05.

[k]The time required to reach the work site by helicopter may be 1 to 2 hours, whereas by boat it may be 10 to 15 hours. Because crews are paid for portal to portal time, the difference becomes economically significant [407].

[l]*Guide to British Offshore Suppliers* as cited in [49*].

[m]The Secy. of State for Scotland, a member of the British Cabinet, is ultimately responsible for making final decisions on most oil-related development programs. This function exists because most petroleum industry projects are not covered under local development plans which require amendments to the original plan, a process depending on the Secretary's approval [285].

[n]For a more detailed look at this complex issue, see [30]:79-94, the primary source of the discussion presented here. A good resume and map showing the location of construction yards and other oil industry facilities are available in [490].

Chapter 5

The United States Offshore

The 48 contiguous states of the United States have a land area of approximately 1.7 million mi.^{2} with a reasonably good geological potential for petroleum strikes. Much of this area, however, has already been intensively explored. Although undiscovered deposits surely remain within this domain, a decline, for the most part, will be its future prospect. Therefore, our hopes and efforts for making large discoveries must be focused upon the continental margin, especially the continental shelf. Although portions of the shelf are geologically unfavorable for hydrocarbon formation, a major part has a potential equal to adjacent productive land areas. Numerous problems confront people in the government who manage offshore federal leasing programs and also those in the petroleum industry who seek to tap new offshore oil and gas reserves. Presently, an area of special concern is the United States' eastern seaboard. While self-sufficiency in energy is a prime national objective, planners and coastal residents of that area face difficult economic and social cost-benefit choices, if present offshore drilling efforts succeed in discovering commercial petroleum reserves.

EARLY VENTURES AND THE TRUMAN PROCLAMATION

United States offshore petroleum production was first associated with onshore discoveries. Working from wooden piers, drillers in 1896 tapped a seaward extension of southern California's Summerland field. This well was the country's first offshore producer.[42] Subsequently, more than 400 shallow pier-based wells (most about 600 ft. deep) were drilled here.[4] During the 1930s, petroleum geologists working on the Gulf of Mexico coast noted that the region's many salt domes indicated a likelihood of oil deposits offshore. Some companies took note of this suggestion, and in 1938, approximately 1 mi. off Louisiana's shore (in 26 ft. of water), wildcatters struck oil. This strike became the famous Creole field. By 1946, nine offshore wells had been drilled along the Gulf Coast—four off Texas and five off Louisiana. These wells, all near the shore, were drilled from rigid, bottom-based platforms.

In 1947, Kerr-McGee Oil engineers designed the first mobile drilling unit; it was composed of two World War II surplus Navy barges (for flotation) and a conventional surplus landing craft (for a platform). The rig's operators spudded in approximately 12 mi. off the Louisiana shore in 16 ft. of water.[42] These pioneering wildcatters struck oil on October 4, 1947.[4] They had (1) verified earlier predictions by USGS officials that the area might contain petroleum; (2) demonstrated the potential for large-scale offshore oil exploration and production; and (3) from a nationalistic viewpoint, vindicated President Harry S. Truman's stance in his September 28, 1945 proclamation, which held that "the exercise of jurisdiction over the natural resources of the sub-soil and seabed on the continental shelf by the contiguous nation is reasonable and just....The continental shelf may be regarded as an extension of the land mass of the coastal nation and thus naturally appurtenant to it."[42] The Truman Proclamation had significant consequences for both federal and state governments and for the development of the offshore petroleum industry.

THE STATES VS. THE FEDERAL GOVERNMENT ON THE CONTINENTAL SHELF

The Truman Proclamation, in effect, preempted coastal states' continental shelf areas.[a] For 6 years, offshore oil developments were delayed while coastal states and the federal government fought a bitter battle in the courts. California in 1947 and Louisiana and Texas in 1950 lost their cases before the U.S. Supreme Court, which ruled that the federal government had paramount authority transcending that of the individual property owner in all offshore areas. Then, under the *Submerged Lands Act of May 22, 1953* (reaffirmed in the *Outer Continental Shelf Lands Act of August 7, 1953*), Congress returned to the states a belt of continental shelf and subsoil extending 3 mi. offshore; the federal government retained control of the seabed and subsoil beyond the 3-mi. limit. The *Submerged Lands Act* also allowed coastal states to claim a territorial offshore limit up to 3-marine leagues (10.4 mi.), if they could provide proof through judicial proceedings that their charter had allowed this claim when they entered the Union or if Congress had approved such an extension prior to the *Submerged Lands Act's* passage. In 1954, the U.S. Supreme Court granted Texas and Florida a 3-league offshore jurisdiction in the Gulf of Mexico.[42] Other coastal states' seaward limits remained at 3 mi.[b]

LEASING PROGRAMS AND THE DEPARTMENT OF THE INTERIOR

The *Outer Continental Shelf Lands Act of 1953* established statutes specifically defining (1) a state's coastline as the average low-tide mark and (2) the authority of state and federal officials over the submerged coastal lands and offshore waters. It also allowed the federal government to begin leasing the OCS for oil exploration, and the DOI proceeded with its first sales in 1954. The first program resulted in 3 sales granting 114 leases (5 for sulphur and 109 for gas and oil). These lands lay off the coasts of Louisiana and Texas, encompassing an area of 486,870 acres. The federal treasury received a total bonus of $141 million and a first-year rental fee of $1.4 million. Two decades later in 1973, two lease sales alone resulted in a total bonus of $3.1 billion and a first year rental of $3.1 million. From 1954 to 1973, a total of 33 offshore lease sales had been made, including areas adjacent to the Gulf Coast, California, Oregon, and Washington. Three of these sales were for sulphur (59 leases), 2 for salt (2 leases), and the remainder for oil and gas. By the end of 1977, the federal government had sold 3,162 OCS leases, covering 14,943,177 acres. Of those, from 1954 through 1977, oil and gas accounted for 3,101 leases for 14,835,557 acres, with Louisiana leading at 1,852 for 8,034,508 acres, followed by Texas (660 for 3,284,510), California (185 for 988,170), Alaska (163 for 904,364 acres).[c] In 1978, 249 oil and gas leases (109 off Louisiana and 62 off Texas) for an acreage of 1,297,280, were added.[223] In only 2½ decades, the administration of oil leasing programs had become an important part of the DOI's functions, frequently of a controversial nature involving a governmental mosaic of half a dozen independent agencies.[d]

Objectives

The DOI's leasing program objectives are to (1) assure that the marine environment is protected, (2) maintain an orderly development of the country's resources, and (3) guarantee a fair monetary return to the people for use of the resources. Unfortunately, these goals sometimes come into conflict. Environmental protection may be interpreted so rigorously that no development can occur, and development that does occur depends on what constitutes a "fair monetary return" at a given time. Establishing a fair market value for offshore leases is difficult, but the DOI attempts to carefully assess the value of each tract. In the past, the task has been especially difficult because in *frontier* areas indirect evidence (based on stratigraphy) was the main information source used to establish the seabed's hydrocarbon potential.[119, 476] Today, the DOI also uses a massive OCS well log data bank, containing information collected over a period of 10 years. The data bank contains logs for more than 13,000 wells drilled by the DOI's USGS and private firms. In addition, application of a statistical simulation model that takes into account the "...uncertainties inherent in hydrocarbon estimating..." helps the DOI set adequate fair market values.[476, e]

Preparation for a Lease Sale

The USGS and BLM administer the DOI's leasing program. The leasing process has several important stages. First, the BLM invites industry to make nominations of OCS areas it would like to see prepared for a lease sale. USGS personnel examine available geophysical data obtained from contract firms specializing in geological and geophysical data collection and from petroleum companies who have agreed to turn over data in return for a permit to do test drilling.[f] The USGS also carries out investigative programs, including bottom sampling, shallow coring, and seismic profiling, usually done from small drill ships. From this data the BLM then makes tentative tract selections. Engineers work out production cost estimates and geologists and geophysicists study individual tracts for potential hazards that might arise during development. State officials and the public are asked to (1) identify anticipated problems associated with the tracts being considered for lease, (2) help prepare a required draft environmental impact statement, and (3) develop lease stipulations for regulating drilling and production. Public hearings are then held and the environmental impact statement is reviewed and issued in its final form. Finally, the Secretary of the Interior examines these materials and decides whether to proceed with the sale as planned, to delete certain tracts, or to include specific regulations with a given tract. A public notice of the proposed lease sale is then made, along with requests for additional state input regarding the development plans for each tract to be leased. If no court injunctions are obtained, the sale takes place in a month or two. When a lease sale has been made, the USGS becomes responsible for all activities relating to exploration, development, and production of oil, gas, sulphur, or any other mineral. The USGS must enforce DOI regulations for conservation, the environment, and safety, and must collect fees for minerals produced, a tax to be payed as money or as payment-in-kind (a portion of the product).[476] Should a commercial oil or gas strike be made, the leaseholder has to present a detailed plan of development and production to the DOI. The company must provide adjacent coastal states with information about its drilling plans, construction program, and time schedule, and the states must certify that these activities are consistent with their Coastal Zone Management Plan. The Secretary of Commerce can override state decisions if he believes that matters of national security emergency are involved.[200]

Bidding

For several years, a debate has centered on the most equitable competitive bidding system for petroleum leases on the OCS. M. Crommelin identified four basic bidding systems (cash bonus, deferred bonus, royalty, and work commitment) and analyzed the advantages and disadvantages of each (Table 9). According to Crommelin, the deferred bonus and royalty systems seem best suited to participation by the small firm with limited capital assets.[110] This point is significant because the BLM's leasing programs have been severely criticized as favoring large corporations.

Under the *OCS Lands Act*, bidding has normally been done in one of two ways—a competitive cash bonus with a fixed royalty rate or a fixed cash bonus

Table 9. United States Competitive Bidding Systems Compared

Type of bidding system	Advantages	Disadvantages
Cash bonus	1. Offers incentive to successful bidders to expedite exploration development to obtain quick return on investment.	1. Difficult for small firms to compete in bidding. 2. Puts even major producers under financial strain. 3. Encourages major producers to join forces in bidding, which may negate "real" competitive bidding.
Deferred bonus	1. Encourages small companies to bid. 2. Bidder can pay sales price in installments and has option of surrendering rights and to discontinue payments.	1. Can encourage excessively high bidding because firm can relinquish its leasing rights without penalty.
Royalty	1. Avoids extensive capital drains.	1. When firms work same structural formation, encourages establishment with highest royalty rate to abandon production before lower royalty operators, which is often too early, creating a waste of economic resources.
Work commitment	1. Government can make the right to bid dependent on an agreement for exploration activities.	1. Can encourage wasteful exploration. 2. Companies can make excessive profits.

Source: Compiled from [110].

with a competitive royalty rate. From the time OCS leasing began in 1954, the competitive cash bonus with a fixed royalty method has been used almost exclusively. Although this system has generated large sums for the national treasury (a total of $28.3 billion through the end of 1978), many mineral economists and specialists in industry agree that, because the competitive bonus system requires large capital reserves, the small operator is often effectively excluded so that the leasing system is not truly competitive. The DOI, in response to this complaint, has made some recent lease sales based upon the fixed bonus system along with a minimum royalty at $33^{1}/_{3}\%$, double the usual rate. This procedure was used in a December 1975 sale of 3 tracts off the coast of southern California and in August

1976 for 15 tracts along the mid-Atlantic. The DOI has also attempted to encourage small-firm bidding by establishing restrictions against joint-bidding of oil enterprises with an annual world production greater than 1.6 million bbl/d, except when the Secretary of the Interior waives the requirement. Joint-bidding was designed to encourage large firms to make bids together with small firms.

These new bidding procedures and strictures are too recent to make firm judgment, but they seem somewhat effective because during 1975 and 1976 alone 100 small companies were participants in leasing sales, accounting for 65% of the acreage involved, and small firms were partners in 13 of the 16 tracts leased under high-royalty provisions.[309, 476] Yet many small companies claim the new leasing programs have not really helped them. Two suggestions have been made that might further encourage small-firm participation—i.e., the use of sequential bidding whereby only part of a block of bids is opened on the first day and the use of oral bidding. The sequential bidding process would allow a company to more quickly know whether it had won or lost a given bid, and oral bidding would tend to hold down bidding prices because companies would attempt only to outbid the highest bidder.[521] In its 1976 and 1977 sessions, Congress attempted to legislate several more leasing options and to require that one-third of the frontier OCS acreage leased each year be offered under bidding systems other than the cash bonus, used in most past sales.[396] These objectives were finally realized in September 1978 when President Carter signed into law several new *OCS Lands Act Amendments*. For a period of 5 years after the amendments' enactment, various alternative bidding systems must be made available, with royalty bidding being required for no less than 20% and no more than 60% of the total area offered for lease each year. Bidding systems are to include:[373]

(1) a cash bonus bid with a royalty of not less than 12.5% in amount or value of the production saved, removed, or sold;

(2) a variable royalty bid based on a percent of the amount of the production saved, removed, or sold with either a fixed work commitment based on the dollar amount for exploration; or a fixed cash bonus as determined by the secretary, or both;

(3) a cash bonus bid or work commitment bid based on a dollar amount for exploration with a fixed cash bonus and a diminishing or sliding royalty based on any formula the Secretary determines equitable but not less than 12.5% at the beginning of the lease period in amount or value of the production saved, removed, or sold;

(4) a cash bonus bid with a fixed share of net profits of not less than 30%;

(5) a fixed cash bonus with the net profit share as the bid variable;

(6) a cash bonus with a royalty of not less than 12.5% fixed by the Secretary in amount or value of the production saved, removed, or sold, and a fixed percent shared of net profits of not less than 30%;

(7) a work commitment based on the amount or value of production saved, removed, or sold.

In 1978, the BLM held the Southeast Georgia Embayment sale, using for the first time a sliding-scale royalty bidding system that "...fixes the percent royalty due the government according to the amount of crude oil, gas or condensate production saved, removed, or sold on the leased tract during each calendar quarter of the year." The royalty due normally varies between 16.7 and 50%; under the amended *OCS Lands Act*, the Secretary of the Interior can set the minimum as low as 12.5%. After an adjustment for inflation, when the quarterly production value is less than $1.5 million, the minimum royalty applies. For every million dollars of increase in production value, a sliding scale of 1% is added to the royalty charge. After the rate reaches 50%, the charge remains constant. Despite this arrangement, the number of bidders in the Georgia Embayment sale were fewer than anticipated, though they included Getty, Tenneco, Shell, Exxon, Trasco Exploration, Pan Canadian Petroleum, Amerada Hess, Sun Oil, and Mobil Oil. Of the 224 tracts offered, only 57 received bids; of those receiving such bids, 31 were singles. Exxon was the heaviest bidder, having placed 14 of the single bids; Getty followed with 6 and Tenneco with 5. Only four tracts obtained more than 4 bids. The 25% bidding ratio was well below most previous sales (Table 10). Several previous disappointments associated with recent sales—as in the Gulf of Alaska, the Destin Dome (off the Florida Panhandle) and the Tanner Banks (off southern California)—gave pause to the oil companies.[500] Perhaps the DOI should look into the stifling effect of its selecting only a small percentage of those tracts suggested by industry. In the October 1978 Eastern Gulf sale, only 89 of the more than 700 tracts suggested by the oil companies were selected for sale by the BLM (see Table 10).[301]

Yet, when oil companies see real potential in the blocks offered for lease, they bid vigorously. In a December 1978 sale in the Gulf of Mexico involving 128 tracts offered, 88 received bids (with 80 companies participating). The DOI again used the sliding-scale royalty system, and of the 59 tracts offered, 38 received bids. This lease sale would have been even larger except that the DOI withdrew 19 tracts because of the likelihood that these tracts and those of an adjacent state might

Table 10. Comparison of Bidding Interest in Selected Past Lease Sales

Area	Sale (month/year)	Tracts offered	Tracts bid on	Percent drawing bids	Total No. of bids	Average bids/tract
Louisiana	12/70	127	127	100	1,043	8.2
Louisiana-Texas	7/74	258	49	18	57	1.2
Louisiana	3/74	206	114	55	402	3.5
Louisiana-Texas	3/75	283	102	36	191	1.9
Gulf of Alaska	4/76	189	81	42	244	3.0
Mid-Atlantic	8/76	154	101	65	410	4.1
Louisiana-Texas	6/77	223	152	68	424	2.8
Southeast Ga. Embayment	3/78	224	57	25	99	1.7
Eastern Gulf[a]	10/78	89	35	39	65	1.9
Gulf of Mexico[b]	12/78	128	88	69	288	3.3

Sources: [500], [a][301], [b][510].

have mutural reservoirs. When this problem seems likely, the DOI, as required under the *OCS Lands Act Amendments*, must work out a method of dividing the oil and gas revenues with the state.[520] Even though the DOI wants OCS developments expedited, the Secretary, in whose office the final decision is made, will sometimes reject bids, if the offer seems too low. For example, of the 223 tracts offered in the June 23, 1977, Louisiana-Texas sale (see Table 10), 152 received bids, 28 of which were rejected.[119] One company, Mitchell Energy & Development Corp. of Houston, has twice been the sole bidder on a tract in Texas' OCS. Its bids have been rejected both times. This kind of experience discourages the small independent firms. Recently, another problem—a $35 million oil-spill liability insurance requirement instituted under the *1978 OCS Lands Act Amendments*—has discouraged small producers from entering the bidding arena.[521]

A Stepped-up Leasing Program

Oil field exploration and development is a lengthy process, even with an aggressive approach by federal authorities and entrepreneurs. The time elapsing from the call for nominations to the sale of tracts takes about 19 months. The time lag from discovery to peak production can range from 9 to 14 years, and even in well-developed areas, the delay may be from 4 to 8 years.[476] Consequently, the DOI has been criticized both for not leasing enough offshore oil prospects and for setting leasing dates and then delaying them.[174]

If compared with other world areas, the United States has indeed been relatively slow to develop its offshore petroleum potential. Under development for little more than a decade, the North Sea had by 1972 55% (77 million acres) of its total area leased, whereas the Gulf Coast, under development for 19 years, had leased only 15% (10.5 million acres) of its total area.[578] In 1977, less than 3% of the United States' OCS had ever been leased, a position far behind the coastal states of Africa (58%), Canada (40%), and Southeast Asia (30%).[476] Despite eight sales held between 1975 and 1977—five in the Gulf of Mexico and one each in the Gulf of Alaska, southern California, and the mid-Atlantic—only 10.2 million acres of the OCS were under lease by December 1977. Although this area was 2.3 times greater than the acreage under lease in 1972, it represented only 2.9% of the total OCS, with a somewhat conflicting though more reliable estimate established at 5.5% by the end of 1978.[223] Prospective lease areas stretch from Alaska's Beaufort Basin off the North Slope to the Kodiak Basin in the south, from nothern Washington to southern California, and from Texas' western Gulf Coast to New England's Georges Bank (Fig. 23). Several of these areas will soon experience the driller's bit. The leasing schedule for 1978 through 1981 calls for 20 sales, including 5 in Alaskan waters, 7 in the Atlantic, 6 in the Gulf of Mexico, and 2 off California.

The accelerated leasing program undertaken by the DOI in the 1970s has experienced heavy opposition not only from environmentalists and the Council on Environmental Quality, but also from Congressmen, the Government Accounting Office,[237] and from within the DOI itself. Conceived in only 27 days by high-ranking officials in the DOI, many lower-level personnal consider the program unrealistic.[479] Those challenging the DOI have felt that the Department is ill

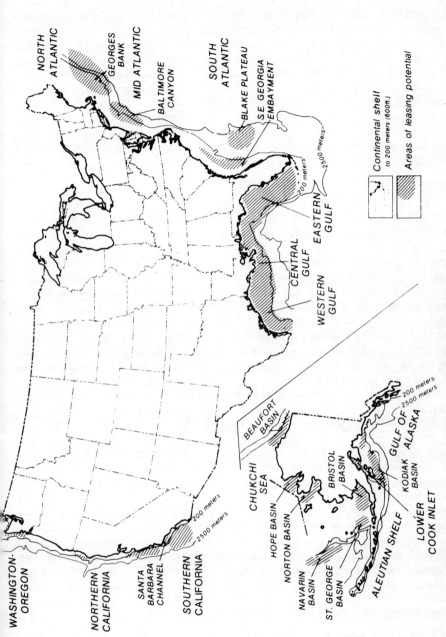

Fig. 23. Potential petroleum-bearing areas—offshore United States. *Source:* [476].

equipped to handle the stepped-up program; environmental data are inadequate; the states are not prepared to deal with onshore impacts; and petroleum companies do not have the resources to develop the 10 million acres of the OCS that will become available. Some fear the public will be cheated by the government's selling leases too cheaply.[237] The accelerated leasing program has been caustically dubbed by South Carolina's Senator Ernest F. Hollings as "drain America first." To quiet some of the criticism, the DOI in 1975 established an OCS Advisory Board, made up of members representing federal agencies (other than the DOI), coastal states, industry, and civic groups. The DOI consults through this Board with the state governors. Although the DOI is not bound by the Board's recommendations, it has pledged to follow them as much as possible.[200]

Conflicting Interests in OCS Leasing Policies

The DOI is not only attempting to stimulate activity on the OCS through accelerated leasing, but is also taking a firmer stand on demanding expeditious development of leases already held by the oil companies. In 1978, DOI began exerting heavy pressure on the oil companies to speed development of long-held leases.[582] Leases are made for 5 years (in compact areas not to exceed 5,700 acres), with extensions being available as long as active tract development occurs. Secretary of the Interior, Cecil Andrus, rejected in mid-1978 all requests for extensions on leases in the Santa Barbara Channel off southern California. Chevron, Exxon, and ARCO lost leases obtained in 1968 for $4.1 million, because the DOI felt that not enough development work was under way. This action may have been intended to let industry know that a stricter application of DOI policies on lease exploration and development is to be applied.[117] Although this new policy may be necessary, some observers point out that oil companies operating in the Santa Barbara Channel area have long been subjected to delays imposed by the DOI itself.[582]

Although the DOI may be able to identify specific areas where the oil companies have been less than diligent in exploratory work, their overall record is good. Between 1972 and 1977, offshore activity doubled in United States waters. In 1977 drillers sank more oil wells, both offshore and onshore, than in any year since 1959, and gas well drilling increased 25% over 1976. Indicative of the effort were changes that occurred since 1972 in the number of seismic crews working and the number of miles of seismic surveys completed. In 1972 only 12 offshore crews were active; these workers averaged 10,206 mi. of seismic lines each month. In 1976, about 25 offshore seismic crews were working and the monthly mileage surveyed was 18,859 mi. While exploration offshore was expanding, onshore activities were relatively stable. There were 239 crews in 1972, and in 1976 almost a like number of 237.[367] Despite the oil companies' efforts, offshore and onshore, United States oil and gas reserves continue to decline, foreboding no relief from OPEC's prospectively predatory policies in the 1980s.

OCS Leasing Revenues

The total cumulative revenue for oil and gas from 1953 through 1978 was $28.3 billion. From a mere $2.4 million in 1953, the annual revenues derived by

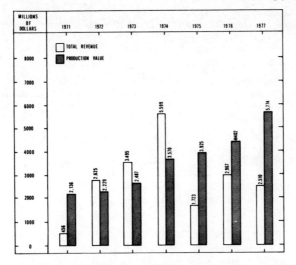

Fig. 24. Outer continental shelf oil and gas revenue and production value, 1971–77.
Source: [223].
Note: Total revenue of the federal government includes bonuses, minimum royalties, rentals, shut-in gas payments, and royalties. Production value is the total average marketed value of old and new oil, less approved allowances for transportation.

the government fluctuated from a low of $11.7 million in 1956 to a high of $3.5 billion in 1973. Total production value, however, rose steadily from $5 million in 1953, reaching a level above $2 billion in 1971 and climbing to $5.774 billion in 1977 (Fig. 24) and $7,096 billion in 1978. Since the Arab oil embargo, the cumulative revenue for 1974–78 amounted to $15.7 billion (Table 11), exceeding the total for the two preceding decades (1953–73) by $3.2 billion. During the corresponding periods, the cumulative production value increased from $15.7 billion for 1953–73 to $24.8 billion for the 5 years thereafter through 1978.[223] The drop in DOI revenues from the peak year of 1974 reflected the corresponding decline in bonuses earned, not enough to be balanced by the consistent increase in royalty income. In 1978, the 5-year trend was somewhat reversed by a rather modest gain of $431 million from both bonuses and royalties. Considering the impact of inflation during these years, the real value of the federal government's income from 1974 through 1978 was considerably less than the total annual revenues would seem to indicate.

Revenues from the OCS leasing and royalties programs go to the Land and Water Conservation Fund, administered by the DOI. These funds provide grant-in-aid assistance to cities, counties, and states for acquiring and developing open space, recreational areas, and public parks; through special authorization, they are

Table 11. Outer Continental Shelf Revenues from Oil and
Gas Operations, 1974-78
(billions of U.S. $)

Year	Bonuses	Royalties	Rentals	Total[a]
1974	5.02	0.56	0.01	5.59
1975	1.09	0.62	0.02	1.72
1976	2.24	0.70	0.02	2.97
1977	1.57	0.92	0.02	2.51
1978	1.76	1.15	0.02	2.94

Source: [223].
Note: [a] Includes minimum royalty and shut-in gas payments.

sometimes used to expand national forests, wildlife preserves, and scenic lands and waters. Revenues accumulated by the DOI declined steadily from 1974, amounting during that year to 94% of the accumulated production value; by 1978 that value had dropped to 70%.

Well Development and Production

At the end of 1978, more than 16,000 oil and gas wells had been completed on the OCS since production began in 1947. The number of active OCS oil wells rose from 50 in 1954 to a peak of 5,704 in 1971, dropping sharply to 3,744 in 1972 and remaining at about the same level through 1978. At the beginning of 1979, there were 3,397 active oil wells and as many as 1,071 that were shut-in. Active OCS gas wells rose rather steadily from 20 in 1954 to 2,615 at the end of 1978.

Drilling of new wells increased after the Arab oil embargo, reaching a high in 1977 then declining in 1978.[g] During 1978, 1,144 new wells were started and 596 producible zones were completed. On December 31, 1978, there were 6,012 active OCS oil and gas zones, with 2,332 production platforms fixed in place.

The share of offshore oil production in the United States' total output rose steadily from 1953, when it amounted to less than .05% to a peak of 17.8% in 1971, whence it began to decline, amounting to 13.2% in 1978. Offshore natural gas production, however, climbed steadily from 0.24% of the U.S. total in 1953 to 26% in 1978, with a significant increase from 1977 to 1978 of 3.3%. From a relatively insignificant annual average of c.68 million bbl. from 1947 to 1953, offshore oil output rose from 49 million bbl. in 1954 to 188 million in 1963, averaging 105 million per year during the course of that decade. During subsequent years (Table 12), an all-time high had been reached in 1971 and 1972, whereupon a year-by-year decline took place even though demand for oil continued to grow, particularly after 1973.

With no less demand in recent years, offshore natural gas fared better. From 20 billion ft.^{3} in 1953, output rose each year during the following 10 years

Table 12. Onshore and Offshore Oil and Gas Production in the United States 1963-78

Year	Crude oil and condensate			Natural gas		
	U.S. onshore (millions bbl.)	U.S. offshore (millions bbl.)	Offshore as % of U.S. total	U.S. onshore (billions ft.^{3})	U.S. offshore (billions ft.^{3})	Offshore as % of U.S. total
1963	2,565	188	6.8	14,027	640	4.4
1964	2,572	215	7.7	14,699	763	4.9
1965	2,607	242	8.5	15,190	850	5.3
1966	2,728	300	9.9	16,268	939	5.5
1967	2,848	368	11.4	16,798	1,373	7.6
1968	2,868	471	14.1	17,484	1,838	9.5
1969	2,846	526	15.6	18,377	2,321	11.2
1970	2,941	576	16.4	19,076	2,845	13.0
1971	2,839	615	17.8	19,275	3,218	14.3
1972	2,841	615	17.8	18,781	3,751	16.6
1973	2,778	583	17.3	18,672	3,975	17.6
1974	2,670	533	16.6	17,371	4,230	19.6
1975	2,562	495	16.2	15,852	4,257	21.2
1976	2,505	463	15.6	15,656	4,296	21.5
1977	2,546	439	14.7	15,485	4,540	22.7
1978	2,733	417	13.2	14,493	5,104	26.0

Source: Calculated from [223].

at an annual average of c.280 billion ft.^{3} reaching 640 billion in 1963. After exceeding the level of 1 trillion in 1967, ever greater amounts were produced, with the most significant gain of 564 million ft.^{3} occurring from 1977 to 1978. Reflecting the trend in federally leased offshore gas output (3.7 trillion ft.^{3} in 1977 and 4.4 trillion in 1978), the gain was also particularly welcome to the national fuel economy in light of the 992 million ft.^{3} decrease in onshore gas production from 1977 to 1978.

Regionally, the total 1978 offshore oil and gas production from both federal and state-leased areas came from 4 states, overwhelmingly led by Louisiana (Table 13). Changes from 1974 to 1978 were generally substantial,[223] particularly with regard to Louisianian gas wells.

With as little as 13.2% of the total oil ouput in 1978 coming from offshore, it is not likely that offshore wells will alleviate the national shortage in 1980 when 55% of the nation's consumption was projected to come from imports (see Fig. 1). Indeed, the offshore oil potential of the United States appears to have been largely discounted for the immediate future by a 1979 task force headed by DOE Deputy Secretary John C. Sawhill, which suggested a wide range of contingency plans such as gasoline rationing, a stiff new tax at the gas pump, reactivation of closed nuclear plants, and even a 4-day work week.

Table 13. Outer Continental Shelf Production of Oil and Natural Gas by State, 1974-1978

State	Crude oil and condensate (millions bbl.)		Natural gas (billions ft.^{3})	
	1974	1978	1974	1978
Alaska	59.9	45.5	73.3	46.5
California	81.6	56.1	30.2	16.7
Louisiana	389.3	310.5	3,871.9	4,909.2
Texas	1.9	4.6	254.3	402.0

Source: [223].

THE NORTHEASTERN SEABOARD: A NEW FRONTIER

The United States government and oilmen have been working to reduce the vulnerability of the country to sudden cutoffs of petroleum supplies, as occurred during the 1973 Arab embargo. Both industry and energy policy officials have placed great hopes in the potential of the United States eastern seaboard for petroleum, especially in the Baltimore Canyon (off New Jersey and Delaware) and Georges Bank (off Massachusetts). In 1973, Assistant DOI Secretary Stephen A. Wakefield forecasted the "...possibility—not a certainty, but a possibility—that significant quantities of oil and gas may be found on the Atlantic OCS close by its major consuming areas. If so, this could measurably reduce the excessive dependence of these areas on imported oil."[562] Although many citizens did not welcome the prospect of offshore petroleum developments in the Atlantic, others have conjured visions of an onshore economic bonanza of oil processing, marketing, and service industries. One optimistic estimate has suggested that offshore oil developments along the entire east coast could potentially support 15,000 workers directly and about 60,000, if all petroleum-based activity is considered.[250] A more conservative estimate has been offered by an independent environmental consulting firm employed by the oil industry. The study shows direct and indirect jobs created can be expected to total 28,000, which is only 2% of the region's total projected population growth during the period of anticipated oil production.[411]

Thus far (by late 1979) only the Baltimore Canyon has experienced exploratory drilling, with more than 15 wells put down; all but 3, that together produced c.50,000 ft.^{3} of gas per day, have been dry or showed only noncommercial quantities of oil and gas. In 1975, USGS officials estimated that the mid-Atlantic area had a 5% probability of containing 14.2 trillion ft.^{3} of undiscovered recoverable gas and, similarly, 4.6 billion bbl. of crude oil. The more likely prospects, however, were 5.3 trillion ft.^{3} of gas and 1.8 billion bbl. of crude oil. As of mid-1979, the USGS was updating these estimates.[402] Whether major oil finds will yet be discovered off the United States' eastern shore has not been resolved, but if oilmen have their way, petroleum soon will be flowing from the Atlantic seabed to refineries and consumers from Maine to Georgia.

Geological Exploration

From December 1975 through March 1976, the first deep stratigraphic test well was drilled in the United States' Atlantic OCS, an area containing sediments and geological structures that have good potential for hydrocarbons. Thirty-one oil companies, the Continental Offshore Stratigraphic Test (COST) Group, in conjunction with the DOI, financed the well. It was put down approximately 90 mi. east of Atlantic City, N.J., on the eastern flank of the Baltimore Canyon Trough in about 235 ft. of water. The well reached a depth of 16,043 ft., passing through thick sand and shale strata that elsewhere may be potentially productive for oil and gas (Fig. 25). The DOI purposely chose a site that was unlikely to encounter hydrocarbon deposits. Various analytical techniques demonstrated that the well area has considerable potential for gas and oil, but enthusiasm should be tempered somewhat. The USGS cautions[469] that:

> Studies of color alteration of visible organic matter, organic geochemistry, and vitrinite reflections show that although many units have high organic-carbon contents, moderately low geothermal gradients may have retarded thermal maturation. This, in conjunction with the scarcity of marine-derived organic matter in the lower part of the section, suggests a reltively low potential for the generation of liquid hydrocarbons. However, the overall combination of source beds, reservoirs, seals, structures, and thermal gradients may be favorable for the generation and entrapment of natural gas. Furthermore, the presence of reservoir rocks, seals, and trapping structures may indicate a significant potential for entrapment of either natural gas or petroleum that was generated deeper in the basin and then migrated either laterally or vertically.

Similarly favorable geological conditions exist in the Georges Bank Basin, lying 125 mi. offshore in 120–540 ft. of water. Two stratigraphic test (COST) wells drilled in preparation for lease sales have indicated source rocks composed of carbonaceous Jurassic limestone and organic-rich Cretaceous shales.[h] These strata, 10,000 to 20,000 ft. thick, lie in an east-west trough with a width of up to 30 mi. and a length of 215 mi. The USGS estimates the basin's indicated reserves at 900 million bbl. of oil and 4.2 trillion ft.^{3} of natural gas.[i]

Leasing

The first Baltimore Canyon sale occurred in August 1976, generating $1.1 billion in bonuses for 93 tracts, totalling 529,466 acres.[109] Exploratory drilling was held up, however, because of court injunctions based on the contention of environmental groups that inadequate environmental impact studies had been done. Drilling finally started in 1978; on August 14 of that year, Texaco announced that it had brought in a promising natural gas well at 14,000 ft. (in 432 ft. of water)

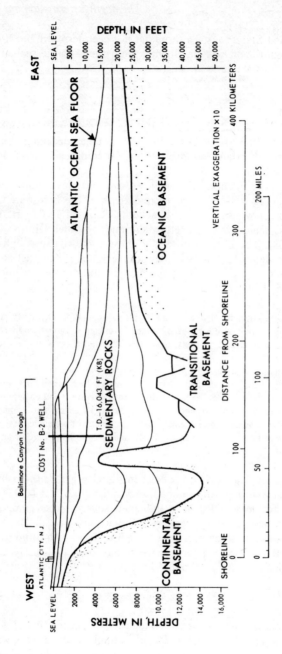

Fig. 25. Generalized cross-section of the United States' Atlantic continental shelf, slope, and rise. Sediments measure as much as 45,000 ft. deep; drilling carried out in the COST No. B-2 Well indicates considerable potential for oil and gas.
Source: [469].

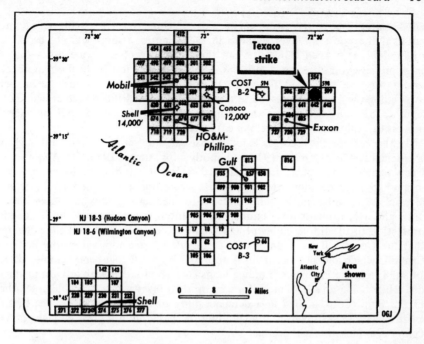

Fig. 26. The Baltimore Canyon's first discovery.
Source: [528]. With permission.

100 mi. east of Atlantic City, N.J. (Fig. 26). Texaco claimed that, to be commercial, its strike would have to yield more than 1 trillion ft.^{3} of gas reserves and produce 200 million ft.^{3}/d, and if the strike proved to be commercial, it would take 7 to 10 years to develop it.[528] To better define the dimensions of its find, Texaco drilled a confirmation well 1.5 mi. to the west of its discovery, but it showed no gas or oil. However, in November 1979, prospects for Texaco's discovery began to look better when a promising new strike, approximately halfway between two wells that have shown signs of gas on a tract leased by Tenneco, reportedly encouraged the company to commence the design of a special production platform.

Texaco, Exxon, Shell, Gulf, Tenneco, Mobil, Chevron, and Houston Oil and Minerals Corp. drilled wildcats throughout the winter of 1978-79 despite severe weather conditions,[535] and by May 1979 had spent a total of $1.5 billion on exploration. The primary exploration targets in the Baltimore Canyon Trough are those areas with normal faults, igneous intrusions, and diapirs.

The BLM held in early 1979 a second lease sale composed of 109 tracts, totalling 649,987 acres, 50 to 100 mi. offshore of Virginia, Delaware, Maryland, New York, and New Jersey, but it did not prove attractive to oilmen, perhaps because of the many dry wells drilled in the area to date and the water depth of some of the leases—as much as 4,700 ft. Secretary Cecil Andrus withdrew 27 tracts from the sale because of potential geological hazards.[156] Numerous researchers have pointed out that the quest for hydrocarbons in the Atlantic offshore will not be easy. A study done at Cornell University and the University of Delaware notes that for each oil reservoir discovered off the Atlantic shore, it will likely require nine exploratory wells,[250] and Berryhill stressed that, because no onshore oil fields exist along the United States' eastern seaboard, it is difficult to predict what petroleum deposits will be found or how long it will take production to begin once the process of exploration and development starts.[42]

In the case of the Georges Bank, a lease sale has yet to be held, although one was scheduled for October 1979 and another for August 1982.[13, j] A sale of 128 tracts had been scheduled for January 1978, but it was blocked by an injunction in an 11th-hour decision when the courts received a lawsuit filed by the State of Massachusetts and several fishing and environmentalist groups to suspend leasing until the Departments of Commerce and Interior could assure both adequate protection to the fishing industry and pollution control programs. Oilmen contend that the Georges Bank injunction is preferable to that imposed on the Baltimore Canyon, which did not occur until after the lease sale had been completed. This situation left the companies holding loans (for 18 months) with interest payments due while being unable to move forward on their investments.[109] William R. Ahern of the Rand Corp. argued in a detailed study of the Georges Bank area that the environmental risks involved in exploiting the basin are not great enough to merit delays in exploration and production.[5]

Onshore Impacts

If major discoveries are made off the United States' eastern seaboard, what does the future hold for the littoral states? One might be inclined to consider the Gulf Coast as a model. Unfortunately, the Gulf Coast's experience does not fit well to the eastern seaboard because its oil industry developments occurred over a long period of time and production began onshore first, only gradually spreading to the offshore. Its communities did not suddenly have to provide all the services, amenities, and infrastructure that will be needed in onshore communities of the Atlantic Coast.[200] A more apposite comparison is Scotland. In their book focused on a comparison with Scotland, the Baldwins have drawn parallels of what is likely to happen to the United States' northeastern seaboard, if new petroleum fields are developed and processing facilities are constructed on adjacent coasts. They see environmental conditions and social and economic structures that may be affected in ways similar to those in Scotland (as discussed in the previous chapter).[30] On the other hand, Thomas A. Grigalunas, a resource economist at the University of Rhode Island, has warned against making specific analogies between these two areas, although he does feel broad regional problems could apply.[213]

The federal government recently seems to have become more aware that the coastal states are bound to absorb a major share of any onland impacts associated with offshore petroleum developments.[286] A step in this direction occurred in 1976 when DOI Secretary Thomas S. Kleppe required lessees' environmental impact statements to provide concerned coastal states with information about their anticipated onshore activities.⁴⁷⁶

If some areas of the eastern seaboard are forced to provide services to businesses or people associated with offshore exploration and development of the oil industry but do not receive any tax income from it, the effect could be fiscally negative. On the other hand, after the exploration phase passes, new jobs and industry are likely to develop in the form of onshore support facilities, as pipeline welding and coating shops, pipeline landfalls, tanker terminals, platform and module construction yards, harbors, and petroleum processing plants. But according to the Office of Technology Assessment (OTA), development of an oil industry in the Baltimore Canyon Trough would not necessarily lead to construction of new onshore oil refineries. Old ones could be used, or they could be expanded. Refining capacity for Delaware, New Jersey, and eastern Pennsylvania for 1985 is projected to total 1.87 million bbl/d which in 1976 was more than 1 million bbl/d. Assuming commercial strikes are made, peak production in the Baltimore Canyon area is expected to be no more than 650,000 bbl/d. Plenty of slack refining capacity should, therefore, exist. Harbor staging requirements for direct support of offshore facilities along the New Jersey and Delaware coasts could range from 55 to 120 acres, depending on the size of the field's total life-time recovery. The best staging sites are at Atlantic City and Cape May, N.J., and at Lewes, Delaware. If a high recovery rate is assumed, a total of 1,645 acres of land may be needed to support activities in the Baltimore Canyon Trough, including support bases in coastal ports (170 acres), pipeline corridors in the coastal zone (150 acres), pipeline corridors outside the coastal zone (550 acres), tank farms (75 acres), and a gas processing plant (700 acres).³⁹⁸ And by 1986, if a total of 10 production platforms (a high-production scenario) were operating in the Baltimore Canyon area, approximately 9,000 people could be directly employed in the oil industry, and worker earnings could total $180 million annually. A median-recovery scenario, with five rigs active, could see 4,700 people directly employed, with an income of $85 million annually (Fig. 27).

What would be the Georges Bank's economic impact on adjacent New England if it proved to have commercial petroleum deposits? Grigalunas foresees no significant regional impact. He notes, for example, that even if Georges Bank proves to have a recoverable oil reserve of 6 billion bbl., 2 billion more than what it is likely to have, the peak annual production rate anticipated would be barely "...equivalent to the region's consumption of products in 1972."²¹³

Grigalunas, through a special application of the Harris regional forecasting model, has produced petroleum-related development and production/employment and income scenarios for New England. His forecasts are based on low-find and high-find production rates and with or without refineries being built.ᵏ Considering that severe inflation has occurred since the publication of his analysis and that no progress has been made in developing the Georges Bank, the *average annual* data he presents must be interpreted only as indicative of potential regional impacts, especially since the base years used (1977-79) have now passed. He suggests that regional employment (without refineries) in the development stage (late 1970s and early 1980s) could vary (for low-find and high-find) between 3,015 and 6,295, with

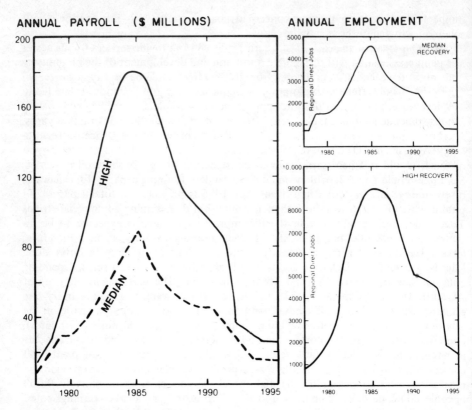

Fig. 27. Annual payroll and direct employment from all OCS activities under high- and median-recovery assumptions. Employment values include exploration, installation and development, and production workers.
Source: [398].

the total stabilizing in the production stage (1985–90) at between 1,375 and 7,575. Development stage payrolls (without refineries) might vary from approximately $33 million (low-find) to $73 million (high-find) with production stage amounts totaling between $18 million (low-find) and $101 million (high-find). Regional income, during the production stage and under a high-find scenario, could be as much as $145 million annually. If petroleum refineries were constructed, they could employ by 1985-90 from 2,650 to 6,825 workers, depending on whether 1 or 3 refineries were built; payrolls might vary from $36.5 million (one refinery) to $96.3 million (three refineries). In sum, Grigalunas concludes, "...the expansion in economic activity described here cannot be expected to substantially increase New England's employment rate or annual per-capita income."[213, 1] Locally, however, the effects of oil-related activity could be significant, as at Davisville, R.I., where oil companies purchased a former Navy staging yard which now houses 30 marine supply and service companies.[407] Ironically, the oil firms picked the site, in part, because they felt the industry's coming and going here would have little effect on the local community.

SUMMARY AND CONCLUSIONS

The United States was one of the early leaders in initiating an offshore petroleum industry. But in recent years, owing to relatively conservative leasing policies of the DOI and more attractive investment regions in foreign areas, as in the North Sea or Southeast Asia, leasing of OCS waters has lagged well behind other offshore oil-producing regions. Although the DOI has been attempting to involve the small producer in OCS petroleum activities, its programs have had only limited success. To improve the situation, the DOI should (1) institute additional incentives for the small firm to join in the bidding process; (2) attempt to adhere to established leasing schedules; and (3) include more of the tracts industry recommends for leasing. Problems of federal vs. state dominion over the continental shelf areas continue to surface, as along the New England seaboard and the Cook Inlet of Alaska, but enough precedent for the 3-mi. state territorial limit has now been established for the problem to diminish, if not to disappear.

If the United States is to reduce its dependence on petroleum imports, it must continue to forge ahead in the exploration of the continental shelf, and where it is technically and economically possible, efforts should be extended onto the continental slope. Although these activities will need to take into account environmental protection measures, lest long-term consequences outweigh short-term benefits, the quest for domestic offshore supplies is bound to continue in the early 1980s. It may not be impossible to close the gap between the costs of offshore extraction and prices in light of OPEC's increase for crude oil by as much as 65% in one year (1979). And, as continuous rounds of OPEC price increases are in prospect, the decline in domestic offshore production might be reversed.

NOTES

[a]The U.S. continental shelf contains 875,000 mi.^{2} or approximately 552 million acres.

[b]As recently as the mid-1970s, the problem of federal and state control on the continental shelf was again before the courts (see Chap. 3, pp. 33). In response to requests by industry, the state of Alaska had scheduled an offshore lease sale (during 1967) beyond the 3-mi. zone within Cook Inlet, a body of water considered by Alaska to be internal. The federal government disallowed the sale, as an infringement on federal territory. The issue was settled when the Supreme Court awarded the federal government all waters beyond the 3-mi. point [452]. F. Bartley has provided a good view of the state and federal territorial limits problem [34].

[c]Data provided by John Duletsky, Chief, Cons. Div., USGS, Reston, Va.

[d]In addition to the DOI, the National Oceanic and Atmospheric Administration, Environmental Protection Agency, Army Corps of Engineers, and the Coast Guard, among others, play a role in managing the OCS.

eFor various methods used to estimate reserves, see [554].

fInformation obtained in test drilling programs must be made public no more than 60 days after the first lease sale takes place within 50 mi. of the test well.

gIn 1978, 80% of the new wells drilled ended up in suspension.

hTwenty-five firms participated in the first well and 19 in the second; the combined cost of the 2 wells was $22.3 million.

iThere is 1 chance in 20 that petroleum reserves in the Georges Bank Basin could be as much as 2.4 billion bbl. of oil and 12.5 trillion ft.^{3} of natural gas. A 40% chance exists that any oil discovery made will be noncommercial [109].

jPending court action delayed the October sale.

kThose attempting to build refineries on the east coast have had great difficulty obtaining permission and buying land. A good example is the failure of the late Aristotle Onassis to obtain a license to construct a $600 million facility at Durham, N.H.

lFor tabular data on income and employment projections, see [213]:14-15.

Chapter 6

Petroleum Exploration and Production Technologies

The offshore quest for crude oil and natural gas has contributed to the development of highly specialized and expensive exploration and production technologies, including geophysical prospecting, drilling, pumping, on-site storage, and rig maintenance. Drilling rigs and production platforms now routinely work in water depths that a few years ago would have been considered inaccessible. New concepts of petroleum production provide more efficient use of capital equipment and labor. Although this chapter's primary focus is upon the technologies of the offshore petroleum industry, several of the exploration techniques described have applications in other ocean mineral industries.

EXPLORATION TECHNIQUES

Only a small portion of the world's continental shelves has received detailed study by geologists and geophysicists. With our increasing need for continental shelf petroleum resources, producers are pushing into ever deeper waters. They have numerous sophisticated equipment items that assist them in their quest.

Seismic, Side-scan Sonar, and Magnetometer Profiling

Seismic studies provide information on the earth's elastic characteristics, earthquake processes, and layering. Scientists obtain needed data by sending energy waves into seabed strata and receiving the waves as they are reflected back toward

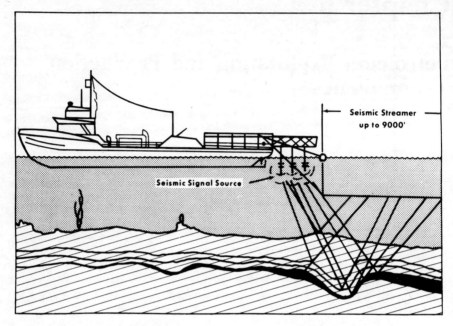

Fig. 28. Schematic diagram of a marine seismic prospecting system. Source: [4].

the ocean surface.[516] Subsea explosives used for seismic work in the past did heavy damage to fish populations.[44] Today, ship-mounted energy emitting devices, as nonexplosive gas or air guns, send seismic waves toward the ocean floor. The reflected waves strike a ship-trailed 4-in. seismic cable (up to 9,000 ft. in length), containing oil for buoyancy and geophones and wires for transferring the seismic wave energy to magnetic tape (digital) recording devices on the ship (Fig. 28). A digital computer rids the data of random energy, "noise," and then presents it in a vertical cross-section format that, when analyzed by the petroleum geologist, can identify potential petroleum-bearing structures.[4] Echo scounding devices, however, do not accurately record all seabed irregularities. Ship movement, for example, is recorded as seabed dips and bumps. Other errors occur due to calibration inaccuracies and because the speed of sound in water changes with varying depths. The error equals 0.5% of the water depth.[346] An added problem of mapping deep ocean areas is a lack of geodetic control points, making ship positioning difficult. Within 50 mi. of the coast, however, radar systems can be used to determine vessel positioning.[436]

Engineers are working on a promising new system for profiling ocean floor sediments and soft rocks. The system consists of a ship-towed vehicle that fires a bullet-like projectile (penetrometer) into the seabed sediments. The projectile, fired by a gas gun primed by compressed air, can penetrate 32 ft. of soft marine

Exploration techniques 115

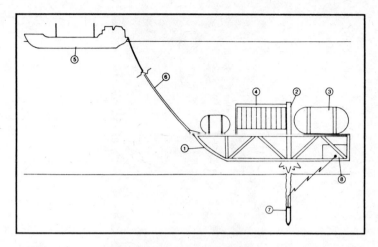

Fig. 29. A proposed bullet penetrometer system. This highly accurate device utilizes a relatively inexpensive disposable seabed projectile unit. It consists of a towed underwater vehicle (1) which has a projectile magazine (4) and a firing tube (2) operated by a gas reservoir (3). A surface vessel (5) tows the underwater unit on or immediately above the seabed, using a towing cable containing gas lines and instrument wiring (6). When the projectile (7) is fired from the vehicle, telemetric accelerometer signals are picked up by a receiver (8).
Source: [63]. With permission.

sands, silts, and clays. By measuring the deceleration/time signatures, sediment types can be determined without retrieving sediment samples (Fig. 29). This system has several advantages. Its operation is as fast and as cheap as other existing systems and it allows continuous surveying without the towing vessel's having to anchor to retrieve samples. It also provides good consistency in the profile depth.[63]

Side-scan sonar devices use transmitters and receivers mounted on a "fish" pulled along at approximately 60 ft. above the seabed and at speeds below 10 knots. These systems provide left- and right-hand profile records resembling pictures. The profile identifies rock outcrops, faults, and sunken objects. Magnetometers are used to find metal objects such as pipelines, cables, and other objects associated with mining operations, and they are also used to detect seabed magnetic anomalies such as are associated with large concentrations of iron ore.[346]

Bottom Sampling

Several devices can provide general geological knowledge of seabed sediments.[375] Gas fired penetrometers, buoyant grab buckets, and weighted tubes dropped to the ocean floor and retrieved by an attached wire are employed to acquire sediment

Fig. 30. Hand-held coring device used to obtain samples of coral. Study of the sample provides information on the petroleum potential of the reservoir rocks. Photo by Shinn, Geologic Division, USGS, Miami, Florida. Courtesy USGS.

samples. Grab bucket samples are limtied to the seabed surface. After the sample has been obtained, the grab bucket ejects its ballast and then surfaces. Seafloor penetration by weighted tubes or drop cores is usually restricted to a few feet,[346] depending on the type of material being sampled. In shallow waters, hand-held coring devices can be used to penetrate several feet into consolidated rock (Fig. 30). For deeper water (as much as 660 ft.), remotely controlled diamond drills can provide core samples of great depth.[139] If the material is sedimentary, the shells contained can be dated to determine the rock's age.

Nuclear Probes

Scientists at the Battelle Institute have developed a nuclear prospecting probe that detects up to 30 elements. This device should prove useful not only in continental shelf areas but in the deep seabed as well. The probe uses a man-made isotope, Californium-252, that emits a beam of low-energy neutrons. The beam is directed onto the ocean bottom; then, depending on the different elements exposed to the isotype, the probe detects the gamma rays given off by the induced radiation. Computer analysis of the short-lived gamma emissions determine the quantity of specific elements contained in a given area. Battelle has experimented with a towable nuclear probe, one that would provide a continuous mineralogical evaluation as the research ship traverses the sea.[137] Some scientists feel the

Fig. 31. Remotely controlled, unmanned submersible, the *RECON III*. An umbilical cord tethers the 330-lb. vessel to the mother ship where, via a TV monitor and control panel, the operator or pilot directs the reconnaissance work. Vehicle equipment includes a TV camera and light mounted on a tilt mechanism, magnetic compass, and depth transducer. Courtesy Perry Oceanographics, Inc.

use of such devices may be unwise because neutron activation creates "...a trail of radioisotopes on the seafloor," a condition potentially detrimental to marine biota.[32]

SUBMERSIBLE VEHICLES

Use of manned and unmanned vehicles is one of the most versatile methods of doing geological exploration and work tasks. These vehicles have various types of instrumentation, including magnetometers, cameras, videotapes (Fig. 31), coring devices, water samplers, and side-scan sonar units.[385] Sediment profiling,[a] topographic surveying,[427] outcrop examination, and drill site selection are some of the tasks performed with these craft.[555] Submersible vehicles may also be fitted with numerous tools such as wire brushes, cable cutters, wrenches, grabs, suction pads[259, 585] and maneuverable arms for using these tools (Fig. 32) in various oil field jobs. In 1975, there were 41 active commercial underwater work vehicles (subs and bells) operating on ocean bottoms throughout the world;[427] by mid-1977 the number grew to 104. Fifty-eight of these craft were manned, with the crews varying from 1 to 4. The high resolution provided by the human eye and the

Fig. 32. *ARMS* bell and life support system. This surface-simulated atmosphere diving system can be used at 3,000 ft. depths. A main observation window and top- and side-view ports provide good visibility for the 2 divers working inside this one-atmosphere bell. There is no need for decompression of the divers because surface atmospheric pressure is maintained inside. The operator of the mechanical arm receives electrical impulses similar to that which would be experienced in direct manual contact. Rotary actuators simulate the human arm's movements. The operator can start nuts, grasp pins, and turn valves. Courtesy Oceaneering International, Inc.

versatility of the human hand and brain are the major assets provided by manned submersibles.[197] Most of the work vehicles are designed to function to a depth of at least 1,000 ft. (some are capable of operating at 3,000 to 4,000 ft.),[327,374] and can remain down for 6 to 14 hours. Because these vehicles can work at great depths while doing heavy tasks, they are often more efficient than divers. Even more important is that the risk is much less than in conventional and saturation diving. They do not, however, except in lockout systems, allow a worker to directly use his manual skills. Some diver-lockout submersibles are equipped to carry 4 men. The lockout compartment "...allows the diver to be compressed to his working depth without affecting the submersible crew" (Fig. 33). The crew can park the submersible within a few feet of the working diver. The work area can be flooded with light from the submersible, and an expert engineer may supervise the diver's work. The mobility enjoyed with the submersible is far superior to the

Fig. 33. A 4-man submersible, *Supersub I*. The vessel is 24 ft. long, weighs 13.5 tons, can work at depths of 984 ft. and travel at 4 knots. The craft has a diver-lockout chamber that permits its use as a submerged work base. When divers have finished their work and return to the pressurized chamber, they can be taken to the surface and transferred to a shipboard living chamber for gradual decompression. Courtesy Perry Oceanographics, Inc.

traditional diving bell systems. Its liabilities are that (1) it is not normally operated during bad weather, because of a susceptibility to damage during launching and retrieval; (2) it is expensive; and (3) it is limited in underwater-time capability to its battery power.[299] A British manufacturer has recently designed a glass-reinforced, plastic-hulled craft valued at $1.3 million that has batteries capable of providing power for 12 hours of continuous work at a depth of 1,500 ft. Two of these craft have been at work in the North Sea for several years. One has been active in the Brent field and the other used in inspecting the Ninian pipeline.[514] Submersibles are also used to inspect wellheads, platforms, and pollution control equipment, or, as in the Santa Barbara oil spill, to find leaks.

DIVING

In most offshore petroleum fields, deep-sea divers perform various, frequently dangerous construction and maintenance tasks on drilling and production rigs, wellheads, and pipelines. As of 1975, about 800 divers were employed in North Sea operations alone, and, as early as 1976, as many as 26 divers had perished in

Fig. 34. This atmospheric diving suit, known as JIM, has been used since 1974 in the North Sea offshore gas and oil industry. Designed for work at depths as great as 1,500 ft., it is pressurized so that no decompression is necessary when the diver surfaces. Both arms and legs are neutrally buoyant, allowing the operator to work with a minimum of friction. The suit weighs 910 lbs., empty in air, and only 60 lbs. in water. Courtesy Oceaneering International, Inc.

North Sea diving accidents.[491] For the most part, divers work at depths below 600 ft., but technicians recently have developed suits that allow work at great depths while remaining at normal atmospheric pressure. A highly effective suit has been used off the coast of Spain at a depth of 1,442 ft. Because no decompression is necessary, the diver has more underwater work time (Fig. 34).[b]

Ths basic problem in deep-sea diving is the relationship of pressure and body absorption and release of gases, such as nitrogen. With increased pressure, the diver's body absorbs nitrogen or helium, that may be used in the diver's air supply to reduce the problem of nitrogen narcosis. Helium is used in place of nitrogen because the body absorbs it at a slower rate. Unfortunately, helium reduces the body's heat retention capacity and contributes to what is called "Donald Duck speech." The garbled speech problem has been overcome recently through the development of a digital recorder that briefly stores the speech and then replays it at a slower speed.[167, c] When returning to normal atmospheric pressure, these gases must diffuse in the reverse direction, causing bubbles to form. The bubbles may block blood vessels, impair the eyes, induce severe muscle and joint pains, and kill. Diving physiologists are now working on ultrasonic devices, monitored from the surface, that detect bubbles in a diver's body tissue.[503] If a diver is stricken by the bends or if a platform is experiencing emergencies (fires or blowouts) while the diver is under decompression, he must be transferred to an onshore hospital. To handle this problem, special mobile compression capsules, ferried by helicopter, have been developed.[534]

Beyond divers' safety, lost work time while in decompression poses a major cost problem.[d] In one work situation off the coast of Malaysia, divers expanded their on-job time by working from a submersible decompression chamber placed at a holding depth of 115 ft. Divers in the chamber were put under a 115-ft.-water-depth pressure and lowered to this depth in water, a point at which a diver's blood

Fig. 35. The *Texas Star* platform, owned by Charles Offshore Ltd., is shown on location in the Gulf of Mexico. The Marathon class 150-44-C self-elevating cantilevered drilling rig is drilling additional wells adjacent to the production platform. Note the helicopter at rest on the helipad. Courtesy Marathon LeTourneau Offshore Co.

becomes relatively saturated with nitrogen so that increased depth has no appreciable effect on his absorption of nitrogen. The divers then left the chamber, proceeded to the working depth at 210 ft., and returned to the chamber to rest, without undergoing decompression. Four-man teams went through 10-day cycles under pressure, spending 5 working hours outside the chamber in a 24-hour period. The program was both a technical and financial success.[281]

DRILLING AND PRODUCTION

The petroleum industry employs two kinds of offshore rigs—drilling and production. These versatile work stations are helping oilmen push exploration and production into ever deeper and more difficult ocean areas. Subsea completion systems also facilitate that objective.

Drilling Rigs

Drilling rigs are either self-propelled or require towing to the drilling site. Most drilling units (jackups, semisubmersibles, or drill ships) have similar equipment—a central drilling aperture or moonpool, drilling mast or derrick, rotary table, hoists, and mud pumps and tanks. They also may be equipped with diving systems, including bells, transfer and decompression chambers, and TV cameras.

The jackup (self-elevating) drilling rig is the most commonly used, especially in shallow waters. It is less expensive than other drilling rigs to build and operate, but presently usable only in water depths of less than about 450 ft.[14] The jackup stands on legs or a monopod that rests on the seabed, with its work area jacked up out of the water to the desired position (Fig. 35). Once in place, the jackup unit is very stable. Its greatest problem is its instability when under tow. The legs may protrude 200 ft. into the air above the water surface, raising its center of gravity. Although the legs can be kept in a lowered position in the water, this procedure slows the transport process because of the water's drag on them. Sometimes the legs are cut off before towing, but rewelding at the new work site each time is an expensive procedure.[78] Moving a drilling rig of any type to a new location is dangerous. The vessel can break loose or founder, if waves and wind cause it to shoal. In the winter of 1973-74, a semisubmersible rig foundered near the North Sea's Piper field,[490] and more recently, Norway lost a unit near Bergen (see Fig. 55 in Chap. 8).

Semisubmersibles and drill ships have the capability of drilling in exceptionally deep water (up to 6,000 ft. for drill ships), but they are expensive and subject to excessive motion. A drill ship must be kept on position by anchors and cables (commonly 6-8) or, more recently, by dynamic stationing (Fig. 36). The column-stabilized floater, or semisubmersible, has a deck supported by columns attached to huge underwater cassions or displacement hulls (pontoons). Because of their size, these rigs are not as subject to seawave motion as the drill ship; they may or may not be self-propelled.[4] Although their towing speeds are only 3 to 4 knots, even when using tugs with 10,000-15,000 hp, those fitted with up to 4 propellers

Fig. 36. The drill ship *Pelerin*. This 16,000-ton vessel can maintain itself within a few yards of the well axis while being subjected to wind gusts of up to 65 knots. Transponder (TR) signals on the wellhead are received by hydrophones (H) that, along with the wind, wave, and current information, are fed into a computer which regulates the thrusters (T). Courtesy Compagnie Francaise des Petroles.

powered by their own motors can reach speeds up to 7 knots (Fig. 37). These propellers, along with thruster systems, allow the rigs to maneuver precisely onto drilling locations and to maintain location. The semisubmersible's water-depth capability is comparable to drill ships, perhaps to 4,000 ft. But they also demand very heavy mooring systems to resist current, wave, and wind loading. Really large units may require combination wire rope/chain mooring systems (3.5 in diameter) with a break strength of 1.2 million lbs.

Fig. 37. The *Aker H-3* semisubmersible exploration drilling rig is self-propelled. The rig is shown here under way to a drilling site. The fjords and islands of offshore Norway provide protection from severe winds and waves. Courtesy AKER Group A/S.

Production Platforms

After a field is determined to be commercially exploitable, a production platform is installed. In 1947, production platforms were operating in waters only 20 ft. deep. Three decades later, fixed (pinned to the sea floor) steel platforms had been installed in waters of more than 1,000 ft., a depth nearly twice the

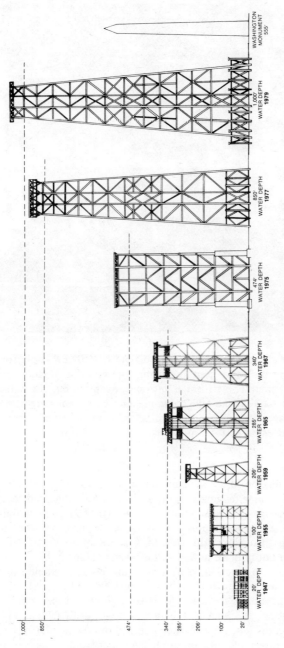

Fig. 38. A schematic diagram showing the evolution of structural changes and depth of operation of oil production platforms over a period of 3 decades.
Source: [269]; redrafted with permission.

Fig. 39. Statoil's and Mobil's Statfjord field *CONDEEP A* platform in the North Sea. The platform contains living quarters for a large number of workers. This concrete platform is movable by pumping the ballast water out of its large basal tanks and columns. The basal tanks are also used for petroleum storage. It requires no pilings and remains stable by resting on the seabed. Courtesy AKER Group A/S.

height of the Washington Monument in Washington, D.C. (Fig. 38). Through controlled directional drilling, as many as 60 wells may be connected to a single production platform. As of late November 1978, a total of 216 production platforms, mostly of the steel type, was reported to be in some phase of planning, construction or installation throughout the world.[332] As of mid-1979, about 1,000 steel platforms had been put in place in the Gulf of Mexico alone.[218]

Another type of production platform is the column stabilized submersible designed for deep waters, up to 600 ft. or more. The submersible platform takes various shapes, but its basic feature is a massive, buoyant base or hull which, after the unit is on station, is flooded or ballasted down until it rests on the seabed (Fig. 39). Once in position, water can be pumped out of the platform's base which is then usable for storage of crude oil produced either from its wells or from nearby satellite wells. Submersible production platforms are usually constructed by casting the base in a graving (dry) dock, then moving it into deeper water where it is allowed to float and then sink, by ballasting, while its height is increased. Norway's North Sea Ekofisk storage unit (capacity 1 million bbl. of crude oil) was the first structure built this way.[308] Another type of submersible production platform utilizes not only its own weight to hold it in place, but also uses the principle of suction or differentiated hydrostatic pressure. The base of the platform, when positioned properly on the seabed, forms a suction seep. Pore water is pumped from the seabed beneath the skirt, keeping the water pressure inside somewhat below that on the outside; this process maintains a suction force on the platform, thus increasing its stability.[186] Sand and clay seabed areas especially lend themselves to the use of this technique;[49] some engineers, however, have been concerned about the possibility of sand liquefaction when a bottom-founded structure such as the Ekofisk storage tank is put under heavy cyclic sheer stresses associated with storm waves.[310]

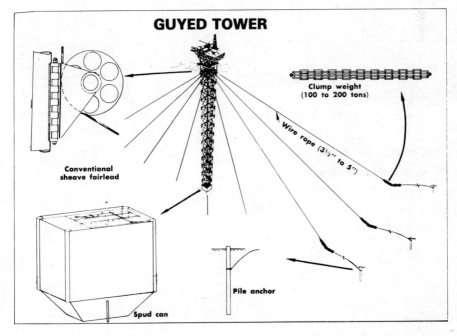

Fig. 40. A schematic diagram of Exxon's guyed tower. The tower has a base called a spud-can that is driven into the seafloor. Water depth, environmental conditions, and the size of the structure determine the number of guylines required. When a storm wave passes the platform, the clumps lift off the sea floor without much increase in tension.
Source: [432]. With permission.

A relatively recent innovation in production platforms is the guyed tower. The first unit of this type was erected off the coast of Louisiana in 1975. Although the guyed tower is bottom-founded, it is compliant, moving in response to wave forces. Studies are in progress to determine the utility of that and other platforms in 3,000 ft. of water (Fig. 40).

Subsea Completion Systems

During recent years, petroleum engineers have been developing wellheads (units that cap or control the flow of oil) that can be installed on the seabed at depths as great as 2,000 ft.[410] (possibly 3,000 ft.)[41] and tied by pipeline networks to central platforms or the shore. These systems are employed in nearly all offshore oil-producing areas, including the Middle East, the North Sea, Southeast Asia, Africa, and the United States. The first unit was installed in 1974. A 1977 estimate predicted that by 1978 as many as 140 subsea units would be operating.[328] Brazil's Campos field off Cabo de Stome (150 mi. northeast of Rio de Janeiro) is a good example of seabed completion engineering. By 1981, a complex system of wells and pipelines will produce and transfer the field's petroleum onshore near Macae (Fig. 41).

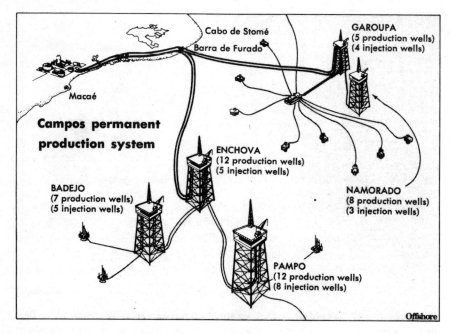

Fig. 41. The Campos permanent subsea and platform production system. It contains 53 production and 25 injection wells (used to maintain pressure). The Campos field has oil reserves estimated at 1.5 billion bbl.
Source: [474]. With permission.

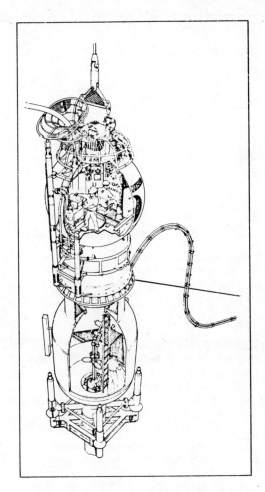

Fig. 42. A wellhead cellar (lower portion) and a one-atmosphere work chamber (upper portion). The work chamber is winched into position above the wellhead cellar and hydraulically sealed to it; after the water is pumped out of the cellar, workmen may enter it for maintenance tasks.
Source: [336]. With permission.

Several factors account for the oil companies' interest in subsea completion systems: (1) as oil production proceeds into deeper waters, well maintenance and monitoring is increasingly difficult, and subsea systems provide easier access; (2) surface platforms are ice, wind, and wave sensitive, whereas most subsea units are

130   *Petroleum technology*

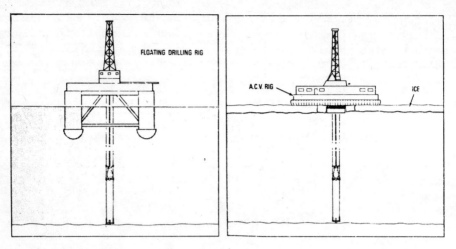

Fig. 43. A semi-submersible exploration rig and an air cushion vehicle (ACV) rigged to lower a dry one-atmosphere wellhead cellar to the seabed for well completion and maintenance work.
Source: [336]. With permission.

not;[463] (3) at times, because of the geological structures of hydrocarbon reservoirs, platform-drilled wells do not work adequately, so that subsea supplementary well systems (including producers and injectors) are used.[38]

Subsea completion systems can be either wet or dry. The dry systems allow maintenance and repair work in a surface atmosphere, shirt-sleeve environment. Nine dry systems had been completed or were in a planning/construction stage in 1978.[81, 328] Dry systems depend on an encapsulated wellhead whereby workers without diving equipment install the necessary hardware. Workers reach the wellhead by using a self-propelled submarine or a transfer capsule (Fig. 42) winched down from a surface support vessel, including a ship, a semi-submersible platform, or, if on ice, an air cushion vehicle (Fig. 43). The wellhead top is covered by a shell. The submarine or transfer capsule attaches to the shell by using a set of lips. Pumps remove the water from the shell, establishing a dry work environment which is maintained by the great hydrostatic pressure exerted on the jointed units.[463] Wet systems are usually diver assisted and, therefore, limited in the depth they can be used. Various safety systems for the transfer capsule workers include a TV-monitoring unit viewed from a support vessel onboard; cable cutters that, in an emergency, can cut the downhaul cable, thus allowing the capsule's positive buoyancy to carry it to the surface; purification cannisters that provide up to 96 man-hours of emergency air supply, in case the umbilical cord fails; and compressed air masks that can provide another 9 man-hours of oxygen supply.[41]

The ultimate objective is to have subsea completion systems maintained and monitored remotely, especially where rough weather frequently makes access difficult.[38, 512, 570] Preliminary design and feasibility studies for close-range monitoring of deep well sites have shown that it is possible to install a semi-permanent seabed home base (50-ft. diameter) from which small manned submarines can emerge to inspect and work on wellheads.[463] Although subsea completion systems hold much promise for a more efficient operation of oil fields, they cannot solve all problems. They cannot, for example, house the large and complex equipment needed in separating crude oil, gas, and water.[255]

SUMMARY AND CONCLUSIONS

Petroleum exploration and production technologies have been developing rapidly, and as they continue to evolve, petroleum geologists and engineers will be able to push farther into the OCS and onto the slope and the deep seabed, areas now thought to hold considerable potential for petroleum reserves. Producers are already seriously considering production work in waters as deep as 2,000 to 3,000 ft., and exploratory drilling is being done at 4,000 ft. Remotely controlled and manned submersibles are likely to become even more important as the "work seahorses" of the offshore as deeper waters come into use. Many tasks can be performed only by saturation or one-atmosphere divers and by men using shirt-sleeve work chambers lowered to the seabed. Drillers are becoming highly skilled in working in once forbidding arctic waters, areas where considerable potential exists for new discoveries.

Advances in petroleum exploration and production technologies are apparently essential to make the quest for offshore oil and gas increasingly more compatible with costs. Greater investments and R&D activities in America and Western Europe are probably in store for the 1980s as areas, that were marginal prior to the surge in world prices after 1973, become increasingly more attractive to explore. Better and safer hardware and vessels should be designed in the not so distant future, leading to more discoveries and likely more economical offshore production.

NOTES

[a] A Canadian firm in Vancouver, B.C. is experimenting with a converted military submarine for making penetrometer measurements [83].

[b] The suit needs little auxiliary equipment or deck space (20 ft.^{2}); a small winch deploys and recovers the diver [266], [574].

[c] Diving specialists are experimenting with breathing an air/oxygen mix (cheaper than helium/oxygen) during saturation diving.

[d] Ear infections cause even more lost time than does the need for decompression [40].

Chapter 7

Transport-Storage-Transfer

The movement of oil and gas within and from offshore fields to onshore refineries or storage facilities is an important part of the marine petroleum industry. Engineers have developed sophisticated systems of marine storage, loading, pipelaying, and shipping that have increased the efficiency of transporting as well as producing oil and gas. Because of their immediate importance in moving oil within petroleum fields and to shore areas proximate to the fields, pipelines will be the primary transport mode examined here, as will special storage and loading facilities designed specifically for the offshore.

PIPELINES

There are two types of offshore oil field pipelines—those gathering oil from several wells or platforms for transfer to a central platform or loading buoy and those connecting a central platform(s) with the coast. For these often relatively short transfer distances, pipelines are more efficient than tankers which must load and unload their cargo. And, although initial capital outlay for purchases of pipe and pipelaying equipment and for the laying of the pipe is high, once the system is installed, maintenance overhead is normally less than for tankers. Consequently, nearly all United States Gulf Coast offshore producers, for example, send oil to onshore refineries by pipeline.[398] The OCS pipeline mileage in the United States increased from 6,700 in 1974 to 7,400 in 1976, and 9,650 in 1977. In 1978, it rose to a total of c.12,400 mi.[119]** A similiar growth situation exists in the North Sea.

North Sea

The North Sea dramatically illustrates the importance of offshore pipelining and its associated problems and costs. For example, as early as 1974-75, the per-mile cost of connecting the Forties field by a 32-in. pipeline to the coast of Scotland was $1.2 million,[30] more than 6 times that for a similar unit buried onshore.[179,255] The line was laid at depths as great as 420 ft., a contributing factor to the cost.

By the end of 1977, more than 2,000 mi. of pipeline, had been laid in the North Sea (Fig. 44); another 1,400 mi. were planned or under construction in 1978.[a] Major operating pipelines were owned in 1978 by Statoil/Phillips Petroleum (494 mi. in length), Total Oil (220 mi.), BP (217 mi.), Occidental Petroleum (124 mi.), and Shell/Esso (93 mi.). These pipelines, mostly in the areas of Scotland (654 mi.), and less in England (274 mi.) and West Germany (220 mi.) where they connected two landfalls with Ekofisk, ranged in diameter from 30 to 36 in.[264,321,b]

A heated debate occurred in 1978 between the United Kingdom's Department of Energy and several oil companies, including BNOC and the British Gas Corp., over the need for additional gas trunklines. The Department wanted to

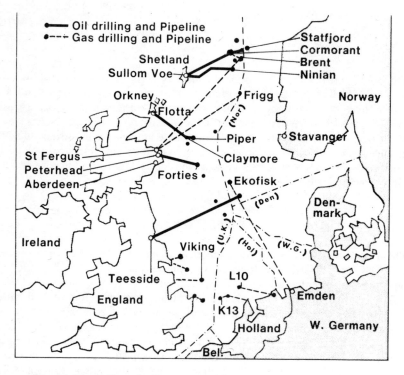

Fig. 44. North Sea pipeline routes, 1977.
Source: [264]; reprinted by author with permission.

see gas flaring ended, but its opponents, who carried out a special study of the problem, insisted the lines were not needed. The two parties also disagreed about the cost estimates for the government's proposed new trunklines, with the government contending that the study groups' estimates were much too high.[540] The Department of Energy is also putting pressure on the oil companies to tie their various fields to existing trunklines. As a result, Shell/Esso recently announced plans to bring its Cormorant oil field's gas ashore via its Brent to St. Fergus line.[363]

Laying of Pipelines

A typical pipeline laying technique involves a 300-ft. lay barge with a 175-man crew which assembles and lays approximately 1 mi. of pipe a day. The pipe is welded and then trailed over the barge's side or stern via a stinger (Fig. 45). Smaller barges follow behind dragging a "jet sled" along the seabed. High-pressure nozzles on the sled force air under the pipe, creating a trench into which the pipe settles.[398]

Getting the pipe safely onto the seabed is a difficult task. The laying down of the pipe at the start of downtime and the picking up of the pipe when work begins once again are especially hazardous periods, because it is then the pipe and stinger are easily damaged. The lowering process demands tremendous pipe strength. To illustrate, the 48-in. diameter trans-Alaska pipeline has a thickness of approximately 0.5 in., whereas North Sea pipelines (30–36 in. in diameter) range in thickness from 0.75 in. to over 1 in.[264] The lowering process is especially difficult during rough seas and when working in great water depths. The barge must be carefully positioned either by dynamic-thruster systems or by anchors. Forward anchor control installations include sophisticated electronic panels with many video screens to facilitate accurate positioning. The best weather in the North Sea occurs between April and mid-October, and even then it permits work during only 65% of the time. Thus, nearly all pipeline contracts contain a weather clause for downtime.[483] A large pipelaying barge recently developed by Brown & Root, Inc. can operate in very rough seas, work in great water depths (up to 1,100 ft. while laying a 36-in. pipe), and store up to 20,000 tons of pipe.[31] The barge is able to project pipe into the sea at an angle which varies with its size, weight, and water depth. A tensioning capacity of 360,000 lbs. that can be exerted on the pipe, a function controlling the radius of the sag bend at the sea floor, allows the barge to work without a stinger. In especially severe climatic areas where the sea may become ice covered, pipelayers have developed methods and equipment to lay pipe under the ice (Fig. 46).

A relatively new pipeline laying technique involves onland preparation of the pipe with subsequent towing to the work site. The technique is called bottom towing. The pipe is welded into sections, coated with concrete, and then the leading end of the pipe is pulled into the sea by a tow vessel. The ship pulls the pipe along the bottom of the seabed. Several projects using this technique have been completed without the concrete and welds having suffered any damage. One major success occurred with the pulling of 7,050-ft. pipeline sections from Tananger, Norway, across the Norwegian Trench, some 244 mi. in waters as deep as 1,260 ft.

Fig. 45. Pipelaying barge, *Semac I*. In March 1977, after trials in Moray Firth, this huge barge moved to the North Sea and began its first operations. The barge, which requires a large crew with many special skills, is laying the Shell/Esso pipeline that will soon carry gas from the Brent oil and gas field to St. Fergus, Scotland. *Semac I*, 433 ft. long and constructed in the United States, was specially designed to work in severe environments. A length of 36-in. pipe is shown passing down the barge's stinger.

Once the pipe is positioned, it usually must be buried in the seabed to prevent its being snagged by ships' fishing nets or punctured by anchors. All pipeline routes in United States coastal waters must be determined in consultation with Department of State and adjacent littoral state officials. The objective is to minimize conflicts among users of ocean space and between local, state, and federal governmental agencies. The hazard of fishing net entanglement is avoided by requiring the burial of all valves and taps, regardless of the water depth. All pipes must be buried to a depth of 3 ft., if the waters are less than 200 ft. deep.[4] The state of Virginia has sponsored a study of problems associated with pipelines connecting the shore with offshore areas. One recommendation resulting from the study was that pipelines should be limited to specific corridors extending from 3 mi. offshore

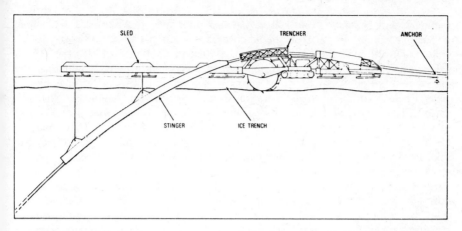

Fig. 46. An example of land-type trencher that can be used in laying pipe under ice. The trencher cuts through the ice and the pipe is fed down through a stinger suspended from the sled.
Source: [336]. With permission.

to the upper limit of the shoreline wetlands, thus reducing their impact upon the land and water interface.[413]

Besides the water jet technique noted above, plowing is another basic excavation method. The trench may be constructed prior to, during, or after pipelaying. A remotely controlled pipelaying device has been developed in Norway that eliminates the need for diver assistance. The device clamps its four wheels onto the pipe from above and then rolls forward on the pipe while digging a trench below. Via an attached control cable, the mother ship monitors the machine's digging head, eight propeller thrusters, and four spherical ballast tanks.[536]

Sea-bottom areas with soft sand or mud are amenable to the use of relatively light-weight tubular skid-mounted plows. Where the seabed is composed of boulder clay, as occurs over much of the North Sea, heavy plows mounted on wheels, on the order of tanks, must be used.[60] If rock is encountered, trenching work becomes especially difficult and expensive. Most sedimentary and conglomeratic rocks, and some metamorphics, can be trenched by using cutter suction and bucket dredges. These dredges, however, cannot operate where wave action is greater than 3 ft. or where water depths are greater than 100 ft. or less than 12 to 15 ft. Consequently, bangalore and shaped charges often must be used to fragment the rock and to heave it out of the trench. Placing of explosives is a diver-assisted job. In some areas it is necessary to remove overburden before the rocks can be blasted. When the rock is several feet thick, numerous passes must be made, perhaps gaining only 3 ft. in a given pass.[421]

In some areas, instead of being buried in a trench, the pipeline can be covered by loose sand or sandbags. Except for a 30-mi. portion, a gas pipeline connecting the North Sea's Ekofisk field with Emden, West Germany, is covered by sandbags.

Denmark and Norway recently have been at odds over the Ekofisk-Emden line. A portion (about 10 mi.) of the uncovered section passes through Denmark's sector, and although the line was commissioned in September 1977, the Danish government, fearful that trawlers and marine construction crews might snag the line, would not allow gas transport across its seabed, until the line was adequately covered; negotiations delayed operation of the pipeline for 7 months. Norway's national oil agency, Statoil, and its partner, the Phillips Group, used 850,000 bags of sand to cover and anchor the pipeline in the Danish sector.[321] The remaining 20 mi. will be covered with dredged sand, using a dynamically positioned drill ship fitted with a specially designed fallpipe and diffuser head.[6]

In some sandy bottom areas, where currents exist, trench-laid pipe can be covered by allowing current-transported sand to fill the trench. There are times, however, when the work of burying pipelines is undone by the sea. Currents and tides may scour away the cover over a pipeline, as has occurred in a section of the Brent pipeline lying near the approach to Scotland's Yell Sound. An uncovered section of the Brent pipeline actually floated to the surface.[490]

Erosion around pipelines can be controlled by placing bunches of polypropylene strands in blocks along the pipeline. Because the polypropylene strands are lighter than the water, they trail vertically from the seabed. Sand particles and other materials become trapped by the swirling strands and fall to the seabed, gradually building up adjacent to the pipe.[21]

Pipeliners are faced with special difficulties when attempting to lay pipe in deep waters. For example, until recently, engineers had considered the construction of a pipeline across the Norwegian Trench (over 1,000 ft. deep in places) as impractical, given existing engineering capabilities. The basic problem of pipelaying in deep waters is position maintenance. With the development of dynamic positioning systems for barges and ships and techniques for preassembled pipelines, deep waters are no longer so formidable.[2] In 1976, pipeliners began to seriously consider laying a gas and an oil pipeline across the Norwegian Trench.[319] In late 1978, engineers determined the feasibility of an oil line to connect the Statfjord field with Sotra, Norway, west of Bergen. The government expects to make a final decision on the line during 1979. The proposed gas line would join the Frigg field with Karmoy, Norway, northwest of Stavanger. Its fate is also to be decided in 1979. Government spokesmen believe it unlikely that this line will be built.[54]

Maintenance of Pipelines

In addition to the hazards of pipeline puncture or scour, oil producers must be alert to various other pipeline problems such as: (1) deformation of a pipe cross section; (2) leaks or failure caused by defects in the pipe; (3) corrosion induced leaks; (4) collapse or buckling; (5) weld separations. These problems must be monitored by saturation divers or manned and unmanned submersibles.[230] Pipeline repair is one of the most difficult tasks in marine pipelining. Recently, divers have been repairing pipes through hyperbaric welding. The system uses a dry work chamber lowered to the seabed around the pipe. The chamber is filled with a breathable helium and oxygen mixture; divers then enter the chamber to make repairs. This technique has been used successfully on the North Sea's Frigg gas line owned by Total Oil.[264]

OFFSHORE STORAGE-TRANSFER SYSTEMS

Problems of petroleum storage for later pipeline or tanker shipment in offshore oil fields present a major challenge to producers. Oilmen use three basic storage-transfer systems: single-point mooring buoys, production-storage platforms (see Chap. 6, pp. 126-27), and independent storage tanks and bladders.

The increasing size of tankers prompted the development of loading systems offshore in deep water. These systems are now being used at oil fields, and one of the most successful is called "single point mooring" (SPM). Construction and operation of SPMs require know-how in mechanical, civil, and marine engineering, and demand knowledge of marine architecture and meteorology, oceanography, and geology.[548] By 1975, some 150 SPMs were operating, and it is expected that by 1980 as many as 200 will be in place, although not all will be associated with offshore oil production. Because tankers moor to the buoy at one point only, the ship can rotate or "weathervane" freely around the mooring buoy (Fig. 47).

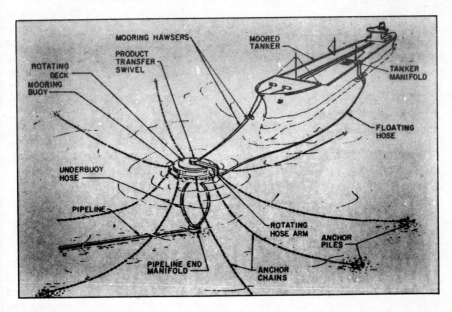

Fig. 47. A catenary anchor leg mooring (CALM) system. This structure contains a mooring buoy as its main feature. Four to eight chains anchor the unit to the seabed.
Source: [149]. With permission.

Oil field loading systems are basically of two types: (1) a permanently stationed tanker receives oil from the loading buoy and then transfers that oil to shuttle tankers, or (2) the oil is loaded directly into the shuttle tankers, which necessitates stopping the pumping system while the loaded tanker moves off to be replaced by an empty one.[149] Waves pose a major hazard in the use of SPM systems not only for the ship but for the loading buoy itself, if the ship should strike it. On the other hand, if the weather becomes rough, the ship can easily disengage from the loading buoy.[548]

A system used since the late 1960s, pioneered by the Chicago Bridge & Iron Co., allows oil storage immediately above the oil field, avoiding the need for piping oil ashore only to be returned to an offshore tanker loading facility. The system consists of a large bottomless shell that rests on the seabed. Because oil floats on water, oil pumped into the shell forces the water out through escape holes in the perimeter of the base. The Chicago firm built a total of three storage shells (500,000 bbl. capacity each) and installed them in the Fateh field for the Dubai Petroleum Co. headquartered in the tiny state of Dubai. The first storage shell, *Khazzan Dubai I*, was installed in 1969 and *Khazzan Dubai II* and *III* were put in place during 1972, all at the same site 60 mi. off the coast of Dubai (Fig. 48). The storage-tank method has also been utilized in the North Sea. Shell/Esso, for example, uses a giant storage tank, *Spar I*, in its Brent field. The tank was towed into position and can take oil directly from producing wells, whereupon, on demand, disgorge it into tankers.

A commonly used storage method depends on using the production platform itself. The Norwegians are leaders in this type of platform construction. The Norwegian "Condeep" platform used successfully in the North Sea's Beryl field has a crude oil storage capacity of 900,000 bbl.[97] Another Norwegian-built unit, the "Conprod," provides storage for 500,000 bbl. of oil while producing 180,000 bbl. a day.[372] Both platforms are constructed of large concrete cells that provide buoyancy as well as storage space (see Fig. 39). A less common but safe and effective method used to store and transport small amounts of oil has been developed recently by a private engineering firm and the U.S. Navy. The system consists of a large bladder (200 ft. long and 11 ft. in diameter) made of a synthetic rubber-coated, nylon fabric and cord material. The bladder can hold 2,762 to 3,000 bbl. of oil, and it is capable of withstanding 15-ft. waves and winds in excess of 40 knots, while being towed.[412] The long-term success of this oil transport technique remains to be proved, and will probably be useful only for small shipments.

SUMMARY AND CONCLUSIONS

One of the most intensively developed petroleum pipeline-usage regions is the North Sea, and it is likely that many additional lines soon will be constructed, one of which may span the Norwegian Trench, long a technically difficult and uneconomic undertaking. New pipelaying methods and equipment have now put all areas of the North Sea, and even ice-covered areas of polar regions, within reach of pipelayers. Pipeline maintenance will continue to be a major task and expense

Fig. 48. The *Khazzan II* oil storage facility. The tank is 205 ft. high, 270 ft. in diameter, and weighs 15,000 tons. After being built on a nearby shore and then floated out to sea on a cushion of air, the tank was submerged in 155 ft. of water; it is shown beginning its descent to the Persian Gulf's seabed. Once on the bottom, the tower extends 45 ft. above the water surface. Courtesy Chicago Bridge & Iron Co.

to the oil companies. The new supertankers can greatly reduce per-unit shipment costs and may make some previously marginal fields more economic. Single-point mooring systems are an especially useful innovation that allows the loading of supertankers without their entering regular ports, which are often unable to accommodate them. Large independent seabed storage facilities, such as the *Khazzan Dubai* units or those built into production platforms, reduce the back-haulage of petroleum from onshore storage terminals. As world petroleum production shifts to the offshore, marine pipelines and storage facilities will become even more important in the decades ahead.

NOTES

[a] Estimated from a map in [321]:132-33.
[b] Also estimated from a map in *Offshore*, **38** (Feb. 1978):149-150.

Chapter 8

Petroleum and the Environment

Offshore development of hydrocarbons is fraught with numerous problems, including natural and man-made hazards to oil field workers and platforms, open water and coastal ecological damage from oil spillage and waste disposal, and conflicts with other users of ocean space and resources, like the fishing and shipping industries. The international dimension of potential environmental damage from the offshore petroleum industry was early recognized under the 1958 *Geneva Convention on the High Seas*. The *Convention* stipulates that "Every state shall draw up regulations to prevent pollution of the seas by the discharge of oil from ships or pipelines or resulting from the exploitation and exploration of the seabed and its subsoil...."[196] Many world states are attempting to abide by this stricture, especially so after several subsequent oil field and tanker accidents focused world attention on the problem.

MAN-MADE HAZARDS

The offshore petroleum industry's pollution record has often been cast in a bad light, but when compared with other sources, its performance is favorable, both in maintaining control over human and industrial wastes and in preventing oil spillage. The control of these problems presents a formidable technological challenge and a major expense to producers.

Human and Industrial Wastes

Numerous federal and state regulations control waste disposal in the United States' offshore petroleum fields. Drilling muds containing toxic materials or additives must be neutralized before being discharged into the ocean or they have to be backhauled to shore. Several studies have shown, however, that dumping drilling muds into the sea has no negative effect on the marine environment.[546] Human wastes must be handled similarly to drilling muds. But on platforms having a limited amount of human waste, electrocatalytic converters, or "electric chairs" as they are known to platform workers, incinerate the waste into a vapor and then pass it through a catalyst. In areas where the weather is hot and humid, it is difficult to store a large volume of galley garbage; consequently, it too must be frequently backhauled.[496]

Oil Spills

A 1975 U.S. National Academy of Sciences study has provided comparative best-estimate data showing that world offshore oil well production in the mid-1970s was annually introducing 0.08 million tons (600,000 bbl.) or 1.3% of the total 6.113 million tons (45.3 million bbl.) of petroleum hydrocarbons entering the oceans. In comparison, transportation sources accounted for 34.9%, river runoff for 26.2%, atmospheric rainout for 9.8%, urban runoff for 4.9%, and natural marine seeps for 9.8%. Translated into percentages and assuming these values remain unchanged, the volume of offshore oil well pollution is 27 times less than that of transportation sources, nearly 4 times less than urban runoff, and 7.5 times less than pollution occurring from natural oil seeps.

Offshore oil fields spillage occurs in various ways, including blowouts, pipeline breakage, and loss during transfer, primarily from supertankers to smaller craft. Spills usually form slicks that can rapidly spread from the point of origin. Oil spill drift is normally a function of wind, current, and wave conditions. Wind accounts for most of a slick's movement, and the rate of movement is usually estimated to be approximately 3% of the wind speed. Subsurface currents and wind-induced surface currents also enter into the problem of predicting speed and direction of slick movement. Wave and slick movement relationships are less well understood; small waves seem to increase slick movement, whereas large waves reduce it. The oil's viscosity, density, and volatility, also, contribute to varying rates of spread.[44]

Several major oil spill disasters have occurred in recent years, and it is these events that account for most of the total volume of spillage. Most spills, however, are small. A 1977 study shows that two-thirds of all reported oil discharges from production facilities in the United States' OCS were caused by equipment malfunctions; 7% stemmed from errors made by personnel.[495] During the years 1971-76, oil spills of 50 bbl. or more occurring on the OCS averaged five events annually. The total spillage for the 5-year period was 51,311 bbl. A few large spills—50 bbl. or more—accounted for most of the spillage (Table 14).

Table 14. Hydrocarbon Spills on the United States' OCS, 1971-76

Year	Spills of 50 bbl.[a] or more			Spills of less than 50 bbl.[a]		
	Number	Bbl. spilled	Ton spilled	Number	Bbl. spilled	Tons spilled
1971	11	1,285	174	1,245	1,492	202
1972	2	150	20	1,159	1,032	139
1973	4	22,175[b]	2,997	1,171	921	124
1974	8	22,721[c]	3,070	1,129	667	90
1975	2	266	36	1,126	711	96
1976	3	4,714[d]	637	949	523	71

Source: [119*].
Notes: [a]50 bbl. = 6.8 metric tons; [b]9,935 bbl. spilled from a ruptured storage tank (Jan. 9), and 7,000 bbl. from a leaking barge (Jan. 26); [c]19,850 bbl. spilled from a pipeline broken when a ship's anchor dragged on the seabed (Apr. 17); [d]4,000 bbl. spilled from a pipeline tie-in broken by a shrimp trawl drag (Dec. 18).

Oil spillages may have significant impacts on marine and coastal ecosystems. Fishermen, marine biologists, and officials at all governmental levels are concerned for coastal biota viability and onshore social and financial costs. In open seas, the major problem is damage to surface organisms such as plankton and pelagic fish. In coastal waters, there may be severe but relatively localized difficulties, especially if the oil is churned into beaches or enters tidal estuaries where it often smothers crustaceans and shell fish or immobilizes and kills birds. Estuaries and beaches may take 2 to 3 years, or more, to recover. On the other hand, studies of Venezuela's Lake Maracaibo and of California's Santa Barbara Channel have demonstrated that many marine organisms' growth rates have not been diminished even when exposed to constant seepage from local oil wells or natural seeps.[44] Under new amendments to the *OCS Lands Act of 1953*, passed in September 1978, a fund of $100-$200 million was established for oil spillage liability; fishermen and others suffering losses from oil spills are eligible for assistance. The fund is maintained by a 3¢/bbl. tax on all oil produced on the OCS. Platform operators are liable for damages and clean-up costs up to $35 million.[373,395]

Small-scale oil spills may go unnoticed or unreported, but major spills have had a significant impact on forming world opinion and prompting tighter offshore oil production regulations and enforcement; the Santa Barbara Channel oil spill of 1969 is a case in point. That spill is attributed to geological conditions, deficiencies in the drilling program, and human error. Without attempting to pinpoint the reason for the spill, one can safely say that it electrified public concern for coastal environments in petroleum-producing regions. Consequently, the U.S. Congress, state legislatures, and local governments and regulatory agencies at all administrative levels moved toward a more vigorous control of the offshore oil industry. The federal *Water Quality Improvement Act of 1970*, containing numerous cleanup and prevention regulations dealing with offshore oil discharges, exemplifies this surge of interest.[293]

The OCS Environment and the DOI

Since 1969, the DOI has instituted much more stringent regulations on petroleum exploration and production safety, including changes in bore hole casing depth specifications and cementing techniques, blowout prevention equipment, well completion methods, and waste disposal practices. All significant pollution accidents must be investigated and the findings made available to the public. And accident contingency plans for oil spill containment and cleanup must be on file with the DOI.[477] Federal regulations passed in 1978 require that before exploratory drilling can begin, oil companies must file environmental reports along with each exploration proposal and obtain an exploration permit from the USGS; and in state waters, a permit from the U.S. Corps of Engineers must also be acquired. The Corps' main concern is unobstructed navigation.[121, 276]

The Conservation Division of DOI's USGS is responsible for administering OCS development plans and environmental affairs. During the fiscal year of 1977, the Division's officials supervised oil and gas production activities of 1,970 leases in OCS waters off the Atlantic, Gulf of Mexico, Pacific, and Alaska coasts, covering some 9,312,000 acres. A total of 540 exploration and development plans were examined and approved. Inspections of drilling rigs (2,655) and production platforms (2,250) resulted in 683 warnings for improper operations.[119]* Although the OTA contends there are no systematically planned government inspections of the construction, installation, and operation of offshore platforms,[398] the work of the DOI seems to have been effective in reducing the number of oil spills. Despite recent increases in OCS petroleum activities, the overall trend for oil spills is downward, as is the total tonnage lost in spills of less than 50 bbl. (see Table 14).

Environmental impact studies prior to leasing. The DOI is responsible for developing very detailed environmental impact statements for all areas it proposes to lease. A good and highly important example of the studies is the BLM's *Environmental Impact Statement* prepared for a proposed 1979 sale of tracts in the mid-Atlantic OCS, the Baltimore Canyon area. The study presents individual probability estimates for the frequencies of oil spills and their potential magnitude from platforms and pipelines (Fig. 49). Notations P1-P6 represent six hypothetical points of origin for petroleum which may be shipped by pipelines passing through points T1-T7 to various places onshore. The BLM calculated that, collectively, the expected number of spills greater than 1,000 bbl. (using pipeline transport) from areas already leased or those proposed for leasing is approximately 0.62 spills during the lease areas' production life. For the proposed lease areas, the probability that no spills will occur during the lifetime of production is 54% (Fig. 50).

Hypothetical trajectories or pathways for oil spills were plotted on a digital map of the area between latitude 35° and 42°N. and between longitude 69° and 76.5°W. Trajectories from 500 hypothetical oil spills were simulated by the Monte Carlo method for each of the four seasons and for each of the 13 points in the mid-Atlantic area, "resulting in a total of 26,000 trajectories that reflect wind

Fig. 49. Potential points of origin for oil spills from existing and proposed lease sites (P1-P6) and transportation routes (T1-T7) offshore in the mid-Atlantic. Shaded tracts are designated for inclusion in the 1979 lease sale; unshaded tracts represent existing leases.
Source: [25]; redrafted.

and current patterns." Figure 51 shows oil spill trajectories for the winter season (which were similar to the trajectories generated for the other three seasons) at site P4 (see Fig. 49), the proposed leasing area's approximate center. The pathways meander, but the net movement is eastward. By combining spill frequency estimates and the probability of a spill following a particular path, the overall oil spill risk posed by oil and gas production in the sale area can be assessed. A similar study done earlier for the same general area for a mid-Atlantic lease sale held in 1976 showed a somewhat different trajectory pattern. The study's designated point of origin for the oil spill was in the vicinity of P2 (see Fig. 49). Trajectories tended to

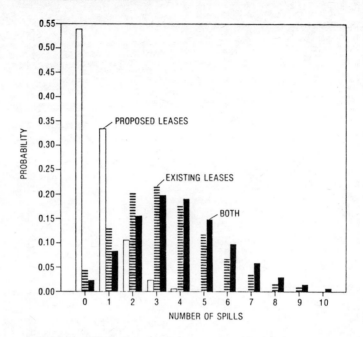

Fig. 50. Frequency distributions for platform and pipeline transportation spills greater than 1,000 bbl. during production lifetime of lease areas.
Source: [25].

be southwesterly for all four seasons. The summer trajectories experienced six landfalls (Fig. 52); the winter season had one landfall. The time elapsing from spillage to landfall varied from 33 to 60 days for summer events to 18 days for the winter's one event. The total time lapse from spill to landfall is important, because the spill's impact on the coast will usually decrease as the length of time the spill is at sea increases.[492] The authors of the 1976 study, stressing the essence of assessing potential environmental damage to coastal zones near offshore oil fields, stated:

> An important fact that stands out when one attempts to evaluate the significance of accidental oil spillage for [the OCS], or any proposed lease area, is that the problem is fundamentally probabilistic.

As in the study conducted for the 1979 sale, the authors calculated frequency probabilities for landings of oil spills larger than 1,000 bbl. during the proposed lease area's 25-year production life. There is a 61% probability that no landings of 1,000-bbl. oil spills will occur, and less than a 1% probability that four landings will occur.[492, a] This study, also, generated probabilities of an oil spill episode's

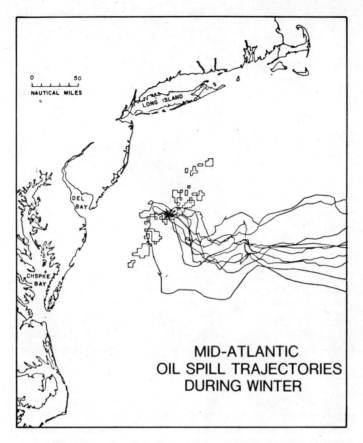

Fig. 51. Example oil spill trajectories during winter for spill site P4 near the center of the proposed mid-Atlantic lease area.
Source: [25]; redrafted.

affecting specific biological resources and recreation areas. The probabilities were relatively low—endangered species (1.5%), migratory birds (4%), shell fisheries (1%), coastal finfish (7%), estuarine and anadromous fish (1%), wildlife refuges and management areas (3%), resort beaches (2%), and park and recreation areas (5%).[492]

The OTA disagrees. The findings of the two studies outlined above seem to indicate that the hazard of oil spills along the mid-Atlantic coast is not excessive. But according to an independent study performed by the U.S. Coast Guard for the OTA, an oil spill occurring in the summer months, when the area is under the influence of a stagnant high-pressure system, has a very good chance of reaching shore,[398] which could virtually wipe out the affected area's seasonal tourist income.[438] In 1976, the OTA pointed out that it is debatable whether oil spills occurring along the Atlantic coast would be handled adequately, and it contended

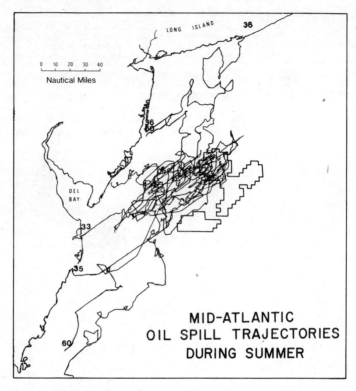

Fig. 52. Example oil spill trajectory results during summer conditions for a spill site in the northwest portion of the proposed mid-Atlantic lease area. Numbers of trajectories reaching the coast give time to land in days.
Source: [492] ; redrafted.

that the USGS did not "...appear to employ the best available system for establishing standards and enforcing regulations dealing with oil spill prevention and cleanup." Although oil spill control is vested in the USGS, the technical knowledge for oil spill detection and containment has been developed primarily in the Coast Guard which, until the *1978 OCS Lands Act Amendments* were instituted, had no statutory authority over oil and gas development activities on the OCS.[398] As of mid-1979, the USGS had an agreement with the Coast Guard to review its oil spillage prevention and cleanup programs.

Blowouts

Most oil spillages are potentially hazardous to marine biota and coastal zones, but those associated with blowouts (an uncontrolled well) pose a special threat—

Table 15. Fires and Explosions on the United States OCS, 1970-76

Year	Number of events	Injuries	Fatalities
1970	12	31	17
1971	19	16	1
1972	31	9	0
1973	30	9	2
1974	22	15	0
1975	27	11	10[a]
1976	32	14	0

Source: [119*].
[a] A tanker collided with a platform. Oil escaping from the tanker ignited and set the ship on fire. Six men died in the blaze (Aug. 15).

explosions and fires that may destroy the rig and endanger personnel. Fortunately, blowout indicence is less than 2 wells for each 1,000 drilled. Blowouts occurring during production operations are nearly always caused by fires on the platform, platform failure, or the collision of a vessel with the platform.[44] From 1970 through 1976, the United States OCS petroleum industry experienced 173 explosions and fires, an average of 24.7 events annually, with a total of 105 injuries and 32 fatalities. Some of these fires and explosions were associated with blowouts (Table 15).

Blowout control. Trapped petroleum deposits are under pressure. Drillers control this pressure by pumping heavy drilling mud into the well to keep the oil and gas from pushing upward through the drill hole. Failure of the mud to restrain the pressurized gas or oil can result in a blowout. All oil drilling and pumping operations have safety backup systems. During drilling from floating platforms, operators attach a blowout preventer (a set of large valves) to the top of the casing on the ocean floor. During production, an elaborate system of valves and fusible plugs, a "Christmas tree," located on the deck or seabed, controls the flow of gas or oil to the surface (Fig. 53); operators also install a safety valve below the seabed. The valve sits within a series of casings set several hundred to thousands of feet into the seabed. The subsurface safety valve is kept open by hydraulic pressure maintained from the platform. If this pressure is lost, the valve closes. Needless to say, these systems sometimes fail, or human error occurs in their operation or installation. Fortunately, within 5 to 15 days, approximately 60% of all blowouts plug, bridging themselves by the collapse of the oil-bearing strata into the drill hole.[44]

Ekofisk. In the North Sea, a major blowout resulted from both mechanical failure and human error on April 22, 1977. The accident occurred on the Bravo Platform of the Ekofisk complex located in the Norwegian sector. Workers, attempting to pull tubing, were replacing a down-hole safety valve when it gave way,

152 Petroleum and the environment

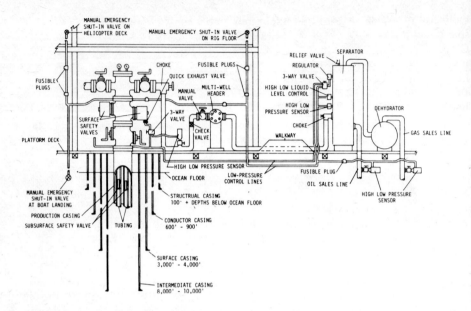

Fig. 53. Schematic diagram showing casing program and production safety system of a typical 12,000-ft. well, in the Gulf of Mexico.
Source: [4].

allowing an oil and gas discharge to reach the platform deck. The discharge became so severe that, after shutting the platform's other 14 wells, all 112 men aboard [563] evacuated the rig, using survival capsules dropped from the deck to the sea surface.[166] To reduce the fire hazard, a fire-fighting vessel sprayed the rig with water.[b] All work in the Ekofisk area stopped for 50 hours, until it was determined that the escaping gas and oil posed no threat to the use of the field.[563] A team from an oil well troubleshooting firm based in Texas was called in to control the blowout.[307] Capping efforts were finally successful on April 30, although the Bravo Platform remained idle until July 19.[563, c]

During the blowout, 15-21,000 tons (110,000 to 155,000 bbl.) of oil discharged into the North Sea. It is estimated that 50% evaporated; only 800 to 1,000 tons (5,900 to 7,400 bbl.) were mechanically recovered during cleanup, primarily because rough seas did not allow recovery equipment to function properly. To reduce the fire hazard, 55 tons of chemical dispersants were distributed, an amount capable of handling about 1,000 tons of oil. Fortunately, because of favorable winds and currents, the slick never reached closer than 70 nautical mi. to any shore, and by June only small (diameter 0.1 in.) scattered tar balls remained. Results of an investigation of the blowout's effects on marine life showed no significant impact, although some birds, fish and plankton were affected. This fortunate outcome may have resulted because of a relatively inactive level of biological processes in that portion of the North Sea at that time of year. Total oil spill cleanup costs were calculated in 1977 as probably between $5-$10 million.[298] Costs of lost production during the field's 8-day closure totaled at least $57 million,[321] and control effort costs were considerable.[563] Investigators found that human error caused the blowout. The down-hole safety valve was blown from the bore hole because it had not been locked correctly and a blowout preventer had been installed in an inverted position.[321, d]

Monitoring North Sea Oil Production and Environment

The Ekofisk accident caused considerable debate within government circles in Norway. Even prior to the blowout, many officials were not happy about Norway's entry into the oil producing business or about the rapid pace of the development effort.[165, 393, e] The blowout cost the Norwegian government 40 million kroner ($7.48 million) in royalty fees and a tax loss of approximately 250 million kroner ($46.75 million).[389]

Although Ekofisk's financial costs were high, some observers view the accident as having had a positive effect because executives in industry and government took a more careful look at safety procedures followed on the North Sea's platforms. Norwegian regulatory agencies have, in fact, been accused of being so obsessed with safety that North Sea oil investment and development are unnecessarily retarded. For example, Norway will not allow oil production on a platform that is also drilling at the same time; Great Britain allows this practice, which some claim increases the accident hazard. Given Norway's strict rules and considering the number of workers injured or killed annually in its oil industry, the government has reason for concern. Injuries during the period 1971-76 averaged 122 per year rising from 21 in 1971 to 328 in 1976; fatalities averaged 2.7 each year. The average number of injuries per 100 workers during 1973-76 was 10.9 per year.[54, f] The United Kingdom, partly as a consequence of the accident, established a team of high-ranking government officials to coordinate control procedures, should a similar event occur in the British sector.[515] In 1969, all states bordering on the North Sea made a pact—the Bonn Agreement—to mutually cooperate on North Sea oil pollution problems. The signatories pledged to exchange information on all oil-related accidents. Thus, within only a few hours after the Bravo blowout, all parties had been informed; they were kept posted on progress in controlling the blowout and in dispersing the oil. In a follow-up, Norway extended an invitation to the governments of Sweden, Ireland, Iceland, France, West Germany, the Netherlands, the United Kingdom, and the United States, as well as representatives of the OECD, EEC, and the Economic Commission of Europe, to meet and discuss the lessons learned from the accident. Out of this meeting came a commitment to convene yet another conference, in early 1978, to consider improved safety and pollution prevention in the North Sea oil fields.[563] By late 1978, Norway and the United Kingdom had worked out a joint program for mutual assistance in emergencies, and they will share in the cost of construction and operation of a special emergency support boat.[539]

The various states bordering the North Sea have different but, for the most part, effective systems of regulating exploratory and development activities of the region's oil producers. Norway and the Netherlands have regulations placing the burden upon the producer to prove they are not damaging the environment; their regulations use words such as "all proper measures," "safe," and "best oil field practice." Norway maintains rigorous governmental inspection programs for platform safety and oil spillage.[g] Although the United Kingdom has specific stipulations for onshore zoning and land use, it holds basically to a "reasonable man" and a less direct governmental monitoring regulatory policy for safety and oil production. This method thus far has been successful in avoiding most conflicts of interest and litigation over conservation problems and among various marine resource users,[578] although not for blowouts and oil spills. The British government has a much less restrictive regulatory scheme than the United States under its *1969 Environmental Protection Act* which requires detailed examination of any likely impacts a given construction program might cause. The United Kingdom does have a *Prevention of Oil Pollution Act* (1971) providing for a fine up to £50,000 for an oil spill. Fines have frequently been levied at a rate less than this amount when the operator has been able to show that "neither the escape nor any delay in discovering it was due to any want of reasonable care, and that as soon as practicable after it was discovered, all reasonable steps were taken to stopping or reducing it."[550] The United Kingdom exerts considerable local control over offshore oil producers by placing supervisory authority in the hands of port city police. Each port's police force is responsible for monitoring all oil rigs and platforms under its jurisdiction. Aberdeen's police, for example, are responsible for the surveillance of operations as far away as the East Shetland Basin.[255]

NATURAL HAZARDS AND THE WORKING ENVIRONMENT

Offshore oil producers and workers must cope not only with man-made hazards associated with platforms—blowouts, fires, explosions, and sanitation—but also face hazards of the marine environment itself, including storms with their winds and waves, currents, scour, and ice. However, elaborate use of technical expertise and sophisticated engineering techniques and safety precautions is allowing producers to venture into areas that only a few years ago would have been economically unproductive and/or technically impossible.

Survival Systems

When hazardous conditions develop on a platform, as occurred in Ekofisk's Bravo Platform blowout, or when in danger of capsizing, drilling and production platform crews need some kind of survival system to reach the water lying more than 100 ft. below. There are several types of survival systems capable of sheltering up to 50 people. These systems are specially built to meet the needs of today's offshore oil industry; constructed of fiberglass, self-righting, self-propelled, and highly fire resistant (Fig. 54) survival vessels have already proven themselves. One

Fig. 54. Rig-mounted 20-man survival capsule system ready for launching. The system is constructed of strong fiberglass and powered by a 40-hp diesel engine. The craft can be launched in waves of 50 ft. or more; it descends from the platform deck to the water at 2 ft./sec. Courtesy Whittaker Corp.

unit, the Harding Safety Lifeboat, was put to a severe test on March 1, 1976, when the semi-submersible rig *Deep Sea Driller* went aground northwest of Bergen, Norway. Winds of 70 knots had whipped the sea into mountainous waves, pushing the drilling rig into shallow waters where it partially capsized (Fig. 55). The crewmen managed to launch one of the rig's survival lifeboats. After launching, a huge wave overturned the lifeboat, but it righted itself and the men inside were rescued after the craft stranded on the shore. Unfortunately, six of seven men who had chosen to cling to the outside of the lifeboat were drowned when it flipped over.[370]

Storms

Exposed to storms moving eastward out of the North Atlantic, the North Sea frequently experiences winds of 50 knots and waves of 60 ft. In severe storms winds may reach hurricane force and waves may attain heights of 80 ft. Consequently, design engineers must plan for the "storm of the century," which might generate winds of 100 knots and waves of 100 ft. The overturning force of

156   *Petroleum and the environment*

Fig. 55. The *Deep Sea Driller* aground off the coast of Norway. Courtesy Harding A/S.

such waves is tremendous, 10 times that of maximum waves in the Gulf of Mexico.[42] Waves of much smaller magnitude can cause considerable downtime during drilling and supply operations. Five times more downtime occurs in the North Sea than in the Gulf of Mexico. Figure 56 provides some notion of the comparative severity of working conditions in the North Sea and other offshore areas noted for severe climatic conditions, as in the Gulf of Alaska and Eastern Canada. Because of these severe conditions, structural steel members used in North Sea platforms need to be 5 to 10 times stronger than those required in the Gulf of Mexico.[359]

In the early period of North Sea development, engineers and government officials did not fully recognize the severity of the area's climatic conditions and the platform design problems they faced. In 1965, because of bad weather, British Petroleum lost a rig, the *Sea Gem*. Forty-five men onboard were killed when the rig went down.[567, h] Heavy weather off the Shetland Islands has caused the loss of a seismic survey ship, a new semi-submersible rig, a rig-standy ship, and a barge.[491] In the same area during rough weather, an articulated tower (owned by Mobil Oil) in the Beryl field broke loose from its concrete base and floated away. The tower was recaptured, secured, and then it broke loose again during a storm

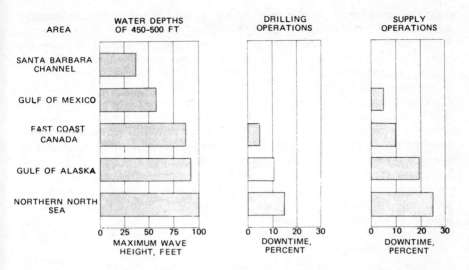

Fig. 56. Sea-state conditions in selected offshore areas and their effect on drilling and supply operations.
Source: [526]. With permission.

on the following day. The tower had to be towed to Stavanger, Norway, for inspection and repairs. This accident caused a 6-month's delay in bringing the Beryl field into production.[3]

Engineering problems of oil rigs are highly varied in relation to weather conditions and currents. Relatively shallow waters in the southern North Sea tend to accentuate wave heights during storms; in the north, not only local storm waves but swells from distant Atlantic storms are an additional problem. Regional coastal configurations and local, fast moving depressions generate wave trains, resulting in highly complicated patterns that create severe and complex stresses on platforms and other offshore oil industry structures.[46,490] Efforts are being made to computerize the stress capabilities of platforms. Instruments mounted on the platform measure stress and movements during storms. Analysis of the data can help identify the approach of the danger threshold.[96,318]

A basic difficulty facing design engineers working with offshore equipment and structures is that long-term historical climatic data are unavailable. To help solve this problem, the United Kingdom's Department of Energy has been working on an oil company-sponsored meteorological and hydrological study, costing approximately £600,000.[255] Ideally, the wave height maximum in a 100-year period should be known for relatively local areas so that the probabilities of the anticipated 20- to 30-year exposure (or lifetime) of the active platform can be calculated. Climatic and wave-height records usually cover a time span too short for statisticians to make reliable predictions. Presently, extrapolation methods (based on a measured 20-year data period) are used for projection and hindcast techniques; in hindcasting, storm conditions are constructed backwards from a known point in time. Monte Carlo simulation methods are applied by using computer calculations based on "...randomly drawn storm variables...," winds, currents, waves and tides.[1]

Although the emphasis here has been on the North Sea, other offshore areas also face difficult climatic problems. During the winter, wildcatters working in the Baltimore Canyon area experience significant downtime, and hurricanes also present a major threat here and in the Gulf of Mexico. In the past, when a hurricane entered the Gulf Coast area, personnel had to be taken off the rigs and operations shut down. This procedure meant at least a 5-day down period even if the hurricane veered away from the oil field. Now, computerized systems have been installed by some producers so that the wells can be left operating when personnel are not on the platform. If it becomes necessary to shut the rig down, this can be done from onshore.[232] Studies made by the European Space Research Organization have recently determined the feasibility of using aerial satellite systems to relay weather and other data instantly to onshore monitoring stations.[456]

Frictional Drag, Currents, and Scour

A problem closely associated with wave impact (slam) as well as with currents is the frictional drag created from movement of the water across marine growth of mussels, sea anemones, and weeds on the platform's legs and cross-members. Engineers designing rigs for the North Sea at first planned for a 4-in. radial growth extending 20 ft. below the water surface. Actual growth patterns more often have an 8-in. thickness, reaching to the seabed. As of the mid-1970s, investigators were uncertain about the precise effects of this additional drag on platforms. The marine growth, however, is removed periodically, and platforms that seem to need it are reinforced.[507]

When oceanic currents or tides flow and ebb, sand grains and other particles may be eroded from the seabed, especially where obstructions create eddies. Depending on local conditions, this erosion process may excavate large holes surrounding the legs of platforms, resulting in a lateral weakness or imbalance in the support base; it can also uncover marine pipelines. Scour is expensive to control, and in conjunction with storms may create extreme danger to platforms. As a result of Hurricane Camille in August 1969, Shell Oil and Gulf Oil each lost a platform off the Louisiana coast near the Mississippi River mouth. Researchers attribute the collapse of the Shell Oil platform to soil failure around its legs, resulting from severe seabed scour.[509]

Scour holes can deepen only so far before reaching an equilibrium point. To allow for this point of equilibrium, engineers must carefully determine the direction and speed of currents before deciding on the depth to set the legs below the seabed. Unfortunately, current direction and speed change make the problem more complicated. Even drilling rigs, remaining stationary only for a short time, may experience scour problems if their legs do not penetrate well below the seabed floor. Sandbags and various leg nets are useful for small scour holes, but they can be expensive in the long run. Adequate protection for an 8-leg platform may require 22,000 sandbags and several divers to arrange the bags after they are dropped overboard.[566] Gravity structures (submersibles) are sometimes ballasted down on a prelaid gravel bed or hinged mats. Both techniques help reduce scour.[38]

Ice Floes and Icebergs

Besides bone-chilling cold and the threat of metal fatigue in platforms and machinery, oil company personnel working in the Arctic and sub-Arctic offshore face the problem of ice. When pack ice and icebergs are moved about by winds and currents, they may be pushed against a rig, either damaging or destroying it.

Conditions off Labrador and Greenland well illustrate the problem of coping with ice, especially icebergs. Icebergs create a hazard when they calve off the Greenland ice cap and move southward in the Labrador Current. In some years, as many as 3,000 icebergs cross the 48th parallel on their journey southward, whereas in other years the number may be fewer than 100. The problems of exploring for oil in this major iceberg region are making costs 10 times higher than in the North Sea and 20 times higher than in Saudi Arabia. Extensive petroleum exploration began during 1976 in the Davis Strait area off Greenland. Icebergs pose such a major threat to the six oil companies with concessions in the area that they use only dynamically positioned drill ships.[i] If danger seems imminent, the ship retrieves its drill string and moves out of the area. Several techniques are currently used or are under study to reduce the iceberg hazard. Experimental work is being done to develop a 100-ft. thick rubber bumper that encircles stationary structures. Presently, drilling and production rigs maintain two radars (one as a backup) that track iceberg movements. Technicians plot an iceberg's hourly position on a master map. If operators decide an iceberg poses a threat, sonar units are used to determine its size, giving them a better indication of how much power will be required to change its course. The iceberg is then lassoed with a floating cable and attached to one or more towing vessels, a hazardous task if it is windy. Icebergs of 3 to 4 million tons have been diverted slightly from their natural trajectory, and success with those of 1 million tons or less has been good. Producers had hoped the iceberg problem would become less hazardous after the United States launched the satellite *Seasat* in June 1978.[208] The satellite would have provided the oil companies with data needed to make more accurate predictions of iceberg movements, but, unfortunately, it developed an unexplained short circuit and quit functioning after only 106 days.[475]

Icebergs and ice ridges can threaten not only platforms, but also damage or destroy seabed pipelines and seabed completion units. After a well has been brought in, it may be tied by pipeline to a central platform, with only a "Christmas tree" completion unit left on the well site. If the deep keels of icebergs or ice ridges scour the bottom, they may sheer off or damage completion units. To reduce the problem, oil field workers in the Beaufort Sea area encase completion systems with steel cylinders.[417] Pipelines laid in the Newfoundland area may have to be buried as much as 50 ft. below the seabed.

Artificial sea-ice islands. The Beaufort Sea is the site of several valuable experiments that might be better termed "working with ice." During the winter of 1976-77, Union Oil Co. drillers at Harrison Bay, 18 mi. west of the Prudhoe Bay field, used a man-made ice island as a temporary drilling base. Construction took place in water 8 to 10 ft. deep. Engineers made a containment ring of snow on the sea ice, and then pumped sea water into the enclosure, allowing it to freeze. As the weight increased, the island sank deeper until it fused with the bottom. The island was designed to have a diameter of 900 ft. and to sit 5 to 6 ft. above the ocean surface; all facilities, including a mess hall, living quarters, and sewage, were contained on the island.[454]

Natural sea ice has also been used successfully as a drilling base. In one experiment, an insulating "sandwich" pad was laid down on the sea ice. The sandwich consisted of two layers of polyethylene film, two 2-in. layers of foam polyurethane, and then two additional layers of plastic film. Finally, a 4-in. timber layer was placed on top. The plastic film was used to provide protection against polluting the ice with drilling mud and other materials.[265]

Landfill islands. The Arctic Ocean's first artificial landfill island was built by Imperial Oil Ltd. of Canada. The island, dubbed Immerk, took two summers (1972-73) to build; it was sited north of the Mackenzie River Delta (Fig. 57) approximately 10 mi. offshore in about 15 ft. of water. Immerk provided the base for drilling the world's first offshore exploration well in the Arctic. In contrast to Union Oil's ice island, Imperial's island was constructed from sediments dredged from the seabed. The sediments were mounded to a depth 15 ft. above sea level, becoming frozen to the seabed during the first winter. The finished work area had a radius of approximately 150 ft. To protect the island from floe ice and wave erosion, engineers built a large erosion beach seaward of the island's work area. The total cost of the Immerk project was $5 million,[450] evidently a worthwhile exploration investment because by 1976 six additional islands had

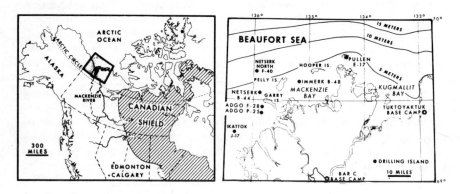

Fig. 57. Beaufort Sea artificial island sites.
Source: [451]. With permission of the Society of Petroleum Engineers.

been constructed in the region's offshore. In 1979, the number of islands constructed reached a total of 15.[275,j]

Offshore tunneling. The petroleum potential of the Beaufort Sea area has led some engineers to suggest the use of an offshore tunneling and chamber system. Two parallel tunnels onshore would be driven seaward—one for air inflow and the other for worker access and rail lines leading to drilling chambers adjacent to the tunnels. A 10-mi.-long tunnel and its network of drilling chambers (containing up to 12 directionally drilled wells) could give access to an area of 30 mi.^{2}. Advantages of this system would be: (1) dry-land drilling technology could be used; (2) hydrostatic head and drilling mud relationships would be easier to manage; (3) surface weather would be avoided; (4) oil spills would present no problem; (5) ocean floor pipelines would be unnecessary; and (6) less financial risk would be involved, because of more stable working conditions. The system would present several special problems in safety, including considerations for the hazards of fire, seismic shock, gas leakage, excessive water inflow, and blowouts.[312,322]

SUMMARY AND CONCLUSIONS

Oil production on the continental margins will become increasingly important in coming decades, requiring significant inputs of capital and human skills to develop environmental impact studies, pollution control devices, and clean-up, when accidents occur, as in oil spills. Considerable differences of opinion, however, prompt disagreements concerning the long-term effects of oil and other types of pollution on marine and coastal biota; clearly, more research must be done as, for example, the type pursued by the Gulf Universities Research Consortium (GURC). Participating scientists have established a comparative set of more than 1 million data points, distributed over an area of 400 mi.,^{2} in the Gulf of Mexico for which geological, biological, and chemical variables and conditions (turbidity, dissolved oxygen, trace elements, species abundance, and hydrocarbon sources) have been measured to determine man's impact on the marine environment.[480]

Greater care must be given to avoiding offshore man-made hazards, including platform accidents, oil spillage, explosions, fires, and blowouts. Natural hazards such as storms, ice, and scour should be further studied to ensure safe platform operation and to avoid unwise siting of rigs which may put them in jeopardy or cause environmental damage. Robert M. Owen has stressed that mining companies should attempt to obtain all previously collected field data (governmental, university, and private) dealing with a proposed mine site, including hydrological, meteorological, biological, and sedimentation information. They should, also, attempt to set up mining schedules that consider such variables as spawning periods of marine biota, establish monitoring and control systems, and develop numerically computerized models of the local ecosystems' interactions and operation.[414]

Although the task of assuring a viable environment and safe working conditions in the offshore petroleum industry is formidable, a cooperative effort of both government and industry should be able to do the job. It is imperative that this commitment continue for, as one prominent author has said, "Ocean space—and its ecology—is one and indivisible."[50]

NOTES

[a] Studies focused on the North and South Atlantic OCS in the U.S. are available in [489] and [493].

[b] After a blowout, an oil well often catches fire. During a fire, spray barges are used to keep the structure cool so it will not collapse.

[c] Efforts to control and kill a blowout are often focused upon drilling relief wells that attempt to intersect the well out of control. Drilling mud and steel balls forced into the runaway well may control the blowout.

[d] The Norwegian government has published a detailed report on the Ekofisk blowout. See [549].

[e] On June 3, 1979, 50 mi. offshore in Mexico's Bay of Campeche, a major oil well blowout began which eclipsed all previous oil spills. The well, *Ixoc I*, spilled 20,000 to 30,000 bbl/d into the sea; currents carried the oil along the Gulf of Mexico's western shore until beaches in Texas were fouled. As of late January 1980, the blowout had yet to be fully contained. Statoil attempted (unsuccessfully) to assist Petroleos Mexicanos in cleaning up the spill. The ripples of this failure were felt in Norway by creating additional opposition to exploration north of the 62nd parallel [82] and [487].

[f] The number of workers in the country's petroleum industry rose from 344 in 1973 to 2,635 in 1976. The Storting provided a detailed listing of Norway's offshore petroleum industry safety regulations [465].

[g] The Norwegian government actually stopped work in the Statfjord field when it determined that living conditions were unsafe; it also requires each well to have four blowout preventers.

[h] For a more detailed discussion of the *Sea Gem* and other accidents in the North Sea, see [103].

[i] The companies include Amoco, Chevron, Arco, Mobil, TGA-Grepco, and Ultamar. One well was spudded 60 mi. north of the Arctic Circle [123].

[j] In one instance, off Long Beach, Calif., oil ompanies use man-made islands as a base in directional drilling to tap reservoirs under the adjacent mainland [158].

Part II

The Deep Seabed

Marine geologists, oceanographers, engineers, and mining companies are now looking to the deep seabed—areas seaward of the continential rise—as a source of both solid minerals occurring as nodular concretions and as metalliferous muds. Mining consortia are gearing up to penetrate this new frontier, having made during the past 15 years tremendous investments in research, capital equipment, and labor to determine the feasibility of extracting minerals in this difficult oceanic environment, lying primarily between 12-20,000 ft. below sea level. Many mineral producers want to proceed with their long anticipated mining ventures; however, they hesistate because of unsettled issues concerning seabed ownership and exploitation. Since 1973, these problems have been before the United Nations-sponsored Third Law of the Sea Conference, but they remain unresolved.

Chapter 9 of Part II examines the physical conditions favoring the formation of oceanic ore deposits, focusing especially on metalliferous muds. Subsequent chapters examine the origin and structure of manganese nodules (Chap. 10); investment and production economics and technology for extracting seabed manganese nodules (Chap. 11); international economics of manganese nodule mining (Chap. 12); and the relationships of deep seabed mining to current efforts to establish an international regime for an equitable sharing of seabed wealth (Chap. 13).

Chapter 9

Plate Tectonics and Mineral Formation

Geological mineral-forming processes that are at work in the seabed are closely associated with sea-floor spreading and global plate tectonics, the movement of earth's lithosphere (crust and upper mantle) from one point to another. Scientists do not agree upon, or fully understand, what processes (thermal uplift plumes, lateral increases in density, or convection cells)[449] are responsible for weaving the fabric of earth's outermost garment; they are, however, working hard to enhance their knowledge. About a dozen large and numerous small plates have been identified by geologists. The plates are traveling at different speeds and in different directions, moving along, as if floating, on the underlying asthenosphere, a plastic-like layer (Fig. 58). The movement of plates creates zones of translation, divergence and convergence, and it is in these three zones (especially in divergence and convergence) that conditions are most favorable for the occurrence of mineralization.[170] In areas of earth's crust where the plates or parts of plates are moving parallel to one another—zones of translation—there is little volcanism present and not much new magma is introduced, although some small-scale injections into faults may occur. The focus here, therefore, will be only on zones of convergence and divergence, especially the latter because the processes of mineral formation and deposition are restricted to the marine environment, whereas convergence mineralization, in part, entails the transfer of minerals from oceanic crust to a continental land mass.

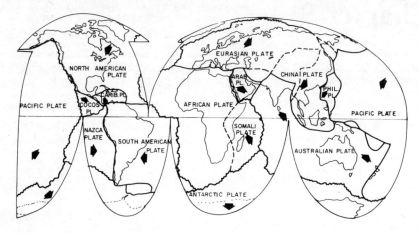

Fig. 58. Major present-day crustal plates. Arrows indicate the general direction of plate movement in relation to the African continent.
Source: [170]. With permission.

DIVERGENCE ZONES

Diverging plates may experience three basic processes or conditions that can contribute to ore body genesis. These include: (1) magmatic segregation at depth in the vicinity of diverging mid-ocean ridges, (2) subsequent intrusions and extrusions of mantle magma into overlying and migrating plates, and (3) hydrothermal solution and precipitation along mid-ocean ridges.

Magmatic Segregation

During the preceding 200 million years (and probably much longer), continental land masses have experienced geological forces that cause them to split and drift apart, creating zones of divergence, whereby the separating plates move laterally, while the rift is continually healed by magma from the deeper mantle. K. O. Emery and B. J. Skinner note that these extrusions carry with them minerals that can be concentrated during the magmatic segregation (differentiation) process, providing the possibility for significant metallic mineral deposits containing chromium, platinum, copper, and nickel. They caution, however, that because the oceanic crust is relatively thin (in comparison to that of the continents), the occurrence of large magmatic chambers is less likely—a prerequisite to the slow cooling needed for "... extreme magmatic differentiation." Because areas of divergence experience pronounced tectonic activity, some of the magma escapes from the chambers—a condition that disrupts the settling and segregation process. The authors conclude that it is thus "... unlikely, although not impossible, that large ore bodies formed by magmatic segregation are hidden in the oceanic crust."[170]

Hot Spots

Local hot spots or plumes (a total of 122 have been identified as having been active within the past 10 million years) exist in the seabed mantle; as the diverging plates move across these hot spots, volcanoes may form, allowing magma from the mantle to reach the seabed surface and form volcanic islands, as in the Hawaiian or Tuamotu chains in the Pacific Ocean.[68] Again, Emery and Skinner are not optimistic that these areas produce significant mineral deposits. They feel that because of the relative homogeneity of the "major-element compositions, and the relatively small volumes of magma that seem to move up the conduits from the mantle for any given phase of eruption, oceanic hot spots do not appear to be materially more promising than spreading belts as a source of primary magmatic mineral deposits."[170]

Hydrothermal Solution and Precipitation

In addition to the processes of magmatic segregation and hot spot mineral formation associated with plates, minerals may be deposited epigenetically in areas immediately along the mid-ocean ridge of the diverging plates. Bonatti contends that the permeability of the seabed in the vicinity of plate formation is high enough to permit seawater to penetrate several kilometers below the seabed. As the water percolates downward, it is heated to several hundred degrees centigrade and then forced to rise in convective cells.[48, a] The mineral deposits form when the hydrothermally circulating ocean water leaches metallic elements from the crust while moving upward and/or laterally from the center of the convecting rift area. Conduits channel the waters through the basalt where it takes into solution manganous manganese (Mn^{+2}) and ferrous iron (Fe^{+2}) along with such elements as zinc and copper. "As the solution continues to react, dissolved sulfur in the S^{+6} state is reduced to the S^{-2} state, causing insoluable sulfide minerals such as pyrite (FeS_2), pyrrhotite (FeS), and chalcopyrite ($CuFeS_2$) to be precipitated on the walls of the solution conduits" (Fig. 59).[170] After the hydrothermal solution again rises to the seabed surface, sulphide minerals may be deposited near the conduits. But most of the dissolved Mn^{+2} and Fe^{+2} is oxidized by the seawater which results in its precipitating out as tiny particles of manganese oxides and ferric hydroxides and oxides. These particles fall to the seabed and form a crust[48, 170] or become a metalliferous component in sea-floor sediments.

Well developed conduit areas could provide deposits of pyrite and chalcopyrite exploitable for copper. Small copper veins occur on the Indian Ocean Ridge, and have been found in oceanic crust in the Atlantic. Geologists think that several intensively mined Troodos Massif copper ores on the island of Cyprus are the result of fossil hydrothermal systems originating on an oceanic ridge which, as a section of the Mediterranean closed, were thrust above the ocean surface onto land (to become ophiolites). By analogy, many deposits like the Troodos should lie in ocean depths, but they will be difficult to locate and exploit (Fig. 60).[460, b] Extensive work by Peter Rona, however, has been directed toward identifying thermal, seismic, geochemical, and acoustic criteria that can help in locating active

168 *Plate tectonics and mineral formation*

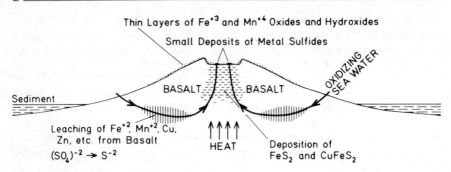

Fig. 59. Schematic diagram of hydrothermal solution circulation, including percolation, leaching, and precipitation processes. Metals leached from contact rocks during the migration of ocean water are precipitated onto the conduit walls and on the ocean floor near the ridge and ocean water interface.
Source: [170]. With permission.

hydrothermal convection discharge areas and mineralization in sea-floor spreading zones. In addition he suggests that petrologic, magnetic, and gravimetric techniques can assist geologists in locating both active and inactive hydrothermally mineralized zones in all areas of the oceanic crust.[458]

The Red Sea Metalliferous Sediments—A Case Study of Origin and Economic Prospects

The Red Sea is an area that well illustrates mineralization associated with diverging plates. It is now being tested for its economic potential as a source for hydrothermally formed metalliferous muds and brines. In the mid-1960s, rich metallic sulphide sediments—65 to 300 ft. thick—were discovered in several small basins (Atlantis II, Chain, Thetis, Nereus, Gypsum, and Vema, among others) along the center of the Red Sea, lying at depths ranging from 6,300 to 7,300 ft. The upper 30 ft. of sediments in these metal-rich basins is estimated to contain 80 million dry weight tons (dwt), which average 29% iron, 3.4% zinc, 1.3% copper, 0.1% lead, 0.005% silver, and 0.00005% gold. Overlying these sediments are salty brines from which it is thought these metals precipitated, after having combined with sulfur,[460] silica, and oxygen. The brines are exceptionally rich in minerals—1,000 to 50,000 times higher than ordinary seawater.[542, c] The most richly endowed Red Sea deeps are Atlantis II, Thetis, Gypsum, Vema, and Nereus.

The Atlantis II Deep was the first to be discovered, with sediment samples having been obtained from it in 1965; it proved to be the premier deep for mineral concentrations. Zinc concentrations in the sediments have been reported to be as high as 5% to 10% and copper between 1% and 4%.[d] In places, the Atlantis II sediments are approximately 65 ft. thick, containing in their southern portion the

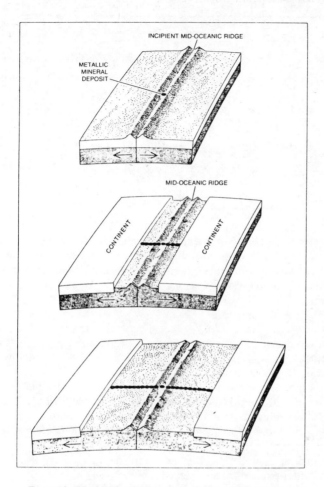

Fig. 60. Hydrothermal mineral deposit formed in a hot-brine pool on the axis of a mid-oceanic ridge. As the ocean basin progressively widens, the mineralized area would be expected to extend from the ridge across the ocean basin to adjacent continents.
Source: [460]. With permission.

equivalent of 150 to 200 million dwt of salf-free mud.[43] One estimate made in 1972 indicates that the upper 30 ft. of the Atlantis II Deep's sediments contain metals valued at $2 billion.[246] Atlantis II is the only one of Red Sea deeps that is presently forming metalliferous deposits.[43]

Geologists believe the Red Sea's metalliferous sediment accumulation process depends largely upon the percolation of waters down through highly fractured and

170 *Plate tectonics and mineral formation*

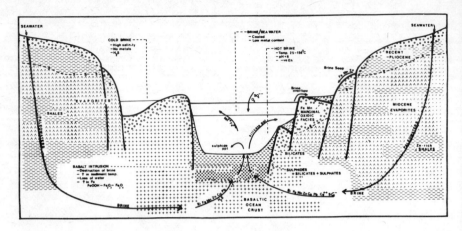

Fig. 61. Cross-section of the Red Sea Trough, showing seawater percolating downward through highly mineralized evaporites and shales and into basaltic ocean crust. The hot hydrothermal mineral solutions finally emerge into brine pools in the central deeps where various precipitates form, creating thick layers of mineral-rich sediments.
Source: [43]. With permission.

mineralized evaporite and shale formations and basaltic oceanic crust lying adjacent to and under much of the sea. As water migrates downward through these fractured rocks, it dissolves minerals that it contacts, becoming a dense and hot brine. Numerous transform faults in the central and northern portions of the Red Sea provide good migration routes, through the otherwise impermeable evaporites, into the axial trough and, at the same time, create natural sediment and brine traps separated by "inter-trough" zones (Fig. 61).[43]

The controlling factors in the precipitation and deposition of the sediments derived from the brines are complex. Basically, however, the process depends upon the brines emerging into deeps and then coming into contact with overlying seawater which results in their cooling and taking up oxygen[43] with subsequent precipitation of iron, copper, and zinc as sulphides, silicates, and oxides, depending on the physicochemical properties of the mixture change from the seabed upward, ". . . as the hydrothermal solutions progressively mix with the water."[112] Precipitates of iron, copper, and zinc sulphides form first within the deep. As the water column becomes adequately oxidizing for Fe^{+3} to be formed from Fe^{+2}, iron silicate precipitates form. Finally, as pH conditions become higher, Mn^{+2} oxidizes to Mn^{+4}. The limits of the precipitate dispersion halo may reach as far as 6 mi. from the hydrothermal vents. Cronan points out that this dispersion characteristic could serve as a useful exploration model in identifying the primary locus of sulphide deposits. The Red Sea should also give geologists clues as to ". . . what other regions on the ocean floor we should look in for such deposits . . .," as in the vicinity of mid-ocean ridges, although conditions there are not fully analogous to the Red Sea deeps.[7,112]

An earlier and alternative (but now generally rejected) hypothesis accounting for the origin of the Red Sea's metalliferous deposits was first offered by Emery and then restated by D. D. Krasov. The primary formation process is attributed to the isolation of the Red Sea from the Indian Ocean, although the hypothesis does not preclude the influx of mineral-rich waters through the lower portions of the rift valley. Radio-carbon dating of littoral sediments in the basin indicate that the Red Sea's threshold (the relatively shallow seabed area in the southern portion) was exposed above sea level at least 20,000 to 14,000 years ago. Excessive evaporation in the isolated basin created lakes that gradually became depleted of water until their salinity and other mineral content became very high. When the sea level rose again and joined with the depression of the rift valley, the lakes and seawater did not mix because of differing densities.[292]

When the mineral potential of the Red Sea sediments was recognized, industrial entrepreneurs and the littoral state governments began to take note, and the economic and political situation soon became clouded. In 1967, the Sudanese government, on the premise that the Red Sea's median line lies east of most of the various deeps, proclaimed sovereignty over them. This claim was reinforced by the government's issuing an Atlantis II Deep exploration license to Sudanese Minerals, Ltd., a subsidiary of the Los Angeles-based firm International Geomarine Corp., which was joined in 1968 by a West German firm, Preussag AG.[343] Despite the Sudanese government's active exploration program, Crawford Marine Specialists, Inc., of San Francisco applied in February 1968 to the United Nations for the exclusive right to explore for mineral brines in the Red Sea. The United Nations claimed it had no jurisdication,[246] which is certainly did not. Still another firm, an international consortium, merely advertized in a major British newspaper that it was claiming the right to explore the brine sinks of the Red Sea.[246] These problems were complicated further when, after a few years, a coup overturned the Sudan government, radically changing the highly favorable terms for Sudanese Minerals and for Preussag. The new Sudanese government demanded that the contract terms previously set at 25% of the net profits be raised to 90%. Sudanese Minerals, unable to cope with the prospect of such unfavorable terms, dropped out of the partnership.[343]

At the time when Sudan was exerting its claims to the Red Sea deeps, Saudi Arabia was announcing its rightful dominion over the same areas, although the Saudi government made no effort to meddle with the exploration programs of Geomarine Corp.[341] In 1975, the Saudis and Sudanese reconciled their differences and joined forces to complete unfinished exploratory work and finance and develop the mining of Red Sea metalliferous sediments, with the Saudis providing most of the capital. A French company, BRGM, was contracted to further explore the dimensions of the sediments in the various deeps while Preussag, still game for action in this potential mineral storehouse, received a 4-year $14 million contract to develop both a mining and a processing system. By 1980, the firm is expected to have a functional pilot mining and processing facility.[343]

CONVERGENCE ZONES

Numerous continental margins, as on the east coast of Asia or the west coast of South and Central America, are experiencing convergence with concomitant subduction, i.e., one plate is being thrust beneath another. The forces generated during the subduction process create a trench and also cause a deformation of continental sediments deposited in the trench. Associated tectonic forces help produce ideal traps for petroleum formation (as described in Chap. 3), and also produce conditions favorable for metallogenesis.

The processes of mineralization in zones of convergence are only partly understood, but Sillitoe and others have stressed the potential of convergence zones for mineralization when mid-ocean-generated mineral-rich crust is carried toward the continents and then subducted. Sillitoe contends there is evidence that ". . . southern Peru-northern Chile and the southwestern United States may be interpreted as regions . . ." which seem to have been mineralized in this way, as in the case of porphyry copper deposits.[484] But how does the mineralization occur? Rona notes that one important model is the "geostill concept," whereby ". . . metals may be distilled from one plate (oceanic lithosphere) and transferred to another plate (continental lithosphere) during subduction." In the early subduction phase, a partial melting of seawater-saturated oceanic crust occurs.[461] During this process, mineralizing gases and solutions are emitted, resulting in ocean waters thermally circulating through the overlying rocks. The solutions may be deposited as sulphide ores, after penetrating upward via subduction-generated faults and fractures. Magma penetration into the overlying continental plate may also occur, creating extremely complex geological formations and rock types that may become mineralized with copper, lead, tin, tungsten, and antimony.[53, 170] The mineral deposits of the eastern Mediterranean, Japan's Kuroko deposits, and those of the Philippines, and the western coastal mountains of North America and South America are good examples of the hydrothermal and magmatic processes associated with zones of convergence (Fig. 62).[e] Rona and Neuman have appropriately suggested that "from a resource point of view, the name 'ring of fire' for the circum-Pacific [convergence] region could well be changed to 'ring of resources'."[462]

Another convergence zone process important to the potential availability of exploitable mineral deposits occurs when (as noted in the case of the Troodos Massif) oceanic crustal deposits may be thrust onto a land mass, i.e., obducted. Theoretically, this process could occur when the abutting continental land mass slices off a section of the oceanic crust which then overrides the adjacent continent to become an ophiolite, instead of being subducted. Ophiolites of various ages occur, among others, in the Appalachian and Caledonian regions of eastern North America, in the Grampians of Scotland, in the Himalayas, in the Kamchatka Peninsula, and in the western Cordillera of North America from Alaska southward.[93]

SUMMARY AND CONCLUSIONS

Although geologists and geophysicists have devoted much effort to study the potential for marine rock and mineral genesis, as they are associated with lithospheric

Notes 173

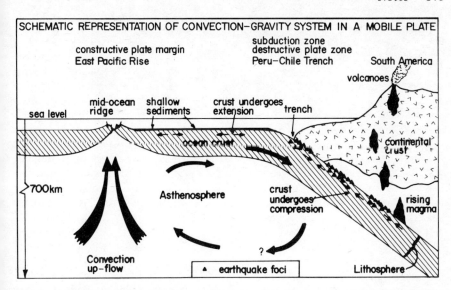

Fig. 62. Sea-floor spreading and subduction of the Nazca Plate. *Source:* [448]. With permission.

formation and movement, much more remains to be done.[66] Current efforts at exploiting prime hydrothermally generated metalliferous deposits should bear fruit, as the problems of mining and processing of the sediments are solved. What is being learned in the Red Sea divergence zone may prove applicable in similar marine environments of the mid-oceanic ridges, as in the East Pacific Rise region.[f] A better understanding of the processes at work in oceanic-crustal and continental-land mass convergence zones should prove useful in locating mineral deposits in both nearshore and onland environments.

NOTES

[a]J. Corliss provided a good discussion of possible explanations for the origin of hydrothermal solutions [104].

[b]One area, the Galapagos spreading center in the eastern Equatorial Pacific Ocean lying at a depth of 8,250 ft. has been studied in considerable detail by scientists. See, for example, [47], [48], and [459].

[c]E. Degens and D. Ross performed a detailed examination of the geological and mineralogical conditions in the Red Sea divergence zone [130].

[d]The lower values were reported in [10]; the higher values were reported more recently in [112].

[e]A useful discussion of ore deposition processes associated with the Kuroko deposits is presented in [553].

[f]An example of recent studies of metalliferous sediments in the East Pacific Rise region is [227].

Chapter 10

Manganese Nodules

In 1872, a 3-masted, square-rigged, wooden ship—the steam corvette H.M.S. *Challenger*—left the British port of Portsmouth on a journey that was to have great significance for the future of the oceans and mankind. The *Challenger*, a ship of the line converted to an oceanographic research vessel, was designed for scientific work with the biological conditions and physical properties of the world's oceans. During its 3½-year expedition, the *Challenger's* dredges frequently brought up black, potato-like concretions at first thought to be phosphorites but which, after chemical analysis, were found to be "almost pure peroxide of manganese."[351] These concretions have come to be called manganese nodules. For more than 80 years after the *Challenger's* discovery of the nodules, little effort was made to determine their industrial value. Then, in the 1950s, scientists began to examine the nodules for their economic potential, and today, they are the center of attention by entrepreneurs and scientists who seek to develop them commercially and try to better understand their distribution, properties, and processes of formation—the primary foci of this chapter.

DISTRIBUTION

Although the amount of ocean floor, sampled as of 1973, totaled only about three dredge hauls for every 1 million km^2,[460] oceanographers had already determined that all the ocean basins, with the exception of the Arctic, contain manganese concretions, nodules (Fig. 63), or pavements (crusts), as in the Blake Plateau

Fig. 63. Manganese nodule deposits in the Antarctic Ocean. Strong currents keep the seabed relatively free of sediments which seems to provide at least one of the requisites for nodule development. Courtesy National Science Foundation and the Smithsonian Institution Oceanographic Sorting Center.

and Tasman Sea areas, where the concretions have become cemented together or have formed initially as a layer covering nearly all of the rocks and seabed, which may extend over an area as large as 5,000 km² (Fig. 64).[433] Nodules occur most frequently at water depths of 12-20,000 ft. in abyssal plains, and it appears that oceanic areas best suited to their information and accumulation are relatively distant from river-borne sediments.[340] Crusts seem to form especially well on exposed submarine elevations, where currents are strong enough for sediments not to accumulate.[241, 283] Nodule distributions are highly variable and appear to depend on the "amount of time that has been available for their accumulation," primarily a function of sea floor spreading rates; on local site factors such as slope; on the availability of nuclei, around which most nodules form; and on altering patterns of seabed-sediment burial and exposure of the nodules.[111]

Distribution 177

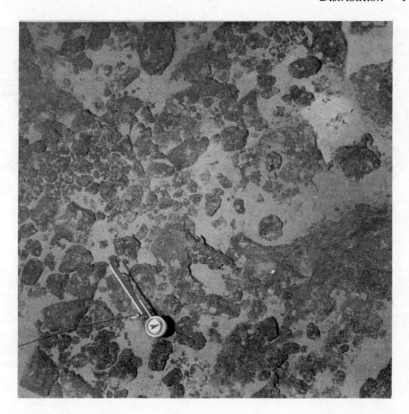

Fig. 64. Slab-like manganese deposits located in the Equatorial Pacific Ocean at 125°56'W., 15°13.8'N., and at depth of approximately 14,000 ft. The compass in the photograph is 13 in. long and 3 in. wide. Courtesy National Oceanic and Atmospheric Administration. Photo by L. Parsons.

Sampling programs have determined that one of the largest nodule concentrations occurs in the mid-Pacific Ocean, lying between the Hawaiian Islands and Central America (Fig. 65) and in an area centered on the Line Islands and the Tuamotu Archipelago (Fig. 66). The Atlantic Ocean, for the most part, stands in sharp contrast to the Pacific in the abundance of its nodules. Although the North Atlantic has been sampled extensively, only limited amounts of manganese nodules and pavements have been found, although the Blake Plateau off the United States' southeast coast and an area east of Cuba do have significant deposits (Fig. 67). This paucity is attributed to a heavy land-derived sediment deposition into the Atlantic and an extensive pelagic carbonate detritus fallout. Manganese deposits are more widespread in South Atlantic waters, because there are fewer rivers dumping sediments into the basin. Deposits are especially abundant south and southwest of South Africa. One significant deposit lies to the southeast of the island of

178 *Manganese nodules*

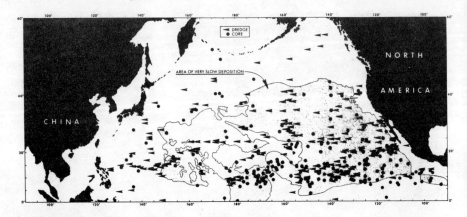

Fig. 65. Manganese nodule deposits of the North Pacific Ocean. The North Pacific receives few sediments from land. The most extensive nodule deposits occur in a sloping area extending from southeast of the Hawaiian Islands (15,000 ft. deep) toward the Central American coast (12,000 ft. deep). These deposits lie in a siliceous ooze region, and have some of the highest metal contents of any nodule deposits discovered up to this time.
Source: [242]. With permission.

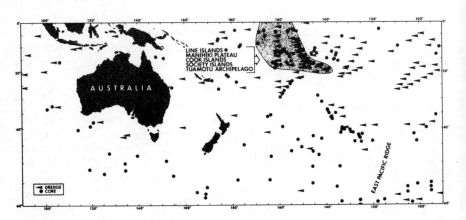

Fig. 66. The South Pacific's nodule distribution. The South Pacific is not as well explored as its North Pacific counterpart, but one especially important nodule area lies between the Line Islands and the Tuamotu Archipelago.
Source: [242]. With permission.

Distribution 179

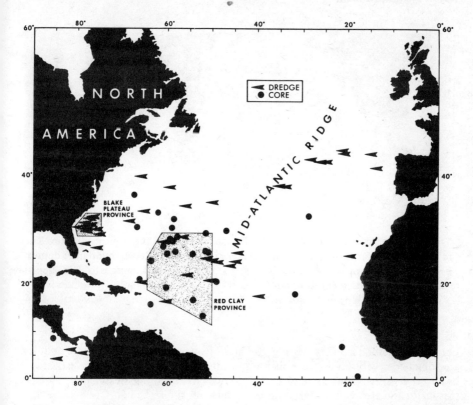

Fig. 67. Manganese deposits of the North Atlantic. The most significant deposits of nodules lie in the Blake Plateau. By discouraging sedimentation, the strong current of the Gulf Stream enhances the development of manganese nodules and crusts. Another major area of deposits is situated east of Cuba in the Red Clay Province.
Source: [242]. With permission.

Madagascar in the Indian Ocean (Fig. 68). Jane Frazer of Scripps Institution of Oceanography (SIO) has developed a useful synoptic world map (based on 5°-square grids) showing sampling stations that reported nodules. Although the map does not reflect nodule abundance, it does give a good idea of their macro-distribution (Fig. 69). According to Frazer, "for some well sampled areas, such as the eastern Pacific, between 0° and 20°N, a finer grid should also give useful results. For most of the world, however, there are not enough nodule analyses available to warrant finer resolution."[189,b]

Coring programs have demonstrated that the preponderance of nodule populations seems to lie either on the seabed or within a few feet of the surface and not at depth. For example in the nodule-rich siliceous ooze area of the northeastern

180 *Manganese nodules*

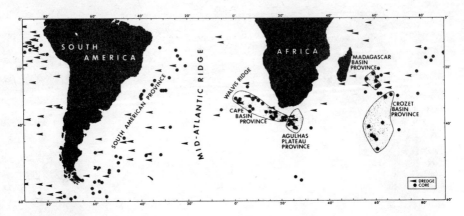

Fig. 68. Manganese deposits of the Indian Ocean and the South Atlantic. The Crozet, Madagascar, Cape, and South American Provinces have abundant nodules. These basins are protected from biogenic and continental sources of sediments, resulting in a low sedimentation rate.
Source: [242]. With permission.

Equatorial Pacific (see Figs. 65 and 76), at 20 ft. below the surface, only 2 nodules are found for every 90 at the seabed-water interface (Fig. 70). Some investigators, however, feel the population of buried nodules has been underestimated.[338] One reason for the greater concentration of nodules on the seabed surface may be that decaying bacteria in the sediments create acids that break down the nodules. The minerals released could migrate back to the seabed/water interface where they can reform as nodules.[409] Another hypothesis offered for the nodules' concentration at the seabed surface is that strong currents keep the nodules from becoming covered. Some have also suggested that marine organisms may cause the nodules to be turned and moved enough to keep them free of sediments.

Most manganese nodules range in size from a few millimeters to larger than tennis balls, although some are much bigger.[399] Nodule surfaces vary from smooth and lustrous to knobby and gritty.[439] Crystal size is sometimes so small that X-ray analysis cannot be used.[399] Fractures and pores that have been partially filled with clay, salts, and manganese material are usually in evidence (Fig. 71). In the past, investigators have frequently described the two main manganese minerals as todorokite and birnessite. More recently, researchers have questioned these designations, because the "... microcrystalline species of manganate minerals occurring in the ocean ..." are not completely known.[c] According to Burnses,[75] the extremely complex internal structure and small crystal size of nodules has "... led to ambiguities ..." in their mineralogical identification, perhaps because of an "uncritical use of X-ray diffraction techniques ...". Geothite is the only iron mineral present in nodules. The other metals contained in nodules seem to be taken up by the iron and manganese oxides either by substitution, absorption,

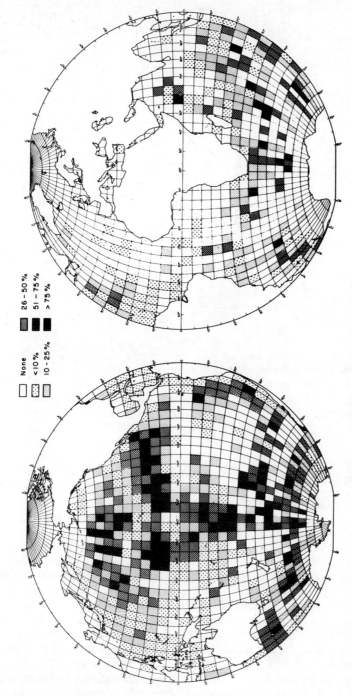

Fig. 69. Percentage of stations reporting manganese nodule occurrences. *Source:* [189]. With permission.

181

182 *Manganese nodules*

Fig. 70. Frequency of nodule occurrence in seabed sediments and at seabed/water interface in the North Pacific. The occurrence of nodules at the seabed/water interface is four times greater than in the top 3 ft. of sediments.
Source: [241]. With permission.

or adsorption, and do not form separate minerals. Some scientists have suggested that, because nodules have no specific mineral formula, they are "... most appropriately classified as rocks."[409] Detrital minerals also occur in the nodules, including very fine grained and widely disseminated barite, montmorillonite, rutile, calcite, illite, quartz, and feldspars. Several researchers believe much more attention should be given to the detrital composition of nodules, which might provide clues to their origin.[498]

There are some 30 metal oxides formed in manganese nodules, with manganese, iron, copper, nickel, and cobalt the most common of the economically desired metals. Zinc, molybdenum, and vanadium also frequently occur,[182] as do calcium, aluminum, silicon, sodium, and magnesium. Table 17 shows the average percent of selected elements contained in ocean samples.[d] The chemical composition of nodules differs greatly in the various oceans and between different areas in the same ocean. In percent of dry weight, nickel ranges from an average of 0.33% in the Atlantic to 0.83% in the Pacific. For copper, the variation ranges from an average of 0.13% in the Atlantic to 0.60% in the Pacific. Average values for cobalt

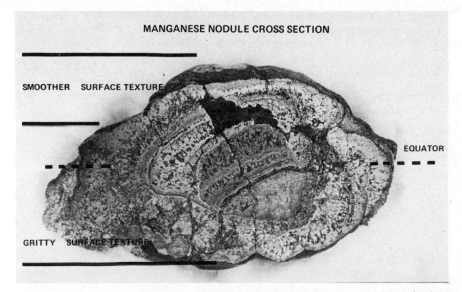

Fig. 71. Scale 1:1.2. Manganese nodule cross-section showing smooth and gritty surfaces, fractures, and layering. Some researchers suggest the laminae may be caused by abrupt changes in the nodule's growth rate, resulting from alterations in bottom current velocities and the amount of suspended particulate matter present in the water. Courtesy Kennecott Copper Corp.

are 0.28%, the same for both the Atlantic and Pacific; manganese averages 13.6% for the Atlantic and 21.3% for the Pacific.

Within the Pacific Basin, the average percentage value for nickel content is 1.25% in the Clarion and Clipperton Fracture Zones (CCFZ), basically a siliceous ooze region, lying southeast of Hawaii—primarily between longitude 155°W and 120°W and latitude 5°N and 22°N (see Fig. 76). In the Pacific outside the CCFZ, the average value for nickel is 0.60%. Similarly, nodule copper content in the CCFZ averages 1.04%, whereas outside the CCFZ it is 0.36%; values for manganese are 25.2% and 18.9%, respectively. Cobalt varies between 0.23% in the CCFZ and 0.30% outside.[e] A generalized view of the metallic distribution within the Pacific shows the eastern portion's nodules as being relatively high in nickel, copper, and manganese. As one proceeds to the west, these values decline, whereas the proportion of iron, lead, titanium, and cobalt increases,[7] and nodules lying in silicious oozes tend to have a much higher copper and nickel content than those located in red clay areas.[243]

Table 16. Average Weight Percent of Selected Elements in Manganese Nodules and Other Ferromanganese Oxide Deposits in the World Ocean

Element	World average	Element	World average
Manganese	16,174	Phosphorus	0.224
Iron	15,608	Barium	0.201
Silica	8.624	Lead	0.087
Aluminum	3.098	Strontium	0.083
Magnesium	1.823	Zinc	0.071
Titanium	0.642	Zirconium	0.065
Nickel	0.489	Vanadium	0.056
Cobalt	0.299	Molybdenum	0.041
Copper	0.256	Yttrium	0.031

Source: [111].

Fig. 72. Scale 1:2. Manganese accretions beginning to form around sharks' teeth. A great variety of materials act as nuclei for nodule formation, including bones of cetacceans and sponge skeletons. Courtesy Kennecott Copper Corp.

Nearly all nodules form around some small object that acts as a nucleus, such as volcanic dust, another manganese nodule or, as sometimes occurs, a shark's tooth (Fig. 72). The oxides form as layers around the nucleus, much like a series of onion skins which can be peeled off. The rate of accretion is exceptionally slow. Nodules have been found to have an age as great as 10 million years. How is this age determined? Because most nodules seem to form around a nucleus, scientists attempt to establish the age of the nucleus, using radioisotope dating techniques such as the potassium-argon method. The data generated "... can be translated to growth rates by dividing the thickness of the [nodule's] Fe-Mn crust by the age of its [nucleus]." Mineralogists have in this way established growth "rates of between 1 and 3 mm/10^6 ...".[57]

ORIGIN

Oceanographers, mineralogists, and chemists have long puzzled over the origin of manganese nodules. Various theories (models) have been offered to account for their formation, but as yet, none has been completely accepted. Debate centers on two fundamental questions: What is the primary source (seabed sediments or seawater) of the metals contained in the nodules, and what is the primary process by which accretion (agglomeration) of the metal oxides occurs? It is widely accepted that a basic source of the elements contained in manganese nodules is mineral enriched seabed sediments, but the sediments' source of mineral enrichment is itself a matter of controversy and should be examined briefly prior to looking at the nodules' origin.

Sources of Mineral-enriched Seabed Sediments

To discuss processes of seabed sediment enrichment is also to ponder seawater enrichment because seawater is the medium by which the mineral material is transported and cycled as solutions, precipitates, and organic matter. For example, the breakdown of seabed basaltic rock and debris itself is a major source of dissolved minerals in seawater which may then be cycled into the seabed sediments.

It has been suggested that seabed sediments are regularly enriched with minerals such as copper and nickel when an abundant supply of metal-concentrating biological organisms is available.[112] Some plankton can concentrate elements contained in seawater at a ratio of 10^3:1, i.e., 1,000 parts in the organism for every unit in its water environment. When they die, settle to the ocean floor, and decompose, their mineral content may be released into the seabed sediment. Researchers of biogenic concentration processes caution that it is difficult to evaluate exactly how much of the metals actually enter the sediment column before being redissolved into the seawater.[362]

Another theory offered to account for the deep seabed's mineral-rich sediments focuses on a continental origin. As granites and basalts weather, elements such as manganese, iron, copper, nickel, and cobalt are carried to the seas by stream systems. Fine particles of these minerals attach themselves to very lightweight particles that are carried fairly equally to all parts of the oceans. Heavier particles, usually free of manganese and iron, tend to settle out nearer the continental margins. Wallace C. Broecker of the Lamont-Doherty Geological Observatory estimated that every 1,000 years an average of 2.4 in. of continental detritus is weathered and carried to the oceans, and that 6% of this material reaches the deep seabed. Assuming that the 6% of the total eroded sediments reaching the deep oceans attaches to 40% of the weathered manganese, the manganese enrichment of the deep seabed could be accounted for. But, because shales make up an estimated 80% of the sediments deposited close to shore and do not show any reduction in their manganese content relative to the continent's "... granite-basalt average ..., this model strains the evidence."[57] In other words, given the large amount of sediments in the deep seabed and the small amount of manganese available for enrichment, the continents cannot be providing enough manganese to account for the 7-times higher then average manganese value common to deep seabed sediments. Others, however, feel that the manganese, and iron added to the oceans from continental sources, are adequate to supply both the seabed sediments and seawater with the manganese they contain.[210]

A third theory holds that the primary source of highly mineralized seabed sediments comes from secondary enrichment. As sediments become buried from subsequent deposition, bacteria act upon organic matter within the sediments, depleting the oxygen until the materials are anaerobic. Sulphate ionizing bacteria then oxidize the manganese from Mn^{+4} to Mn^{+2}, thus changing it from an insoluble to a soluble form. The buried sediments finally begin to reduce and the manganese can move toward the seabed surface where it is reoxidized and deposited as precipitates. The entire sedimentary column's manganese is then associated with the uppermost sediment layer. But, according to Broecker, a flaw in the hypothesis has been shown by scientists having measured the lower sediments for manganese depletion and finding none, even at depths of more than 30 ft.[57]

A final theory relates to discussion of the previous chapter, "Plate Tectonics and Mineral Formation." Hot basalt extrusions in mid-ocean ridge areas are associated with hydrothermal emissions that bring dissolved minerals to the seawater/ seabed interface where precipitation occurs. Theoretically, any seabed surface rock should have received a similar layer of iron-manganese enriched material as it emerged from the ridge area. With time, as the emerging layer moves from the crest, its iron-manganese content should be depleted, i.e., there should be an inverse relationship of proximity to the ridge and manganese accumulation. Such is not the case, as sediments well removed from the ridges have far more manganese than would be normally expected. Broecker notes that this unexpected manganese content of the sediments may come from the mid-ocean ridges' mineralized hydrothermal solutions which escape immediate precipitation and are carried throughout the deep sea.[57]

Nodule Formation Processes

Any model explaining the formation of nodules must account for their morphological structure, their chemical composition, and their density of distribution at the seabed/water interface. The model must also demonstrate the processes by which "... nodule components are extracted from the seabed sediments and transported to the interface"[439] and/or by which they are precipitated or adsorbed from seawater. There is appreciable agreement that nodules are most plentiful where (1) currents are strong, (2) sedimentation rates are low, and (3) oxygen levels are high.[439] Beyond this concensus, considerable debate exists about the primary processes responsible for nodule formation. For example, until recently many investigators felt that when nodules contained "... mineralogical phases able to incorporate high concentrations of Ni and Cu ...," a significant scavenging of nickel and copper into the nodule matter would occur.[112] Now, there is considerable doubt that a prior existence of these mineral phases is necessary for the accretion of nodules' high nickel and copper values.[188]

Biological. Several researchers feel that numerous microorganisms contribute to nodule growth through mineral adsorption. Under the leadership of H. L. Ehrlich of Rensselaer Polytechnic Institute, investigators have done detailed and controlled laboratory experiments to determine the significance of microbial flora (contained in manganese nodules) in the process of nodule formation. They have found that oxidizers promote accretion by catalyzing Mn^{+2} (already contained in the nodule) to Mn^{+4}. The Mn^{+4} then absorbes more Mn^{+2}, providing "... more substrate for biological oxidation and thus contributes to progressive growth of nodules." Reducing bacteria convert Mn^{+4} to Mn^{+2} by utilizing "... reduced carbon ... as the source of reducing power," and this process also makes soluble the seawater trace elements of copper, nickel, and cobalt, thus contributing to the degradation of the nodule structure.[164,f]

In addition to the bacteria, organisms such as benthic foraminifera of the genus *Saccorhiza* "... attach Mn micronodules to the surface of macronodules as part of their tube-building activities." Foraminifera that absorb the minerals from the seawater and then secrete them in building their tubes (0.02 in. diameter) have been found preserved (not replaced with clay or other materials) inside nodules[212] (Fig. 73).

Precipitation. One theory holds that nodules grow electrochemically. A nucleus such as a shark's tooth attracts hydrous oxides of iron, manganese, and other metals which are precipitated onto the nucleus surface so that it builds outward.[409] It has been suggested that the iron oxides must precipitate first before manganite and trace elements can be deposited.[74]

Debate has focused on whether the seabed or the seawater surface of nodules is the principal zone of accretion. Werner Raab of Kennecott Exploration Corp. has presented arguments that opt for an underside accretion of the nodule[439]—an idea now generally agreed upon, at least for the accretion of manganese, copper, and nickel.[188] This process is related to the presence of sea-floor igneous activity. The intrusion of sills and dikes could increase the temperature of seabed sediments, a process that increases "the solubility of metals in the interstitial water (leaching of metals from the sediments)." The increased heat within the sediment column would encourage the interstitial solutions to move convectively to the seabed/water interface. Because the mineral "... concentration and temperature gradients at the interface are steep and the oxidation potential of overlying sea water is high

Manganese nodules

Fig. 73. An example of a fossil embedded in nodule materials. The magnification is 25,000 times. Many scientists think that various benthic organisms may play a role in nodule growth. Courtesy Deepsea Ventures, Inc.

relative to interstitial water ...," rapid precipitation of iron, copper, nickel, and manganese will occur. If nuclei are present, nodules may form as the solutions precipitate on their surface. If no precipitation sites are available, crusts may form or the solutions may be dissipated into the seawater.[439] Since the 1880s, scientists have suggested that nodules may form when waters containing sulphate and carbonate react with seabed rocks to release manganese, primarily as a bicarbonate. The seawater's dissolved oxygen causes the bicarbonate to form into a colloidal manganese-peroxide that, upon suspension in seawater, becomes hydrated and then accretes on various hard surfaces available.[361] Mero notes that this and several other theories do not take into account the presence of metals other than iron and manganese in the nodules.[342]

Direct adsorption. One of the most frequently proposed mechanisms for the removal of metals such as chromium, lead, zinc, zirconium, and cobalt from seawater is direct adsorption onto iron and manganese oxides.[429] Murray and Brewer have cautioned that adsorption on iron and manganese oxides has not been proved, "although circumstantial evidence ... is abundant." The models so far proposed are based on very limited experimental work or under conditions that were only partially analogous to the seawater environment. There is, however, a strong argument for the seawater as the basic source for the metals contained in nodules, because when nodules become covered with sediments, they stop growing.[362]

SUMMARY AND CONCLUSIONS

Although manganese nodules have been known for 100 years, only recently have scientists and industrialists focused on them as a possible source for minerals. Nodules have a nearly worldwide oceanic distribution, being most common in the eastern equatorial zone of the Pacific, least common in the North Atlantic, and absent in the Arctic. Most nodules form around a nucleus, but they are highly varied in shape, surface texture, and chemical composition. Nodule populations in a given sediment column are normally densest at the seabed/water interface. Although the highest metal contents have been found in nodules located primarily in the siliceous ooze region of the northeastern Equatorial Pacific, there are oceanic areas which have higher densities of nodule populations. They are most numerous in sloping, abyssal siliceous clay areas with significant currents, whereas manganese crusts are most common in raised seabed relief areas, as on seamounts, and where currents are even stronger, a condition that minimizes sediment deposition. Theories attempting to explain nodule formation center on variations of three (biological, precipitation, and adsorption) basic mineral concentrating and accreting processes. The mystery of the manganese nodule has not yet been solved, but as the drama of seabed resource exploitation intensifies in coming decades, scientists will further part the curtain of the unknown, ready for the next act, the mining of the deep seabed.

NOTES

[a]As of mid-1979, core and grab sampling programs have provided about 100 samples per 1 million km^2 of ocean floor [188].

[b]Data presented in the map were extracted from a comprehensive sediment data bank which has been developed at SIO. For a useful description of the data bank and instructions for retrieval, see [190].

[c]They suggest that a better procedure is to identify the crystalline structure of the elements contained. See [20].

[d]Elements of relatively negligible weight percent included tungsten (0.006%), tellurium (0.005), chromium (0.001) and gallium (0.001). Also present in even more negligible proportions were bismuth (0.0008%), cadmium (0.0008); silver (0.0006) and tin (0.0003).

[e]Data for the CCFZ are based on approximately 1100 assays; for the Pacific outside of the CCFZ, the number of assays was approximately 2050; in the Atlantic, the total number of assays was about 440. See [188].

[f]Ehrlich's earlier work (1963) was criticized as inconclusive because the experimental conditions were "unnatural" and may have been influenced by contamination [499]. But Ehrlich made no claim that his findings were definitive, noting that "it remains to be determined whether the mechanism of Mn incorporation into, or removal from, nodules actually occurs in the sea" [163].

Chapter 11

Manganese Nodule Investment-Production Economics and Technology

During the last decade, numerous companies have invested large amounts of capital, gambling in the hope that they can soon put to sea for the harvest of a crop of metal-rich manganese nodules. Several variables are encouraging mineral producers and others to venture into the deep seabed mining industry. These include (1) an anticipated long-term increase in demand, along with potentially depleting high-grade onland sources, for nickel, copper, cobalt, and manganese; (2) an expectation of profits equal to or greater than onland mine sites; and (3) a desire of states to assure themselves a steady and *safe* supply of minerals.[45] Mero stressed this last point by noting that the $250 million invested by industry in nodule exploration and development (as of 1975) was not primarily a function of imminent shortages of metals from onland mines, but rather indicative of a concern by the mining industry for assured supplies coming from many world states that are establishing restrictive political or economic policies.[345] The primary driving force behind the mining companies' investments, however, is the hope for profits under a system of free market competition.

Although no commercial mining of the deep seabed has begun, several consortia are in advanced stages of testing equipment and processing methods. When production does begin, there may be significant impacts on certain onland areas, as at ports or near processing sites; on world metal markets and prices and the competitive position of onland mining industries producing the same or competing minerals; and on the economic-political relations among world states. These topics are the concern of this and subsequent chapters.

PREMATURE PREDICTIONS AND REALISTIC ASSESSMENTS

For nearly a decade, many lay authors have implied that mining of the deep seabed was imminent and that mining companies would be reaping excessive profits, i.e., amounts well beyond the average corporate returns on investment within the mining industry. As of the mid-1970s, mining firms in the United States have been reported to have a pre-tax return on investment (ROI) of about 27% annually,[311] but those in industry would probably be willing to debate this figure. Where does the truth lie?

First, the matter of profits. Mero suggested that, initially, seabed mining companies may make very high profits relative to total capital investments, because onland producers will continue to dominate metal market prices. Then, as large onland producers "rush" to the sea, prices will decline, creating a highly competitive situation and making minerals more readily available at a lower cost.[345] Those in industry and the financial community might criticize Mero's scenario as not taking into account the enormous research and development requirements that are necessary before a viable seabed mining enterprise could begin operating. Also, taking the long view, Moncrieff and Smale-Adams, using investment and operational cost ranges, contend that pre-tax ROIs may be only 10–15%.[352] Moncrieff's and Smale-Adams' pre-tax figures have been criticized as being too low, because of their assuming exceptionally large capital investment costs and overly high operational costs. Critics have used Moncrieff's and Smale-Adams' low operational cost estimates and high revenue estimates to calculate a pre-tax ROI of approximately 50%.[311] If a 50% royalty fee is assumed, a 3-million tons-per year (tpy) plant would yield an ROI between 27% and 47% annually.[541] Two more recent investigations have also addressed the ROI problem. One study by the consulting firms Dames & Moore and EIC Corporation came up with much more modest ROI after-tax figures—between 9.2% and 15.8% for a 3-million tpy establishment.[135] Finally, a detailed modeling study done at MIT suggests the prospective seabed nodule mining investor might realistically expect a ROI of 15% to 22%.[394]

Given these informed but varying estimates of potential ROI by deep seabed mining companies, it seems prudent to give only a tentative answer to the original questions regarding the mining companies' prospects for reaping excessive profits. Overall, it does appear that there could be some "economic rent," i.e., profits may be "... higher than the normal profit for industries with similar risks ...,"[105] but it is unlikely that massive profits will accrue indefinitely; in the long run, the mining companies' ROI after taxes may measure between 10% and 18%.

When will mining begin? One so-called "expert group," the United Nations Conference for Trade and Development, which should have used more caution, predicted in 1974 that deep seabed mining would begin that same year.[45] Such statements mislead the public, because mining firms pursuing nodule deposit mapping and sampling programs, equipment efficiency analyses, and economic feasibility studies are still a long way from commercial production. Even those working within the industry have been led to make premature predictions about the mining industry's progress toward commercial seabed mineral extraction. In 1974, A. H. Rothstein and R. Kaufman predicted, "we foresee with great assurance

Predictions and assessments 193

that a few [deep seabed mining] units will be in production in the late 1970s." They saw mining beginning in 1978 with two establishments becoming active, two more in 1983, one in 1984, and two in 1986.[464] Given the international political situation at that time and the progress of preliminary preparation for mining, their prediction was not entirely unrealistic.

Others in industry have been more conservative in their optimism for the start of seabed mining. Blissenbach expressed his conviction in 1975 that full-scale mining operations could not be carried out prior to the mid-1980s.[45] In 1976, Moncrieff took the view that "the time when the first nodule operation will come into production and the rate at which further operations will come in, are still uncertain factors. However, it is unlikely that the first will start before 1980, or that on the average, more than one new one will start every two years between then and 1990."[351] Tinsley, looking at the seabed mining industry from a financiers' viewpoint and from a more recent vantage point, also took a conservative view of the time schedule for nodule mining's start-up. He was quoted in 1977 as projecting 1984 as the earliest possible year for a major nodule mining enterprise, one likely to require an investment sufficient for a 3-million tpy establishment.[381]

A 1978 Library of Congress study has produced two estimated (hypothetical) timetable scenarios for the development of seabed mining. But because no seabed mining yet exists and because future economic and political events are always uncertain, the researchers stress that these estimated schedules are no more precise than the assumptions they are based upon—namely:

(1) Each mine site or nodule deposit would have the following average percentage composition: manganese, 28; nickel, 1.25; copper, 1.1; and cobalt, 0.25.

(2) A single operation would produce 3 million metric tons (dry weight) of nodules per year. However, for manganese, only 3 to 4 million metric tons of nodules would be processed annually by the year 2000.

(3) Metallurgical processing efficiency would be 95%.

(4) Through the year 2000, world demand would increase at an average annual rate as follows: nickel, 4%; copper, 4%; cobalt, 3.5%; and manganese, 3%.

Other assumptions include a total investment per mine site of $500 million to $700 million; each of five consortia would initially develop only 1 site (in sequential years) and an additional site at approximately 5-year intervals up to the year 2000; a return on investment of 13-29%; Law of the Sea negotiations will not affect international metal market relationships, as production controls. Given all these assumptions, the researchers see a total of 17 mine sites operating in the year 2000 if mining begins in 1985, and 12 mine sites if mining begins in 1990.[128]

Considering the diverse sources of opinion presented here, the consensus seems to be that the mid-1980s will be the earliest mining can begin; and when it does start, it will develop very slowly. Because of the capital needed and the long-term commitment necessary to a rather hazardous economic and political future, only a limited number of those in the mining industry have attempted to enter the field of deep seabed exploitation. Who are these intrepid few?

MINING CONSORTIA

In the early 1960s, a small number of corporations, including Kennecott Copper Corp. and Newport News Shipbuilding and Drydock Co. became interested in the commercial potential of seabed nodules. The latter firm's ocean mining interests were vested in a subsidiary, Deepsea Ventures, Inc. When Newport News Shipbuilding and Drydock Co. was purchased by Tenneco, Inc., it made Deepsea Ventures a wholly owned subsidiary. Deepsea Ventures continues to be one of the leaders in seabed mining. Various firms during 1962–68 individually invested a collective total of between $10 and $20 million on research and development that looked at the potential of nodules as a source of metals. From 1969 to 1974, much more detailed engineering work took place, and additional companies, including Metallgesellschaft AG, Centre National pour l'Exploitation des Oceans, and various Japanese firms under the corporate umbrella of the Sumitomo Group, joined in with the seabed mineral exploration and development work. During this period, these firms invested another collective total of $50 to $80 million in investigative programs.

In 1974, a significant series of events began to take place. Many of the previously independently operating seabed mining companies decided to join forces and form consortia. Three basic factors account for this development: (1) The costs of generating full-scale technical and economic investigations demanded more financial strength than most of the individual firms could command; (2) the uncertain international political situation for the regulation and ownership of seabed mining enterprises made such large individual-firm investments hazardous; and (3) the technical and personnel capabilities of the various firms could be used more efficiently in collective efforts.[128] By 1977, industrial firms from the United States, Canada, United Kingdom, Netherlands, France, West Germany, Belgium, and Japan were members of five different consortia. An additional enterprise, the CLB Group (composed of about 20 companies) banded together to work out one specific type of production technology—namely, a Continuous Line Bucket nodule mining and lifting system. The five other groups were organized under a broader relationship: Kennecott Consortium was the first group formed (1974); Ocean Mining Associates followed soon thereafter (1974) with Deepsea Ventures, Inc. acting as its service contractor; next came Ocean Management, Inc. (1975); 2 years later the newly organized Ocean Minerals (1977); and finally Afernod (1977), an acronym for the French Association for Nodule Exploration. Because of the importance of these consortia for seabed mining's future and their diverse approaches to developing seabed mineral deposits, a brief examination of each is in order (Table 17).

Kennecott Consortium

The Kennecott group has six members. The most important of the six is the Kennecott Copper Corp. with a broad experience in marine exploration and development work since 1962. Kennecott's preeminence in expertise is reflected by its 50% ownership and management of the consortium. Prior to the consortium's

Table 17. Ocean Mining Consortia and Joint Ventures

Name	Percent ownership	Country
Kennecott Consortium:		
Kennecott Copper Corp.	50	United States
British Petroleum Co.	10	Great Britain
Rio Tinto Zinc Corp., Ltd.	10	Great Britain
Consolidated Gold Fields, Ltd.	10	Great Britain
Noranda Mines, Ltd.	10	Canada
Mitsubishi Corp.	10	Japan
Ocean Management, Inc.:		
Inco, Ltd.	25	Canada
Arbeitsgemeinschaft Meerestechnische Gewinnbare Rohstoffe (AMR)	25	Fed. Repub. of Germany
Sedco, Inc.	25	United States
Deep Ocean Mining Co. (DOMCO)	25	Japan
Ocean Mining Associates:		
United States Steel Corp.	$33^{1}/_3$	United States
Union Minere, S.A.	$33^{1}/_3$	Belgium
Sun Co.	$33^{1}/_3$	United States
Ocean Minerals Co.:		
Lockheed Missiles & Space Co.	n.a.	United States
Standard Oil of Indiana	n.a.	United States
Royal Dutch Shell	n.a.	Netherlands
Bos Kalis Westminster	n.a.	Netherlands
French Association For Nodule Exploration (Afernod):[a]		
Centre National pour l'Exploitation des Oceans	20	France
Commissariat a l'Energie Atomique (CEA)	20	France
Bureau de Recherche Geologiques et Minieres (BRGM)	20	France
Societe Metallurgie Nouvelle/Societe Le Nickel (SMN/SLN)	20	France
Chantiers de France Dunkerque	20	France
Continuous Line Bucket (CLB) Group[b]	—	—

Source: [128]; [a][530].
Notes: n.a. = unknown; [b]About 20 companies in 6 countries, including Australia, Canada, France, Federal Republic of Germany, Japan, and the United States Consortium for development of mining system but not commercialization.

formation, Kennecott had completed extensive nodule resource delineation surveys which prompted other members to focus their efforts on developing mining technology.[420] Much onland work has been done with a mining system composed of a single, towed mining head with an in-line pumping system that should be able to dredge about 3 million tons (dry weight) of nodules annually. The consortium's technicians are reported to be also working on robot collectors. Much processing work seems to have been focused on an ammonia-based process, the Cuprion, designed to recover nickel, copper, cobalt, and molybdenum.[530]

Ocean Mining Associates

The Ocean Mining Associates (OMA) consortium's linchpin is Deepsea Ventures, which has been working with marine mining for more than 15 years and performing extensive experimentation with a prototype mining system, a towed mining head. Preliminary work with air-lift nodule recovery techniques on the Blake Plateau has encouraged them to go ahead with larger scale development and testing of equipment. This work was to have been carried out during 1978 in the Pacific Ocean, southeast of the Hawaiian Islands. OMA has been reported to be developing its mining systems for an output of 1 million tpy (dry weight) recovering nickel, copper, cobalt, and manganese, the only consortium to make a commitment to the production of the latter. Some have speculated that OMA may eventually recover as many as seven of the nodules' metal components.[530] Metals such as vanadium, silver, zinc, molybdenum, and cadmium could be extracted by the mining companies, if economic conditions are favorable.

Ocean Management, Inc.

Ocean Management, Inc. (OMI) is composed of members who have all had experience in marine mining and in developing ocean resources technology, making it a unique group among the various consortia. This organization has thus progressed rapidly in its development and testing programs. It has already completed major pilot-testing programs in the Pacific where it successfully recovered nodules in water depths of 15,000 ft. According to Tinsley, the group's recovery method may be a hydrolift and/or airlift system(s), which, if a 2-ship operation were used, could produce 3 million dry weight tons annually.[530] Partly because of the unsettled Law of the Sea negotiations, OMI drastically cut back on its programs in 1978, closing its central office in Bellevue, Washington, and reducing the number of personnel involved. Contrary to rumors within the industry, the consortium has not folded, but is in a "holding pattern."[506]

Afernod

Composed of five firms, with an equally shared ownership, this group has been active in exploratory work since 1970, but did not formally declare its commitment to an active nodule mining development program until 1977. It has mapped extensive areas of the Pacific that appear to be prime mining sites. This group, under the leadership of Centre National pour l'Exploitation des Oceans, is continuing extensive equipment testing work, especially with 2-ship CLB systems.[128, 530]

Ocean Minerals Company

The leadership of Ocean Minerals Company (OMC) reports to Lockheed Corp.; since the mid-1960s, Lockheed Missiles and Space Co., a subsidiary of Lockheed Corp., has been pursuing ocean mining research, especially as it relates to nodule recovery systems. Extensive economic studies completed by OMC have shown that seabed manganese nodules should be able to compete effectively with onland producers.[572] OMC was scheduled (as of early 1979) to carry out extensive deep sea tests of equipment, using a technically advanced deep sea mining vessel leased from the United States government, the *Glomar Explorer*, a remnant of a deep seabed mining "dropout," the Summa Corporation.

The Howard Hughes Episode

During the early 1970s, widespread news and trade reports indicated that the Howard Hughes-controlled Summa Corp., with its newly built 36,000-ton ship, the *Glomar Explorer*, would soon become the world's first commercial nodule mining enterprise. When the *Explorer* with a crew of 40 "mining staff" and 130 operational personnel sailed on its first mission in June 1974,[437] it appeared the eccentric Howard Hughes was ready to undertake yet a new venture. Indeed, he was.

The *Explorer's* "mining staff" (CIA employees), with the help of the ship's technicians, succeeded in raising not nodules but a Soviet nuclear submarine lost in the depths of the North Pacific Ocean.[a] Government officials thought the submarine might contain nuclear missiles, nuclear-tipped torpedoes, and commuications code books. Project Jennifer, the code name for the program, estimated by some sources to have cost as much as $350 million,[211] was considered a waste of time and money by many in the federal government and the U.S. Senate as exemplified by the criticism of Senator Frank Church. Many legitimate deep seabed mining firms were chagrined that the federal government would spend such a large sum on this clandestine operation but be reluctant to give them investment guarantees in case adverse international decisions might emerge from the United Nations' Third Law of the Sea Conference (discussed in Chap. 13).[437] As a result of the Howard Hughes episode, numerous underdeveloped states participating in the Conference expressed a deep suspicion about all scientific research and mining exploration,[399] and many continue to hold this view.

Other Interested Groups

Among the other interested parties engaged in considerable work of assessing the possibilities of manganese nodule production were Union Carbide Exploration Corp., Bethlehem Steel Corp.,[399] and Tenneco, Deepsea Venture's former controlling corporation. However, their efforts appear to have been discontinued. The research organization Battelle Memorial Institute has also done extensive work with nodule mining technology.[183]

Various government agencies in the United States have been active in working with manganese nodules. These include the USGS, the Bureau of Mines (BM), the Naval Research Laboratory (NRL), NOAA, and NSF. The USGS' and BM's efforts have been focused on determining nodule origin, composition, and mineral potential. NOAA has been mainly concerned with environmental problems associated with exploitation, and the NRL has been investigating methods for improving photography in exploration programs.[399]

INVESTMENTS AND PRODUCTION TECHNOLOGY

As the preceding discussion tends to imply, the deep seabed mining industry is not a business for the timid, the impatient, or the poor. Tremendous sums of capital are required in the slow, frustrating process of developing mining and processing systems. The costs are so great that investment funds often must be generated from industrial sectors outside of individual mining companies,[147] as is evident from the consortia membership that includes several steel and oil companies. Some 1977 estimates placed the "Deep Seabed Club" entry fee for one mine site at as much as $700 million.[442]

At present rates of investment, ocean mining analysts expect that more than $1 billion may be spent by 1985 by the deep seabed mining industry. If adequate financial investment protection (government guarantees or insurance programs) from international usurpation were available, this figure might reach $3 billion.[381] Richard Tinsley recently provided useful, if necessarily tentative, estimates of likely future investments to be made by the various consortia in their pre-mining research and development programs from 1980 to 1985 (Table 18), given a favorable international political climate and an aggressive investment policy by consortia. His estimates are premised on a range of ±30%. Although Tinsley (from a mid-1978 perspective) believed that Kennecott might invest a total of $450 million between 1981 and 1984, a Kennecott officer indicated in August 1979 that only a maximum of $150 million[153] would be allocated, probably due to lack of significant progress in the UNCLOS III negotiations in 1978. Similarly, Ocean Management's estimated investment of $450 million during 1982-85 may be much lower because it has temporarily suspended all additional major investments and development.

Tinsley's estimates include projections in so-called escalated dollars, which provide for inflation in cost factors.[b] In such escalated values, the grand total for 1980-85 could reach $2.27 billion, of which Ocean Management would account for $780 million, Kennecott for $738 million, Ocean Minerals for $408 million, and Ocean Management, Inc. for $340 million. Much of the projected technology development and research work must be geared to specific conditions at anticipated mine sites, an important consideration in the overall plans formulated by the various consortia.

Table 18. Estimated R&D Investments in Seabed Mining, 1980-85
(millions of 1977 U.S. $)

Consortium	1980	1981	1982	1983	1984	1985	Total
Kennecott		50	150	150	100		450
Ocean Mining Assoc.	40	60	60	65			225
Ocean Management, Inc.			60	160	220	10	450
Ocean Minerals Co.	30	40	50	50		30	250
Grand total	70	150	220	425	320	40	1,375

Source: Compiled from [530].

MINE SITE EXPLOITABILITY

Factors which affect the exploitability potential of a given mine site can be classified into three categories: (1) relative locational relationships, (2) site characteristics, and (3) characteristics of the nodules. Each of these variables is related to world metal market conditions, to the size of the mining and processing facility, and to the type of technology systems used in gathering, lifting, and processing operations.

Relative Location

The distance to service base ports and to onshore processing facilities could be critical to the economic efficiency of a mining site, depending on the needs of mining and auxiliary vessels and work crews and the requirements for human and industrial waste disposal. Ships will need fuels and equipment, and, if nodule processing occurs aboard ship, chemicals used in beneficiation. Food and other essentials must be available to workers on the ships which may be on station for a period of several months. The problem of local environmental risks and hazard prevention will, by necessity, enter into the calculation of costs.

Site Characteristics

Various local physical conditions bear on the exploitability of specific mine sites. Mero notes that the slope of the seabed should not exceed approximately 10%, and weather conditions in the area should allow at least 250 working days per year of operation.[340] Although weather conditions may have some significance, more important is the *mean* time between equipment failures.[153] Water depth will determine the time and expense necessary to reach the seabed with mining equipment and to send nodules to the surface. Speed of currents throughout the water column, consistency of the seabed sediments, and general relief of the seabed terrain must be reckoned with, because they influence the handling of mining and lifting systems. If the density of the nodule population is too sparse or the areal extent too small, a company will be unable to operate profitably. Both Mero and Kaufman have cited a seabed surface nodule density threshold of at least 5 kg/m^2 or 1 lb./ft.^{2} (this threshold has been set by a United Nations group of experts from industry and government at 10 kg/m^2), if a site is to be economically exploitable.[278, 340] Supplying adequate nodules for a 3-million tpy processing establishment over a 15–20 year lifetime may require a seabed area of 7,700 to 15,450 mi.^{2}, equivalent in its upper range to the size of the Netherlands (15,058 mi.^{2}) or Switzerland (15,830 mi.^{2}).[45, 351]

The Nodules

An important factor for an economic mine site is access to nodules that have relatively consistent dimensions. A uniform nodule size will reduce engineering problems associated with collection, transport, and processing. The type of nucleus in the nodule could be important, as will the total and specific metal contents and the consistency of grade. The grade of nodules that can be exploited will depend on the general economic situation and on specific site and situational relationships. Some of Mero's economic analyses have been based on a total nickel, copper, and cobalt content of 2.8% of the dry weight tonnage produced.[340]

The most crucial variable in the exploitability of most nodule deposits is the marriage of metal grade and population density. The ideal marriage of high grade and adequately dense populations of nodules has been an important research focus of several investigators. Estimates of the number of economically viable mining sites that are available for first-generation nodule-producing establishments have varied greatly. On the high side, a range of 80-185 has been offered, and on the low side, a range of 14-56 has been determined.[c] The wide range of these estimates is, in part, related to the basic assumptions made about what constitutes economic exploitability and the methodology used for estimation.

In the recent past, some researchers have estimated the number of "prime" mine sites throughout the world's oceans by using the "Global Estimator Approach." This methodology assumes a randomness of nodule sampling stations that provide data on metal grade and density of population, with both being considered as independent variables.[17,d] Jane Frazer has cautioned that these variables have not been proved to be independent, also noting that recent estimates of "prime" mining locations premised on a random distribution of sampling sites cannot be justified because sampling programs are no longer truly random. They are skewed toward the northeastern Equatorial Pacific (the main area of research interest during the past several years), where both nodule populations and their metal content are relatively high.[189]

Frazer suggested an alternative methodology, the "Grid Estimator Approach," for identifying "prime" mine site areas. Her methodology uses an arbitrary grid system to show the distribution "... of nodule grade and frequency of nodule occurrence," using data values generated from SIO's Sediment Data Bank. A computer analysis determines whether a grid unit should be classified as "prime" or "not prime." Based on the average grade of $Cu + 2.2\ Ni \geq 3.76\%$ per unit of dry weight, "prime" nodule mining sites are centered in the northeastern Equatorial Pacific (Fig. 74). Frazer concludes that, although the Grid Estimator methodology, as applied at Scripps Institution of Oceanography, should provide some notion of the total "prime" first-generation mine sites and their location (given available data), it must be looked upon only as a rough estimate, with the best one being 28 sites.[189]

The Northeastern Equatorial Pacific

It is clear from all available public data and from the areal focus of the mining consortia that the northeastern Equatorial Pacific stands out as the region most likely to first experience commercial nodule mining, as illustrated by Deepsea Venture's formal claim in late 1974 to a mine site at approximately 15°N., 126°W., about 1100 mi. southwest of San Diego, Calif.[e] Figures 75 and 76 demonstrate the preeminence of the region in relation to worldwide distribution of copper contained

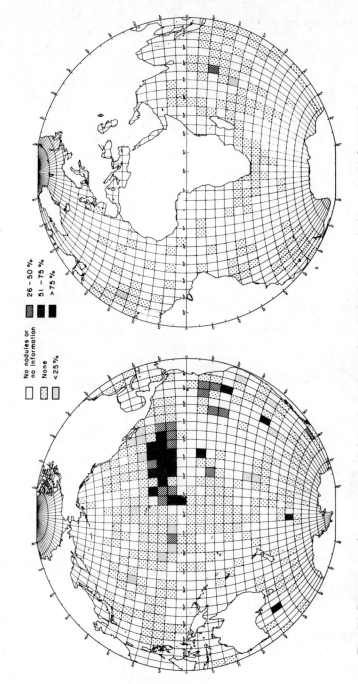

Fig. 74. Percentage of manganese nodule analyses above the required mine grade (Cu + 2.2 Ni ≥ 3.76%). *Source:* [189]. With permission.

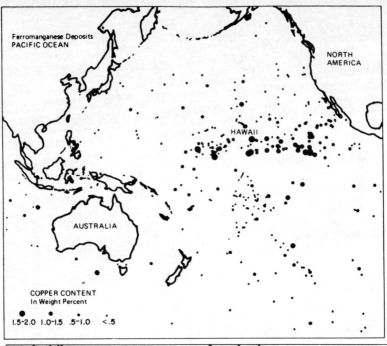
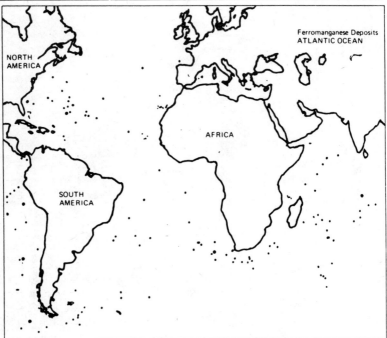

Fig. 75. Copper content of ferromanganese nodules of Pacific and Atlantic Ocean.
Source: [244]. With permission.

Mine site exploitability 203

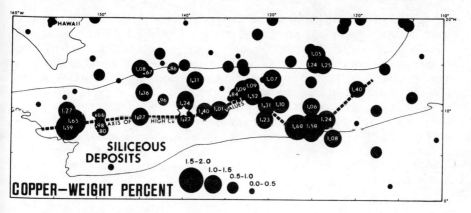

Fig. 76. Copper content of manganese nodules southeast of Hawaii.
Source: [244]. With permission.

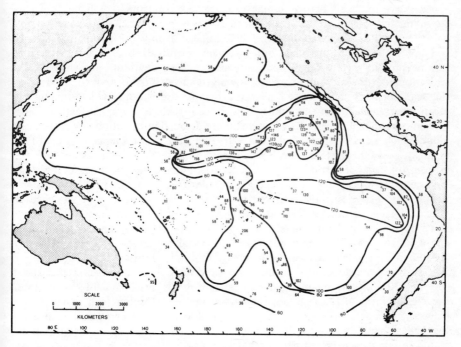

Fig. 77. Dollar value of manganese deposits per dry weight ton in the Pacific Ocean. Manganese nodules vary greatly in the total value of metals contained in them. Values assigned are $2.20 for each percent of manganese; $7.90/kg (2.2 lbs.) of cobalt; $1.80/kg of copper; and $4.00/kg of nickel. The high value areas are considered to be those enclosed by the $120 isoline.
Source: [340]. With permission.

in nodules. Nickel content is also relatively high in this region. The area's importance becomes vividly recognizable when these distributions are translated into potential dollar values. The collective dollar value (per dry-weight ton) for nickel, copper, cobalt, and manganese along the region's east-west axis (bounded by a $120 per-ton isoline) ranges from $109 in the east to $156 in the west (Fig. 77). Considering the inflationary pressures on the dollar since 1975-76 when these values were assigned, and taking into account the metal price fluctuations in the late 1970s, the values would be different if a similar study were made today. For example, the producer price of cobalt increased in one year (from December 1977) by as much as 213%. In addition, while producer prices in late 1978 were around $20.00 per lb. of cobalt, occasional spot prices on the "free market" were as high as $40.00 per lb.[547**]

EXPLORATION

Identifying exploitable mine sites is a slow and expensive task, but it is obviously a necessary first step before mining can begin. Mero estimated that it can cost approximately 5 million 1976 dollars to survey a nodule deposit covering 3,860 mi.^{2}.[340] Exploration programs must determine not only the population density and ore grade of the nodules, but must also identify topographic obstructions and measure the average relief, slope, sediment structure, currents, and water depth.

Various devices are employed to provide the needed data. Acoustic devices, as side-scan sonar systems, towed near the bottom are helpful in determining the location of areas barren of nodules and in identifying topographic features, with TV and photographic cameras providing more detailed information. TV sets can be mounted on sleds or suspended from a drifting ship. Still cameras, that may take as many as 1,000 shots in one lowering, are kept at the proper distance off the bottom by using acoustic devices. Tethered sampling units, including dredge buckets designed specifically for collecting nodules (Fig. 78), provide sediments and nodules for analysis, as do sediment corers and spades that may be designed to obtain rectangular cores of about 7 x 7 in. to 10 x 13 in. for lateral dimensions and 7 to 16 in. for depth; the spade devices provide relatively undisturbed nodule and sediment samples, often needed for precise analyses. Free-fall devices, relatively cheap and easily recovered, also can be used to obtain seabed samples. The free-fall unit is dropped overboard, whereupon it sinks to the bottom at a rate of about 115 to 230 ft./min., depending on the shape of the device and the "... ratio of the displacement of the flotation element to that of the submerged weight of the ballasting elements." Once on the bottom, the device secures a grab sample, releases the ballast, and returns to the surface; attached signalling systems help the research vessel recover the sampling unit.[340]

Battelle Research Institute's Northwest Laboratories have been developing a nuclear probe, using the man-made element Californium-252, that can determine concentrations of up to 30 different elements contained in nodules. A beam of neutrons causes exposed materials to give off gamma rays that can be measured and analyzed by computers on board ship. This technique may help avoid

Fig. 78. A nodule dredge unloading its cargo onto the deck of an exploration vessel. Courtesy Kennecott Copper Corp.

some future costly and slow grab and core sampling programs.[106, 137] But according to Marne A. Dubs of Kennecott Copper Corp., the nuclear probe technique is also somewhat slow to be truly useful.[153]

MINING AND PRODUCTION TECHNOLOGY: COST AND BENEFITS

After mine sites have been identified and mapped, a new phase of technology must be applied, i.e., mining—an expensive and complex operation when the ore lies 2 to 4 mi. underwater. The mining process is composed of three steps—gathering, lifting, and beneficiating, i.e., the separation of ores from the gangue. The five mining consortia, and others, are developing numerous techniques and devices for each phase of the mining process.

Gathering Systems

Producers will likely depend on one or more of three basic gathering devices to collect nodules from the seabed. These systems are either towed or self-propelled, and can be classified as (1) continuous line bucket (CLB), (2) fixed area, and (3) continuous path.

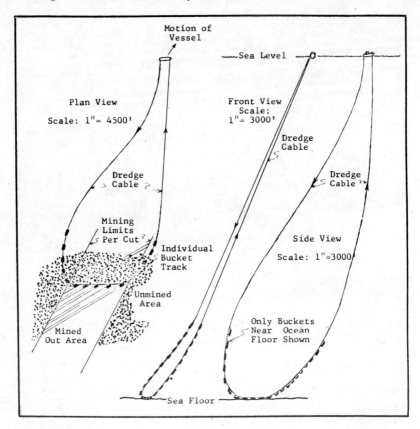

Fig. 79. Schematic drawing showing the operation and design of the 1-ship CLB system of nodule mining. The dredge cable passes over one side of the vessel, its attached buckets picking up nodules on their way across the seabed; the buckets ascend to the surface and then deposit the nodules on the mining vessel, before making another circuit. Cable diameter and buckets are not drawn to scale.

Source: [344]. With permission.

Engineers are working on two types of CLB systems—1-ship (Fig. 79) and 2-ship (Fig. 80). The 1-ship system has been extensively experimented with by a group of companies (the CLB Group noted earlier) working together for the specific objective of developing this mining system. Early tests in 1972, using wire cables and focusing on the feasibility of CLB nodule extraction (up to 250 tons/d),[106] determined that if the descending and ascending portions of the cable are not kept well apart, the two portions tend to become entangled. More recently, a French organization (CNEXO) and several Japanese firms (the Sumitomo Group) have experimented with a 52,000 ft. polypropylene rope in conjunction with a 2-ship

Mining and production technology 207

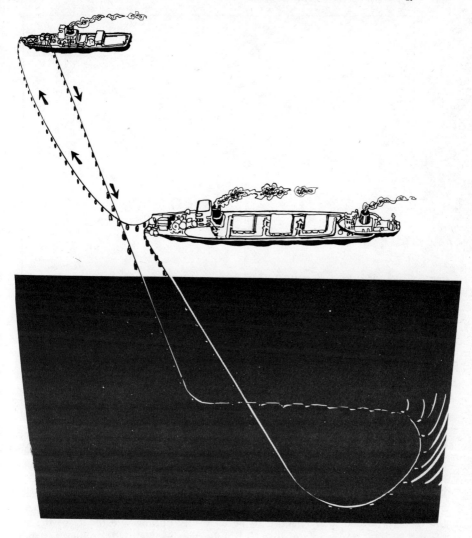

Fig. 80. A 2-ship continuous line bucket for nodule mining developed in France and Japan.
Source: [126]. With permission.

system, accomplishing considerable success. If full-scale tests prove that the buckets can be filled to 50% of their capacity, it is likely the system will be used to mine nodules commercially, because it is relatively inexpensive and uncomplicated.[295] In addition to these two assets, CLB systems do not have to be designed for a specific depth, nodule size, or sediment type, and they are easily retrieved. According to

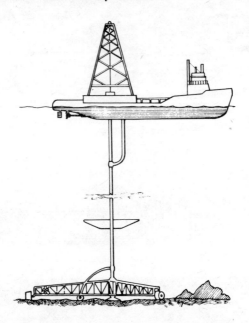

Fig. 81. A fixed area sweep method of mining nodules. This sytem was developed by N. Koot and R. N. Nold (US Patent 3,433,531).
Source: [126]. With permission.

Mero, a CLB system takes only 6 hours to assemble or disassemble, and it can load a 100,000-ton carrier in 7 days.[340] Some disadvantages are that the polypropylene cable may have to be sharkproofed, and the buckets cannot be controlled as they move across the seabed.[126] In the process, they have a tendency to pick up too much sediment, although much of this material may be washed away during the bucket's return to the surface.[217] Mero has given what people in the industry would likely believe to be an overly optimistic estimate of the cost of CLB systems. He estimates that CLB units may be able to lift nodules for about $5/ton (in mid-1970s dollars), with a capital investment of $10 million for a 2-million tpy plant.[340]

Engineers have designed some mining head systems to collect nodules at one site and then to move on to another. These are the fixed systems. A ship or barge can support these gathering devices, as illustrated in Fig. 81. Another group of mining heads, continuous path collectors, is being developed by marine engineers. Several continuous path collectors use the seabed as the support base, a method that may present problems when the terrain is rough, if obstacles block the passage of the vehicle or dredge head, or if the sediments' load-bearing capacity is poor (Fig. 82). If the load capacity of the sediments is low, the collecting vehicle or

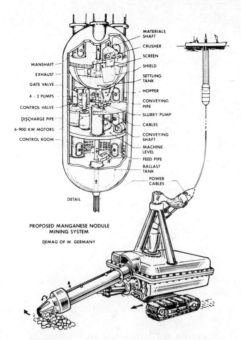

Fig. 82. A crawler-mounted gathering arm loader for nodule mining, proposed by Demag of West Germany.
Source: [126]. With permission.

mechanism may sink into them, as in siliceous ooze zones whose porosity may reach as much as 90%. This problem may be overcome, in part, by making the collection vehicle buoyant (Fig. 83)^f to reduce its weight or by distributing the weight in a way to allow the vehicle to remain on the seabed surface.

A basic problem with several of the gathering devices is to determine the proper width of the collecting surface relative to the speed of the ship. These two variables must be carefully adjusted. According to Krutein, if the lift pipe is moved too fast through the water mass, vortex effects may vibrate the pipe, causing fatigue problems. On the other hand, as the forward speed decreases, the required mining rate will demand a wider nodule collector.[295] People in industry have estimated that the overall efficiency of various collection systems may be anywhere from 20% to 70%, a crucial problem if a given mining site is to be economically viable. Collecting efficiency is likely to be in the 10%–20% range when seabed mining first begins, but it should improve to perhaps 50% by the year 2000.[153] The French have developed a shuttle-dredge system (composed of 14 dredges) *said to be* 70% efficient. The system utilizes a slightly negative buoyancy to assist the dredges in their downward journey and a positive buoyancy to help the loaded dredges surface.[313]

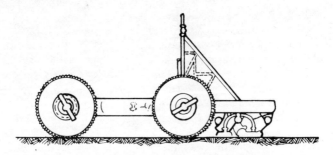

Fig. 83. A semi-buoyant crawler that utilizes an array of vacuum cleaner dredge heads (US Patent 3,504,943).
Source: [126]. With permission.

Lifting Systems

Various lifting systems are under development or have been proposed. Besides the CLB cable and rope system, most units function either through pneumatic-hydraulic or pump-hydraulic mechanisms, suspended from a ship (Fig. 84). Both lifting methods are technically more sophisticated than the CLB system. The pneumatic-hydraulic systems depend upon the principle of expanding air bubbles

Fig. 84. The 20,000-ton, 560 ft., converted ore carrier *Deepsea Miner II* owned by Deepsea Ventures, Inc. In late 1977, this ship was used to carry out hydraulic pipe air-lift tests at the owner's claim site to determine the optimal depth for injecting air into the lift-pipe system. The 70-ft. high, 52-ft.-diameter geodesic dome provides an all-weather shelter that houses a gimbal-mounted derrick above a 27-ft. wide, 34-ft. long "moon pool" through which a pipe weighing up to a total of 1 million lbs. can be lowered. Courtesy of Deepsea Ventures, Inc.

that create the lift required to establish a flow of water from the ocean floor to the surface, which carries the nodules and sediments along with it.[182] To function properly, the volume of air, the points along the pipe at which the air is injected, the diameter of the pipe, and the size of the nodules must be mutually calibrated. The dredge head must be designed to screen the nodules for proper size, and the air injection process must be maintained at a level that carries the nodules to the surface at desired rate.[399] Deepsea Ventures plans to use an air-injection dredge system in its mining operations and has successfully tested its equipment on the Blake Plateau at depths of 2,500 ft. in the early 1970s and off the southern California coast in 1977. The Blake Plateau tests brought up some 60,000 tons of nodules at a recovery rate of 10–50 tons/hour. Nodule recovery costs (in mid-1970s prices) were estimated at between $10–$20/ton, with the equipment valued at between $30 and $100 million.[340]

Other lifting systems may use conventional hydraulic dredge pumps. This system type functions similarly to the air-lift method, but its operation depends upon pumped water to induce upward flow in the pipe. The technology of this method has been well refined in the field of pipeline coal slurries and in mud pumping in oil fields. Both the air-lift and pump-hydraulic lifting devices may have difficulties operating in seabed areas with highly irregular topography or where there are frequent obstructions. A recent government study contends that, given numerous unfavorable working conditions and a sparse nodule population, these systems could function at only 9% of their designed capacity or, if conditions are all ideal, the rate might reach 58%. Many in industry expect an average efficiency rate of about 25%.[399]

The pump and pneumatic systems will put a considerable volume of sediments into the upper water column, because both the nodules and seabed sediments will be carried to the surface vessel. The CLB systems will create more disturbance on the seabed, because of their intense scraping action. Both CLB and hydraulic lift systems will be very heavy, weighing up to several million lbs. The complexity and the time required to assemble and disassemble hydraulic units will demand that they operate continuously, whereas CLB units should have more operational-timing flexibility.[340]

Nodule Mining Costs

The nodule tonnage produced by individual mining establishments will have a major bearing on the relative investment requirements and profits gained from seabed mining. In 1977, the consulting firms of Dames & Moore (D&M) and EIC Corp. completed a joint analysis (at various scales) of the estimated costs and profits from mining and processing nodules.[135]

Most operations were expected to be scaled at between 1 and 3 million tpy. The D&M and EIC study shows that fixed capital for a 1-million-ton, 1-ship enterprise (including expenditures for the ship, air-lift and pipe system, bottom miner, and port facilities) totaled 48 million 1976 dollars. If R&D and working capital costs are added, the total is $102 million; the 3-million tpy enterprise operating with two ships requires fixed capital amounting to $112.5 million, and R&D and

Table 19. Nodule Mining Capital and Operating Cost Estimates

Capital and cost component	Scale of operations (millions of dtpy)				
	1 M tons[a]	2 M tons[b]		3 M tons[b]	
	One ship[c]	Two ships[c]	One ship[d]	Two ships[e]	One ship[f]
Capital investment (millions of 1976 dollars)					
Fixed capital (total)	48	92	66	112	82
Ships	36	72	45	84	51
Airlift and pipe[g]	6	12	10	16	14
Bottom miner[g]	2	4	3	6	5
Port facilities	4	4	8	6	12
Exploration and R&D	50	50	50	50	50
Working capital[h]	4	7	6	9	7
Total	102	149	122	171	139
Operating costs (1976 dollars per wet ton)					
Mining (total)[i]	15.1	14.7	12.1	13.0	10.9
Power[j]	4.6	4.6	4.6	4.6	4.6
Bunker fuel	0.4	0.4	0.4	0.4	0.4
Wages[k]	1.5	1.5	1.1	1.3	0.9
Insurance[l]	0.8	0.7	0.5	0.6	0.4
Supplies	0.1	0.1	0.1	0.1	0.1
Overhead and misc.	0.3	0.3	0.3	0.3	0.3
Capital charges[m]	7.4	7.1	5.1	5.8	4.2
Transportation and handling	6.8	6.8	4.6	5.1	4.1
Total (per wet ton)	21.9	21.5	16.7	18.1	15.0
Total (per dry ton)	23.4	28.0	21.7	23.5	19.5

Source: [a][533], [b][135].
Notes: Mining ship sizes [c]50,000 dwt, [d]85,000 dwt, [e]65,000 dwt, [f]120,000 dwt; [g]Scaled with an 0.8 capacity exponent; [h]Equals 8% of fixed capital; [i]Mining of 1.3 to 3.9 million wet metric tpy, 1.0 to 3.0 million dry; [j]Power at 1.5 hp/ton/day for hoisting, 3.0 hp/ton/day for ore transfer; [k]Wages scaled with 0.5 capacity exponent; [l]At 2.1% of fixed capital per year; [m]20% of fixed capital, covering depreciation, maintenance, taxes, interest, etc.

working capital of $59 million—a total investment of $171.5 million. The 3-million-ton operation thus represents a significant improvement in the investment ratio in comparison to the 1-million-ton establishment. Operating costs also show an economy-of-scale; the larger the annual tonnage capacity of the mining plant, the better the operating cost ratio, with the 1-ship operation showing the best cost structure—$21.70/dry ton for the 2-million-ton plant and $19.50/dry ton for the 3-million-ton plant (Table 19).

Processing

Once the nodules have been lifted to the surface, the last phase of the mining process (beneficiating) can begin, either on a ship or at an onland site believed to be the more likely one by most analysts. Two problems with at-sea processing are that waste handling costs will be higher than at onland sites,[399] and motion of the vessel will make difficult, if not hazardous, the use of some processing techniques. A large investment will be required for both onland or at-sea processing plants. Mero estimated that an investment of 50 1975-76 dollars will be needed to extract, process, and transport each ton of raw nodule capacity of a large-scale operation. And if the metals can be removed at approximately $.20/lb., each ton of nodules might produce $50-$150 in revenues.[345]

The main task in processing manganese nodules is to "... break down the nodule matrix and liberate the contained value metals."[133] Since a detailed discussion of chemical and physical processes of reducing and selectively leaching the metal oxides in manganese nodules is beyond the scope of this presentation, only a brief overview will be presented.

The most likely processing techniques will be based upon leaching[g] (hydrometallurgy) and smelting or heating (pyrometallurgy) methods dependent on sulphate, ammoniacal, and chloride systems (Table 20). Depending on the process used, probably three or four metals will be recovered. Three-metal processes extract cobalt, copper, and nickel while 4-metal processes also capture manganese. Among the hydrometallurgical processes, sulphuric acid leaching techniques may be the primary methods used in 3-metal processes, because the unwanted manganese

Table 20. Classification of Extractive Metallurgical Process Systems

Systems type	Process
Sulphate	High Temperature Sulphuric Acid Leach[a]
	Smelting[a]
	Sulphuric Acid Reduction Leach
	Reduction Roast/Sulphuric Acid Leach
	Sulphation Roast
Ammoniacal	Reduction/Ammonia Leach[a]
	Cuprion/Ammonia Leach[a]
	High Temperature Ammonia Leach
Chloride	Hydrochloric Acid Reduction Leach[a]
	Hydrogen Chloride Reduction Roast/Acid Leach
	Segregation Roast
	Molten Salt Chloridation

Source: [135].
Note: [a]Process likely to be used in first-generation plants.

dioxide is insoluble in weak sulphuric acid[409] and because sulphuric acid is usually readily available. High temperature sulphuric acid leaching has been reported to have a better metal recovery rate than the low temperature process. But some researchers contend that low temperature hydrometallurgical processes are best suited to the extraction of copper, nickel, cobalt, molybdenum, and manganese.[77] Reduction ammonia leach processes are already well proven from use with lateritic nickel ores and should be economically feasible in nodule metal production.[133]

At least one firm, Deepsea Ventures, is expecting to extract manganese with a hydrometallurgical method, using a chloride process, that is based upon "... a leach stream subject to a separation of the liquids and solids (mostly clay) followed by ion exchange separation of the metal bearing streams and the subsequent electrolytic deposition of metals." The resulting product is a very pure manganese metal.[182]

Pyrometallurgy will also likely be used in the beneficiation process, if manganese is desired as an end product. A liability of smelting techniques, however, is the high energy demand, a costly prospect now and in the future. On the other hand, smelting provides a high usable-metal recovery rate.[133] Kennecott Copper Corp. has done extensive work with a pyrometallurgical process using sulfur-dioxide roasting to convert the nodules' metallic oxides into a more workable sulphate form. A leach liquor of the soluble sulphates is then processed to obtain precipitates of copper directly and, through the use of heat and pressure, cobalt and nickel. Manganese sulphates are processed to a dry, crystalline manganese sulphate that can be decomposed to provide manganese oxide that with further processing results in commercial manganese.[311]

Recovery rates of the copper and nickel in the nodules may be 90 to 95%, whether hydrometallurgical or pyrometallurgical techniques are used. Cobalt recovery is more dependent on the process used, but will probably be at least 40% and, perhaps, as much as 80%. These recovery rates would result in a 3-million tpy plant's producing 30,000 tons of copper, 35,000 tons of nickel, and 3,000 to 6,000 tons of cobalt annually.[351]

Nodule Processing Costs

The D&M and EIC study presents an analysis of estimated capital investment and operating costs for 3-metal manganese nodule processing plants of 1, 2, and 3 million dry tpy. The 1976 findings show that the 3-million-ton plant has an investment economy-of-scale when compared with the 1-million-ton plant ($349 vs. $174 million, respectively). Operating costs for the 1- and 3-million dtpy-processing establishments show relationships similar to those for capital investment. Operational costs for each dry ton of nodules in the 3-million-ton plant costs 17.3% less than its 1-million-ton counterpart (Table 21). According to the D&M and EIC report, even larger nodule mining and processing facilities could provide greater economies-of-scale, but "... it may be difficult to satisfy capital requirements much greater than half a billion dollars, even by a consortium, on a first generation operation."[135] The study also indicates that a 4-metal (including manganese) nodule plant would be more economical, assuming a stable world

Table 21. Three-metal Nodule Processing Cost Estimates

Scale of operations (millions of dtpy)	Capital investment[a] (millions of 1976 U.S. $)			
	Fixed: plant	R&D	Working capital	Total
1 M tons	115	50	9	174
2 M tons	200	50	16	266
3 M tons	277	50	22	349

Scale of operations (millions of dtpy)	Operating costs[b] (per dry ton in 1976 U.S. $)			
	Utilities and supplies[c]	Labor[d]	Capital charges[e]	Total
1 M tons	17.9	1.90	23.0	42.8
2 M tons	17.9	1.25	20.0	39.2
3 M tons	17.9	0.98	18.5	37.4

Sources: [a][135], [b][533].
Notes: [c]Supplies = $4.90; [d]Wages scaled with a 0.4 capacity exponent; [e]20% of fixed capital.

market existed for manganese. Even though total capital and investment costs would be higher than for the 3-metal plant, the actual metal recovery rate would increase approximately 10 times, because manganese is such a large portion of nodules' total metallic content. The additional revenues provided by manganese recovery, and the fact that world manganese markets might be saturated, may contribute to first-generation, 4-metal plants being smaller than 3-metal plants (Table 22).

Potential Revenues

The scale of the mine and processing plant, metal content of the nodules, costs of capital and operations, and world market conditions will each have an impact on the profitability of a given enterprise. Table 23 illustrates the range of revenues that may be available to the nodule producer, depending on a given percentage of metal content in the nodules mined. Revenues for nickel, for example, could range from $74/ton for high-grade ore to $64 for low-grade. Collectively, a 3-metal (nickel, copper, and cobalt) operation might have an income of $106 per/ton of high-grade nodules and $91 for nodules in the low-grade range. A 4-metal plant, including manganese, could provide a net revenue value of $230/ton of high-grade nodule ore and $125 for low-grade.

A comparison of financial elements of investment and operating costs, taxes, and revenues, shows that a 3-metal processing operation (using 2 ships) with a

Table 22. Four-metal Nodule Processing Cost Estimates[a]

Investments and costs	Scale of operation (millions of dtpy)	
	0.5 M. tons	1.0 M. tons
Capital investment (millions)[b]		
Fixed: plant	135	235
R&D	50	50
Working capital	11	19
Total investment	*196*	*304*
Operating costs (per dry ton)[b]		
Supplies	5.7	5.7
Utilities	26.0	26.0
Labor	5.7	4.3
Capital charges	54.0	47.0
Total	*91.4*	*83.0*

Source: [135].
Notes: [a]See notes to Table 19. [b]1976 U.S. dollars.

3-million dtpy capacity might expect an after-tax ROI value of 15.8%, if high grade nodules are available. If low-grade ores are used in the calculation, then the ROI would be 11.1%. D&M and EIC contend that the "... three-metal operation becomes attractive only at a 2-3 million tpy scale ..., and the low range ROIs suggest that 2 million tpy may be the cutoff point below which investments will be made" (Table 24). A similar analysis for a 4-metal operation demonstrates that for a 1-million dtpy plant the ROI is adequate, but very marginal for a 0.5-million dtpy processing plant. The ROI is 14.8% for high grade ore used in a 1-million dtpy facility and only 9.1% in a 0.5-million dtpy unit (Table 25).

Table 23. Range of Nodule Operation Revenues

Range, price and value	Nickel	Copper	Cobalt	Manganese
High range grade (%)	1.5	1.3	0.3	25
Low range grade (%)	1.3	1.1	0.2	25
Recovery efficiency (%)	90	90	60	90
Price per lb. (1976 U.S. $)[a]	2.5	0.8	3.0	0.3
Net value per ton (1976 U.S. $)[a]				
High range[b]	74.0	22.0	10.0	124
Low range[b]	64.0	19.0	8.0	124

Source: [135].
Notes: [a]Assumed future prices based on common factors. [b]The price of cobalt (c.$25/lb. in December, 1979) is grossly underestimated.

Table 24. Potential Profits of a Three-metal Nodule Processing Operation
(millions of 1976 U.S. $)

	Scale of operation (millions of dtpy)				
Range	1 M tons	2 M tons		3 M tons	
	One ship	Two ships	One ship	Two ships	One ship
High					
Total investment	276	415	388	521	488
Revenue	106	212	212	318	318
Ore cost	28	56	44	71	59
Process cost	43	78	78	106	106
Margin	35	78	90	141	153
After tax profit	18	39	45	72	77
Low					
Revenue	91	182	182	273	273
Margin	20	48	60	96	108
After tax profit	10	24	30	48	54
Return on investment (percent)					
High	6.5	9.4	11.6	13.8	15.8
Low	3.6	5.8	7.7	9.2	11.1

Source: [135].

Transportation

Whether the nodules are processed at sea (which seems unlikely) or moved directly to land, their movement will entail the need for extensive transport linkages between mine site and port. The nodules can be transported whole, ground up for dry shipment, or slurried. Slurries will likely be the most common method used. Crushers on the mining vessel can prepare the nodules for transfer to cargo vessels by recycled-seawater slurry systems. Once in the cargo vessel's holds (smooth sided and hopper like), the slurry will be dewatered at both top and bottom— dewatered ores are much more stable and cheaper to transport. After the cargo ship reaches port or a single-point mooring buoy, the ore can be reslurried and pumped off the ship. Whole nodules are cumbersome to handle and would require conveyor systems because without grinding they are too heavy for slurry pumping. The use of clamshells for loading or unloading would be exceedingly slow and require a large deck space. Overall, the liabilities of handling whole nodules dim the prospects for this shipment method. Drying and grinding nodules would have the advantage of reducing their bulk by at least 30%, by ridding them of both surface and capillary waters. Pneumatic devices are rather slow but they

Table 25. Potential Profits of a Four-metal Nodule Processing Operation (millions of 1976 U.S. $)

Range	Scale of operation (millions of dtpy)	
	0.5 M tons	1.0 M tons
	One ship	One ship
High		
Total investment	287	406
Revenue	115	230
Ore cost	19	28
Process cost	45	83
Margin	51	119
After tax profit	26	60
Low		
Revenue	108	215
Margin	44	104
After tax profit	22	52
Return on investment (percent)		
High	9.1	14.8
Low	7.7	12.8

Source: [135].

could be used in loading and unloading dry ores. However, high humidity on board ship could possibly create problems of caking, or if dry conditions prevail, dust may be a nuisance on the transport vessel and in port.

Distance to port processing sites will be an important variable in the economics of the manganese nodule industry. Four areas of the United States are likely prospects for onshore processing: Hawaiian Islands, southern California, the Pacific Northwest, and the Gulf Coast. Whichever area's ports are used, the ore carriers will have to be rather limited in size (probably no more than about 65,000 dwt) because of the inadequate depths in most United States ports, although offshore single-point mooring buoys could avoid this problem. Limited draft (39 ft.) and beam (105 ft.) capabilities of the Panama Canal and an entryway draft limitation (70,000 dwt ships) off the Columbia River in the Pacific Northwest will pose additional constraints.[134]

The relationships of distance and vessel size are important because the fewer ships that are needed the greater the savings in transport costs. The distance to Texas' Gulf Coast from the prospective Equatorial Pacific mining region via the Panama Canal, for example, is approximately 5,000 to 7,000 mi. The economies-of-scale provided by large carriers will be sorely needed if mining companies are to be competitive with onland producers nearer markets. Mero has calculated that bulk ore carriers could move nodules to ports at rates from $0.0003 to $0.0005/ton-km. (in 1975-76 dollars), depending on carrier size and the availability of automated loading and unloading equipment. Mero, assuming a situation where ships must return to the mine site empty, estimated that transport costs will range from $3 to $11 per ton (including a $1/ton handling charge).[340] It has been suggested, however, that overall transportation costs may be reduced if vessels capable of carrying fuels and other supplies are used for return voyages to mine sites.[134]

The Environment

The overall effects of manganese nodule mining and processing on environmental systems in the sea or onland are still unknown. Nor are the operational problems at sea completely predictable. But scientists and specialists in industry and government have already done considerable work in identifying potential problems for the marine environment. And economic and social planners in some areas that have favorable locations for nodule processing have been weighing the assets and liabilities of siting facilities in their area.

The weather hazard. Work at sea always poses a certain hazard,[h] especially in relation to weather. The most likely mining sites lying between Hawaii and the Central American coast should not be frequently subjected to extreme weather conditions, but some severe oceanic tropical storms may be spawned during late summer off Central America which then move westward into the prime mining area. During these disturbances, mining ships will have difficulty in remaining on station and in transferring nodules from the mining ship to transport vessels. The most hazardous problem stemming from heavy seas would be associated with processing operations aboard ship. The yawing, pitching, heaving, and surging could cause chemical spillage and fire hazards.

Biotic and physical systems disruption. Much more important in the realm of deep seabed/environmental relationships are the potential hazards created for marine and onland environmental systems and coastal communities that may be near the mining and processing sites. In recent years, NOAA has initiated a program to assess these environmental and socioeconomic impacts. The United States government expects nodule extraction to become a major mining activity in coming decades, and by having environmental impact information available, seeks to assure producers of a minimal delay in beginning mining operations. Delays will create added expense to industry and it is possible that nodule mining will enhance the United States' strategic mineral position.[133] In late 1978, preparations were under way for much more extensive studies of the environmental relationships of deep seabed mining. The program—Deep Ocean Mining Environmental Study (DOMES II)—is now under way, and it will generate data that should assist federal agencies in establishing deep ocean mining regulations. Problems associated with benthic ecosystems, heavy metal entry into the water column, and tailings disposal onland and at sea will be some of the topics investigated.[571]

Benthic organism destruction: Roughly 77% of the world's ocean basins lie at depths greater than 10,000 ft.; this vast area includes only about 0.8% of the

Table 26. Amount of Sediment Disturbed at the Ocean Floor by Dredging

Type of dredge	Separation of track lines (meters)	Nodules mined per day (metric tons)	Suspended solids in bottom 10 m of water (ppm)
Towed	0	4920	9460
	50	4920	2850
	100	4920	1430
	500	4920	290
Self-propelled[a]	0	4600	18440
	50	4600	6500
	100	4600	3200
	500	4600	660

Source: After [368].
Note: [a]Assuming a track with a width of 17.6 m (3.6 m for propulsion device which interacts with ocean floor) and a calculted width of 14 m to mine 4600 metric tons per day.

total faunal biomass.[561] Abyssal fauna dwell on or very near the bottom and also in the upper few inches of sediments.[i] These organisms will experience heavy destruction during the mining process.[457] But eventual repopulation of mined-out areas should occur, assuming adequate zones of undisturbed seabed are left.[409]

Sediments: A relatively recent study (1978) by Ocean Management, Inc. monitored by NOAA involved a small-scale testing program of hydraulic pump mining of manganese nodules over a period of several months. The project was carried out about 865 nautical mi. southeast of Hawaii in waters 3-mi. deep. Preliminary analyses indicate little persistent effect from the use of a collector head on the sea floor.[180] Still, researchers may question these findings, if applied to long-term environmental relationships. According to a study by Amos and Roels, for each ton of material removed from the seabed, probably 2.5 tons of bottom sediments will be disturbed, if the dredge head is towed, and 5.5 times if it is a self-propelled mechanism. Generally, the closer the track lines or passes of the mining device the more suspended solids will be generated within the bottom 10 m (33 ft.) of water (Table 26). Sediment plumes near the seabed and fallout during lifting will be especially severe with CLB systems. As the sediments settle out, benthic organisms' habitat will be covered with debris that may adversely affect their feeding and reproductive patterns as well as their gas exchange mechanisms. Some researchers are concerned that scientific work will be harmed when the sediment records are disturbed.[217] Another potential consequence of dredging the seabed is that some organisms (fish and mobile and sessile invertebrates) may be transported from one location to another, possibly affecting ecosystem balance.[399]

If hydraulic mining systems are used, a large volume of sediments will be carried to the surface where it will be dumped back into the ocean. Washing or beneficiating the nodules on the mining or processing vessels, may help create a persistent surface water sediment plume that could alter photosynthesis processes. But several research experiments have shown that surface plumes usually dissipate in a relatively short period. Ocean Management's study findings indicate that there is little effect from nodule and fine particulate separation operations on the surface.[180] And Amos and Roels have calculated that a mining area with a long axis of 90 mi. and a width of 18 mi. in the prime mine site Equatorial Pacific area will experience a complete flushing by the North Equatorial Current (speed—4.9 in. s^{-1}) in only 11 days. A test done by the *Deepsea Miner* on the Blake Plateau showed that a dye plume "... diluted to approximately 0.05% of its original concentration 8 minutes after discharge."[11]

More study is needed to answer many crucial questions about particulate pollution created by seabed mining and surface processing of manganese nodules. For example, will particulates tend to accumulate at the pycnocline, i.e., where the lighter and highly mixed surface waters meet denser and poorly mixed waters, usually at about the 180-ft. depth?[253] Can the mineral enrichment of local surface ecosystems be put to work, as in aquaculture?[45] Or, will the nutrients stimulate excessive algal blooms that upon dying create excessive demands for oxygen during decomposition processes?[11]

Onland Impact of Manganese Nodule Processing: A Case Study

Because most nodule processing will occur at onland sites, it is important that industry, local communities, and government begin to examine in detail the potential impacts of these activities that, in some areas, may begin within the next decade. The state of Hawaii has recently completed a scenario study (abbreviated FPI) which provides many useful insights and serves as the source for most of the following discussion.[508]

FPI researchers addressed two basic questions: (1) What is the feasibility of Hawaii, relative to "Mainland" sites, for providing a viable location for a manganese nodule processing plant? And (2), Is this industry socioeconomically and environmentally desirable for Hawaii? Discussion will address only the second topic.

To date, Hawaii has been involved with seabed mining of nodules primarily as a support base for exploration vessels and as an important locus for deep seabed research projects, primarily at the University of Hawaii. According to the FPI study, none of the consortia have decided on a processing plant site. If Hawaii is to persuade entrepreneurs to locate a nodule processing plant in the state, it must provide for the establishment's various needs—one of these needs is energy and another the proximity to the prime mine sites of the Equatorial Pacific. Hawaii's "Big Island" can provide both requisites, as it lies only 500 to 1,500 mi. from the prime mining area and possesses a potentially usable geothermal energy source, only recently tapped, that could provide for the development of the 25–30 megawatt power plant needed in processing operations.

Socioeconomic impacts: Estimates of the economic impacts on the state and the local plant siting area must be considered only tentative, because the size of plant, type of beneficiation process, and the date of start-up cannot be predicted

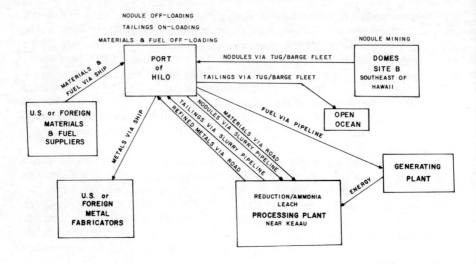

Fig. 85. Scenario adopted for study of a prospective nodule processing plant located near Hilo, Hawaii.
Source: [508]. With permission.

with certainty. Two primary scenario assumptions (Fig. 85) are that the facility will be a 3-million wet (2.25 million dry) tpy plant, using a reduction/ammonia leach process. Other assumptions include a mining operation composed of two 55,000 dwt ships mining nodules at 11°42′N. latitude and about 138°24′W. longitude some 1,085 nautical mi. from the port of Hilo, which will receive the nodules transported by a fleet of tugs and barges (each with a 35,000 dwt capacity). Special facilities at Hilo will unload the barges; then, the nodules will be pumped as a slurry to an assumed plant 9 mi. south of Hilo near the city of Keaau.

A summary of estimated investment costs for the various phases of the mining and processing operation is presented in Table 27. Total capital investment requirement is placed at approximately $521 million and the annual cost of overall operation at $235 million. This large investment will have a significant impact on the economic and social conditions of the "Big Island" and the state as a whole. Gross revenues generated by the processing facility's overall metal output of 72,700 short tons of nickel, copper, cobalt, and zinc and copper sulphate total approximately $262 million annually, an amount slightly larger than the state's total crop sales of $260 million in 1976.

Using an input-output modeling technique, the researchers determined the possible effects of this establishment on employment and household income. Construction of the processing plant and other facilities would likely take 3 years. Construction expenditures would be approximately $91 million per year, enough to increase by 34% the total output of goods and services in Hawaii County and

Table 27. Cost Estimates for Mining, Transporting, and Processing Manganese Nodules in the Hawaiian Island Scenario (millions of 1977 U. S. $)

Category	Capital investment	Annual cost
Mining	123.9[a]	43.31
Ocean transportation	52.4[b]	17.35
Port facilities	6.6[c]	1.64
Onshore handling, transportation	13.0[d]	3.84
Processing	325.0[e]	168.61
Total	520.9	234.75

Source: [508].
Notes: [a]Mining ships and system, exploration, working capital; [b]Tugs, barges, pumps; [c]Approach trestle, unloading pier, mooring system, ponds, miscellaneous facilities; [d]Pumping stations, pipeline system, land; [e]Plant, land, R&D, working capital. A generating plant for electricity would cost an additional $20–24 million.

5% statewide. Personal income would grow "... in the various sectors (of the economy) in the same proportions as total output (gross product)." For Hawaii County, the total growth in personal income would be $42.3 million and for the state $50.5 million at the construction phase.[j] New jobs during the construction phase would total about 5,000 for Hawaii County with an additional 1,000 distributed elsewhere in the state, but a major share of these jobs will be lost when construction ends and the production phase begins.[k] On the other hand, gross product will increase by nearly 84% in Hawaii County, rising from $169.4 to $311.5 million. For the state, the gross product increase would be 66%, increasing from $201.9 to $335.9 million.

The social impact of the nodule processing plant will be significant because the area is primarily rural. As of April 1977, some 9.5% of the labor force in Hawaii County was unemployed. Many of the management and engineering personnel needed for the construction and operation phases of the plant will come from outside Hawaii County; other jobs should go primarily to the local populace, many of whom might change their place of residence to be nearer the work site, a process which would likely stimulate home building construction and need for services. The small town of Keaau would be changed radically, with the most drastic impacts coming during the beginning of the project. In short, "the rural mood and pace of Keaau would be lost."[508]

Physical environment impacts: The Keaau area has abundant open land so that space requirements and land use conflicts should not pose a problem to the construction of the processing facility. Plant site facilities will need 200 acres; in addition, the nodule slurry pipeline connecting Hilo with Keaau will require a 50-ft. right-of-way.

Processing operations will demand about 1.9 million gallons of water per day or nearly 700 million gallons annually. Fresh water should be available from established supply systems, but a well could be drilled without damaging the area's groundwater resource. Local water quality should not prove a problem, because either tailings basin or at-sea dumping (the scenario method) will be used to handle wastes. Handling wastes, of course, does not necessarily dispose of them. If the facility is not designed to process manganese, waste in excess of 2 million tons/year will be generated. The waste will contain significant quantities of toxic elements. The problems of waste disposal still need much detailed study, especially in relation to surface water plumes, if, as projected by the scenario, ocean dumping is practiced. To reduce the sediment plume problem for pellagic biota, the FPI study notes that one possibility might be to pump the tailings several hundred feet below the water surface, a technique that could be as economical as surface dumping.

Air quality should not present a problem, especially given the region's rather steady northeast trade winds. Leakages or spills of processing chemicals, as chlorine, hydrogen, and hydrogen sulphide gases or, perhaps, nitric and sulphuric acids might pose ephemeral air quality problems. Noise levels can be readily controlled and should not, even in the interior of the operating plant, exceed the level common to an automobile freeway.

Overall, based on the findings of the FPI study, environmental problems associated with the processing plant and transport facilities should not present major difficulties. But the waste disposal problem still looms as the crucial issue to any prospective operation on the island. How it is resolved legally and environmentally may be the fulcrum for the economic viability and social acceptance of manganese nodule mining and processing operations.

SUMMARY AND CONCLUSIONS

Entrepreneurs and scientists from many branches of industry have joined forces to mine the deep seabed. Their investments over the past 15 years have still to bear fruit. The public has been misled by careless prophets who have predicted that the start of deep seabed mining was, at most, a matter of only 1 or 2 years away when in fact it is unlikely to occur until the mid-1980s, if then.

Mining companies still have much work to do in testing mining, lifting, and processing equipment and techniques, an expensive and slow process. Numerous variables enter into the exploitability of mine sites, including their relative location, their physical characteristics, and the quality and extent of their nodule resources. Potential revenues from seabed mining should be adequate for producers to compete with their onland counterparts, but excessive returns on investments, though considered unlikely, would depend on the price of metals. An especially important area of concern centers on the effects of mining and processing on the oceans' natural systems, but the present general consensus seems to be that no extremely detrimental effects will result. Many investigations are under way to more comprehensively measure the potential at-sea and onland physical and socioeconomic impacts of the manganese nodule mining industry.

NOTES

[a]Initial survey and prototype-testing work did involve the recovery of nodules, and much of the unique hardware developed is directly applicable to nodule mining.

[b]Escalated values refer to increases in construction costs due to inflation, supply and demand interactions, and changes in technology.

[c]For the low estimate, see [189]; for the high estimate, see [239].

[d]H. Menard and J. Frazer have demonstrated through the use of regression analysis that nodule metallic content and abundance are not independent, but rather are negatively correlated. For example, analyses of samples taken at 350 nodule locations in the Pacific Ocean showed that for a nodule metal content between 0.25% and 0.50%, some 23% of the seabed was covered; between 2.25% and 2.50% grade, only 2% of the seabed contained nodules [339].

[e]Deepsea Ventures filed a notice of discovery and placed its claim on then Secretary of State Henry Kissinger's desk on Nov. 15, 1974. The U.S. government never formally recognized the claim, and it will not likely do so until the Law of the Sea Conference establishes a treaty.

[f]A West German firm has been working on developing a buoyant magnetic collector [45].

[g]Leaching takes advantage of the nodule's porosity.

[h]Insurance may be a major item in the production costs of nodules and must be carefully evaluated in costing studies. Because marine mining firms have only a limited safety experience, insurance companies will base charges on claim rates from established at-sea operations. If claim rates differ from these, the marine mining industry may quickly see radical adjustments. See [295] and [379].

[i]Researchers in 1975 on the ship *Oceanographer* made 25 box cores in the siliceous ooze zone of the Equatorial Pacific midway between Hawaii and southern Mexico. Analysis of the cores, covering a total area of 6.25 m^2, had a total organic weight of 5.39g, equivalent to 860 mg/m^2. See [11].

[j]At the production phase, personal income growth would be $20.6 million for the County and $26.7 for the State.

[k]County employment would decline from 5,029 at the construction phase to 1,707 at the operational, and State from 6,080 to 2,415, respectively.

Chapter 12

Manganese Nodules and the International Marketplace

Much recent debate has centered upon whether deep seabed nodule mining will have a significant impact on world metal markets and on the economies of individual states. These arguments are highly speculative, because until 1985, or later, no deep seabed mining establishments are likely to be active which might provide a guide to (1) the actual output that can be expected and (2) the influence this output may have on world metal supplies and prices. Many within the United States government and industry would like to have available an abundant and an assured supply of several minerals that seabed nodules can provide. But a few developed and many LDCs dependent upon domestic production and export income from the same metals that will be extracted from nodules are concerned that their mineral industries will be adversely affected and their foreign exchange earnings jeopardized. While national governments worry about these possibilities, prospective nodule miners are speculating about their ability to compete with onland producers and about international market conditions.

THE BIG FOUR: RESERVES, DEMAND, AND MARKET RELATIONSHIPS

Nickel, copper, cobalt, and manganese—the "Big Four"—will be the primary interest of the nodule miners. Relationships of these metals to the international marketplace will be a function of reserves and demand which are, in turn, dependent upon technology and general economic conditions. The market problems

228 *The international market place*

of each of the "Big Four" will vary, as will the states affected by their production. Although onland mining is not within the scope of this book, a brief review of world resources, reserves (which, at a given time, can be economically exploited), and mine output of the "Big Four" metals is included to enhance an understanding of the international marketplace and its future influences on deep seabed mining.

Nickel

Although manganese constitutes the largest share of the metallic content of nodules, nickel is the most important economic incentive to production. Identified world onland resources with an average nickel content of 1% total approximately 130 million tons of which 80% are in laterites and the balance in sulphide deposits. The United States has more than 6.3 billion tons of sulphide deposits that average 0.2% in nickel content. Onland world nickel reserves totaled over 54 million tons of metal in 1978 (Table 28) which at forecasted consumption rates for 1985 would be barely sufficient to last 47 years.[128] This estimate is probably low because it is based on fragmentary information,[337] and because mining companies have a tendency for proving reserves only to meet short-term needs rather than attempting to identify their full reserve ownership. The world's mine output decreased by 24% from 1977 to 1978, and the United States extracted 19% less in 1978 than in 1977 even though total domestic nickel consumption increased by 18% during the year (of which about 36,000 tons, or 17%, came from recycling scrap).

Some mineral industry specialists believe that during the next 2 decades world demand for nickel will increase at least 6% annually,[486] continuing the 6.5% trend of the last 25 years. A 6% rate seems too optimistic. A recent United States government study has made a more conservative (and probably more realistic)

Table 28. Onland Nickel Mine Output and Reserves in 1978
(thousand metric tons of metal)

Country	Output	Reserves	
		Ore grade (%)	Quantity
New Caledonia	63.5	1.0-3.0	13,605
Canada	164.6	1.5-3.0	7,800
Cuba[a]	34.7	1.4	3,084
United States	10.5	0.8-1.3	181
Other non-Communist	179.2	0.2-4.0	25,336
Other Communist	136.1	0.4-4.0	4,353
World total[b]	588.4		54,420

Source: Calculated from [547**].
Notes: [a]Estimate. [b]Discrepancy due to rounding.

world demand estimate of 4% annually, up through the year 2000. With 1976 as the base year, a 4% rate "... would essentially triple the amount of nickel ... consumed annually and would result in a cumulative consumption of 31.8 million metric tons ..." by the year 2000.[128] The probable annual world demand at the turn of the century for primary (mined) and secondary (recycled) nickel will approximate 2 million tons.[337]

The U.S. Bureau of Mines has made projections for domestic demand of both primary and secondary nickel, and it expects the annual growth rate between the years 1977 and 2000 to be 3.6%, which is comparable to the recent past. United States consumption of primary and secondary nickel in 1977 was approximately 209,600 tons. In 1978, it rose to 232,600 tons,[547**] and in the year 2000, it is projected to be 508,000 tons. In 1978, United States produced only about 6.3% of the primary nickel it consumed, and in the year 2000 the figure is expected to be only 8.5%. Although secondary sources are expected to supply an increasing proportion of domestic consumption (perhaps 141,000 tons in the year 2000), the balance of 367,000 tons remains to be obtained from primary ores, 91.5% of which must be supplied by imports. In 1979, the economically exploitable nickel reserves of the United States amounted to only 181,000 tons. If, however, new mining and processing technologies can be developed, it is estimated that as much as 245,000 tons (70% of the United States' annual primary ore demand) could be produced from low-grade sulphide deposits such as the Duluth gabbro of northeastern Minnesota, currently considered only a potenital reserve.[123]

As noted earlier, the economic feasibility of manganese nodule mining depends primarily on the competitiveness of the nickel recovery phase of production. But prospective nodule producers should be cautious about solely focusing their production on nickel for at least three reasons: (1) In the future, more sophisticated technology may increase the use of secondary recovery techniques in recycling plants, a situation that might intensify competition in the nickel market; (2) if prices should rise too high, other metals such as molybdenum, vanadium, chromium, columbium, or titanium can become substitute alloys in some steel products, while enamel and paint can replace nickel in many plating processes; and (3) nodule entrepreneurs will be competing with onland sulphide and laterite producers whose operations also require a large capital investment and a long time to deliver. Investments in time and capital by large onland firms render them relatively unresponsive to changes in market conditions, making it potentially difficult for the nodule producer to compete.[128] This problem is especially true in the case of the nickel industry because it is oligopolistic. International Nickel Company of Canada, Ltd. (Inco), and Societe LeNickel of New Caledonia (controlled by France) presently account for about two-thirds of the non-Communist economies' nickel output and 50% of the world's output. One important factor in the nickel marketplace is that Inco is both a major producer of nickel and a member of the "seabed mining club," which might allow it to adjust its onland production to accommodate the output of deep seabed mining establishments, without nickel prices dropping appreciably.[313] It is unlikely, however, that Inco could politically afford cutbacks in Canadian mines, especially considering the current situation in the country's nickel industry. Given current market conditions, some mineral specialists within the Canadian government feel their country's nickel industry could be hurt by seabed mining.[420] Their concern is reflected in the government's recent formation of a group of experts to monitor deep seabed mining developments as they relate to Canadian interests and also to advise national governmental agencies and the country's delegation to the United Nations Law of the Sea Conference.[343] Analysts in the United States DOI's Ocean Mining Administration, however, feel that established nickel operations, as Inco's sulphide mines at Sudbury, Ontario, should be able to produce more cheaply than the deep seabed companies, but nickel recovered from nodules should be competitive with newly opened onland lateritic operations.[476] But land-based producers might counter this point by noting that a domestic seabed nickel mining industry, as in the United States, might receive preferential treatment through the establishment of import quotas and other non-tariff trade barriers.

Blissenbach feels that nickel production in such states as Canada, New Caledonia, and the Soviet Union should be unaffected by seabed mining of nickel, because growth in demand will offset increases in nickel production and may actually be so strong that additional land sites will be required.[45] A good example of entrepreneurs forging ahead in onland nickel production is the current development of rich lateritic deposits in the Department of Cordoba, Colombia. These deposits should begin to enter the world market about 1982. Geologists think that they may contain 25 million tons of ore with a 1.5% to 3.2% nickel content; another 40 million tons contain 1.0% to 1.5% metal, and there are additional reserves of marginal deposits.[86] In the future, as more nickel becomes available, prices may decline until nickel can capture other metal markets such as aluminum,[345] although some observers, such as Tinsley, are skeptical that this will occur.[532]

The price of nickel, which did not increase appreciably from 1974 to 1977, declined during 1978 because of excessive producer stocks which dominated the market. However, world prices increased from an average $2.06/lb. in 1978 to $3.20 in December 1979—an additional burden on the U.S. balance of payments. During 1974-77, 58% of United States nickel imports were supplied by Canada followed by New Caledonia and the Dominican Republic with 7% each; the import reliance of the United States increased from 70% in 1977 to 77% in 1978. Refining from domestic ore and foreign matter yielded over 33,000 tons of metal in 1978, roughly seven times less than the 230,000 tons estimated by the CIA as USSR's total for that year. The nickel resources contained in seabed nodules (estimated in 1975 at 690 million tons)[337] indeed would seem to hold a potentially important place for the United States and other countries dependent upon sulphide and laterite ore imports. But whether nodule nickel can compete with onland laterite ores remains a debatable issue.

A highly significant development that has and will affect the relative competitive position of lateritic and nodule nickel production is the increasing price of energy. Nickel production from laterites is highly energy intensive. An analysis done in the mid-1970s of the comparative operating costs of laterite and nodule nickel producers showed that with costs of oil at $10-$11/bbl. the nodule miner could compete very well. A crude oil price of $24-$30/bbl. (as of late December 1979) should have put potential hydrometallurgical (but not pyrometallurgical)

Table 29. Comparative Costs of Laterite and Nodule Nickel Production
(1976 U.S. $ per lb. of metal)

Cost	Laterites	Nodules[a]
Operating	0.96–2.87[b]	1.01–1.39
Capital investment	6.07–8.00[c]	3.89–5.92

Source: [589].
Notes: [a]Assumed to be a 3 million tons/year establishment. [b]Universal Oil Products Co. ($0.96), Falconbridge Dominicana ($1.39), LeNickel ($1.67), Japanese Producer ($2.87). [c]Indonesian Inco ($6.07–7.66), Guatemala Inco ($7.67–8.00).

nodule producers in a favorable position, relative to laterite nickel producers. On the other hand, energy needed in transporting nodules to processing sites and markets will be significant enough to offset some of this advantage. But overall capital investment costs per pound of nickel produced will probably be higher for the laterite producer (Table 29).

Given the various conditions outlined here, how much of the future nickel market expansion might seabed mining capture? One hypothetical projection, assuming production begins in 1985, indicates that the annual seabed output of nickel could increase from 36,000 tons in 1985 to 612,000 tons in the year 2000. During 1985–2000, seabed mining is projected to supply 5,004,000 tons or 20% of the cumulative demand of 24,540,000 tons (Table 30). If this projection can be considered reasonably accurate, it would mean that present onland nickel producers will not lack for markets.

Copper

Because of technological advances during the past 30 years, world copper reserves have increased faster than consumption. Onland reserves in 1978 were put at about 498 million tons (an amount that, at current consumption rates, could last perhaps 68 years); another 726 million tons are estimated to exist in seabed nodules. Both the United States and Chile led the world in 1978 onland reserves with an estimated 97 million tons each, followed by the USSR and Zambia. Estimated world onland production in 1978 totaled approximately 7.6 million tons (Table 31). Several major new onland projects were under construction, including the La Caridad in Mexico and the Carr Fork in Utah,[a] all planned before a severe fall in world copper prices occurred in 1974.[547]**

World copper consumption in the late 1970s was slowly rising; significantly, this trend has been coupled with a decline in output in many countries. Cumulative demand through the year 2000 should total approximately 346 million tons with copper markets expected to grow by some 4% annually over the next two

232 The international market place

Table 30. Projected Primary Nickel Demand and Hypothetical Seabed Nickel Production Beginning in 1985

Year	Annual demand	Annual increase in demand	Annual seabed production	Percent[a]
	(thousand metric tons)			
1985	1,124	45	36	80
1986	1,169	47	72	77
1987	1,216	49	108	73
1988	1,265	51	144	71
1989	1,315	53	180	68
1990	1,368	55	216	65
1991	1,423	57	252	63
1992	1,480	59	288	61
1993	1,539	62	324	58
1994	1,600	64	360	56
1995	1,664	67	396	54
1996	1,731	69	432	52
1997	1,800	72	468	50
1998	1,872	75	540	96
1999	1,947	78	576	46
2000	2,025	81	612	44
Cumulative	24,540		5,004	

Source: [128].
Note: [a] Annual increase in demand supplied from annual increase in seabed production.

decades. This annual growth rate will be slightly lower than that experienced during the previous 20 years, which averaged about 4.5%.[128] The United States has about 20% of the world's total copper reserves, and it is the world's largest producer, accounting in 1978 for an estimated 17.7% (1,347,000 tons) of that year's world output (7,630,000 tons). Despite this large production, the United States imported about 25% (620,000 tons) of its 1978 copper consumption (2,480,000 tons), and it will likely continue to do so, although this situation is due in large part to price advantages and trade needs. During the 1974-77 period, imports came primarily from four important suppliers—Canada (28%), Chile (21%), Peru (13%), and Zambia (10%); the remaining 28% was distributed among several minor exporters.[547**]

Although demand in the late 1970s has been slowly rising while production declined slightly, large stocks have created a glut on the world copper market resulting in depressed prices. Because of this situation, which may persist for several years, producers extracting copper from manganese nodules probably cannot depend on the copper market to sustain their operations,[128] although their relatively low operational costs could allow effective competition with onland producers. The magnitude of world copper production and consumption precludes

Table 31. Onland Copper Mine Output and Reserves in 1978 (million metric tons)

Country	Output	Reserves
United States	1.347	97.0
Chile	1.061	97.0
USSR[a]	0.853	36.3
Zambia	0.635	33.6
Canada	0.789	31.7
Peru	0.336	31.7
Zaire	0.435	23.6
Philippines	0.263	18.1
Papua New Guinea	0.190	14.5
Poland	0.254	12.7
Australia	0.190	8.2
Republic of South Africa	0.190	5.4
Other non-Communist	0.687	77.1
Other Communist[a]	0.399	10.9
World total[b]	7.630	497.9

Source: Calculated from [547**].
Notes: [a]Estimate. [b]Discrepancy due to rounding.

any important effect on world market conditions by seabed copper mining. One estimate indicates that even by 1990 seabed mining will not account for more than 2% of the world's annual production.[53, 399] For example, a 10-million ton/year nodule producing facility could supply only about 1.3% of the world's annual copper demand.[135] Several copper-producing units of this size still will not upset world market conditions to any degree, and they will pose little threat to onland producers. A few states, such as Peru, Chile, Zaire, and Zambia, do not agree with this conclusion (perhaps rightfully so) because of their significant dependence on foreign exchange earned by copper exports. But a 1974 United Nations Conference on Trade and Development (UNCTAD) made a strong point when it noted that seabed mining's annual output would not equal many of the annual fluctuations of the world copper industry.[543]

Cobalt

Mineral specialists in 1978 placed identified world cobalt resources at 4,536,000 tons with economic reserves pegged at 1,451,200 tons. If the metal content of seabed nodules is added, very large potential resources are available. Zaire leads in cobalt reserves and significant amounts are also present in New Caledonia, the Philippines, and Zambia. Lacking reserves, the United States is dependent on imports; in 1978 its consumption of 9,433 tons came primarily from Zaire (42%), Belgium-Luxembourg (23%), Zambia (7%), and Finland (6%). The world's top

Table 32. Onland Cobalt Mine Output and Reserves in 1978
(metric tons of metal)

Country	Output	Reserves	
		Ore grade (%)	Quantity
Zaire	10,884	0.3-2.0	453,500
New Caledonia	4,172	0.1	272,100
Philippines	1,088	0.03-0.12	190,470
Zambia	2,267	0.1-0.2	113,375
Australia	3,447	0.08-0.12	48,976
Canada	1,814	0.03-0.1	29,931
Botswana	181	0.06	26,303
Finland	1,361	0.2-0.7	18,140
Morocco	1,814	1.2	12,698
USSR and other Communist	3,628	0.01-0.3	317,450
World total[a]	30,838		1,451,200

Source: Calculated from [547**].
Note: [a]Discrepancy due to rounding.

1978 producers were Zaire (10,884 tons), New Caledonia, Australia and Zambia (Table 32).

Produced mainly as a by-product of copper, nickel, iron, zinc, chromium, and lead mining, cobalt was once used primarily as a coloring additive, but functions today as an alloy for increasing the corrosion, heat, and abrasion resistance of metals. It also improves metal strength and is especially useful in electrical magnets, machine tools, missiles, jet engines, ceramics, paint, and glass.[236]

Despite cobalt's wide application in industry, total world demand is relatively limited. World consumption in 1973 was only 28,163 tons, and in the United States—10,023 tons. The U.S. Bureau of Mines preliminary data for 1977 place the estimated total annual domestic demand for cobalt at 8,299 tons. By the year 2000, annual world demand is anticipated to reach 80,000 tons with United States consuming 20,272 tons from both primary and secondary sources. The annual demand growth rate anticipated up to the year 2000 is approximately 3.3% for the world and 3.1% for the United States.

Because cobalt is produced as a by-product, its availability depends in large part upon the status of production in other mining industries, especially copper and nickel which are the ore sources for most cobalt. If production in these industries declines, cobalt could become in short supply, prompting price increases or substitutions (e.g., nickel, iron, and platinum in magnets; copper, chromium, and manganese in paints; and nickel in ceramics and jet engines). Recent events provide a good illustration of cobalt's supply, demand, price, and by-product relationships. From 1970 through mid-1977, cobalt prices inflated from $2.20/lb. to $6.00/lb., an increase of 272%. At this price, according to a DOC study,

Table 33. Projected Primary Cobalt Demand and Hypothetical Seabed Cobalt Production Beginning in 1985

Year	Annual demand	Annual increase in demand	Annual seabed production	Percent[a]
	(thousand metric tons)			
1985	48	1.7	7	412
1986	50	1.7	14	412
1987	51	1.8	21	389
1988	53	1.9	28	368
1989	55	1.9	35	368
1990	57	2.0	42	350
1991	59	2.1	49	333
1992	61	2.1	56	333
1993	63	2.2	63	318
1994	65	2.3	70	304
1995	68	2.4	77	292
1996	70	2.4	84	292
1997	72	2.5	91	280
1998	75	2.6	105	538
1999	77	2.7	112	259
2000	80	2.8	119	250

Source: [128].
Note: [a] Annual increase in demand supplied from annual increase in seabed production.

nodule-contained cobalt could be produced at a profit.[135] From mid-1977 to April 1978, prices advanced to $6.85/lb. whereupon a major crisis struck the industry when Zaire failed to supply the world market. Internal strife in Zaire, primarily centered in Shaba Province which was invaded by insurgents from Angola, shattered the country's copper and cobalt industry, and the price of cobalt skyrocketted, reaching $25.00/lb. by March 1979[565] where it remained throughout the year. This abrupt market change encouraged Canada's Falconbridge Nickel and Zambia's copper producers to announce plans to increase their by-product cobalt output. However, because both nickel and copper were in oversupply at the end of 1979, such plans could exert downward pressure on the price of these metals even though the copper market strengthened considerably during the year.[b]

If manganese nodule mining begins on a large scale in the decades ahead and if onland mining enterprises continue to produce, it is likely that cobalt prices will decline (with due allowances for inflation), at least to the level of nickel.[378] Table 33 illustrates the impact seabed mining could have on total annual supply relationships of cobalt from 1985 to 2000. The annual increase in cobalt produced by seabed mining could supply 412% of the increase in demand for 1985 and

250% in the year 2000, an impressive oversupply of the market. Considering cobalt's limited market, its pronounced inelasticity of demand, and substitute availability, prospective nodule producers cannot rely on cobalt as the primary metal produced. The most significant problem associated with this potential oversupply is that several states, especially Zaire, depend on cobalt to generate a portion of their foreign exchange. In 1977, Zaire produced an estimated 30 million lbs. (13,605 tons) of cobalt, which at mid-1977 prices of $6.00/lb. was valued at $180 million.[547**] If cobalt prices were to fall to the level of nickel ($2.20 in mid-1977), as Nigrelli suggests they may, Zaire could lose $114 million in export income—a 63% decline from its 1978 cobalt export revenue. Nigrelli, of course, was writing from a pre-1978 invasion perspective, before prices rose precipitously. If manganese nodule mining does adversely affect the cobalt market, it may be that Zaire will have to diversify its export program,[311] not an easy task to contemplate for that country.

Manganese

Manganese is vital to the steel industry which consumes about 90% of the world's total annual production. This dependence and the uncertainty of supplies coming from some countries that are at times politically unstable account, in part, for the interest of some steel producers in seabed nodule mining.[214] Onland world manganese ore reserves are estimated by the U.S. Bureau of Mines to total about 5,442 million tons (Table 34) or approximately 1,814 million tons of metal. Again, as with nickel, this important metal's real reserves may be somewhat underestimated, and certainly expanded somewhat by lowering the ore grade considered commercially exploitable (normally 35% metal content). Current world onland reserves are adequate to meet demand well beyond the year 2000[254*] (official

Table 34. Onland Manganese Mine Output and Reserves in 1978
(thousand metric tons)

Country	Output	Reserves
Republic of South Africa	5,261	1,995,400
Australia	1,542	299,300
Gabon	1,905	149,700
Brazil	1,088	86,200
India	1,814	58,900
Other non-Communist	1,361	53,500
USSR and other Communist	9,705	2,721,000
World total[a]	22,675	5,442,000

Source: Calculated from [547**].
Note: [a]Discrepancy due to rounding.

estimates place reserves at 25 years). At present, the United States has no commercially exploitable reserves and no domestic production since 1973, although total potential resources could provide 66.8 million tons of manganese metal.[254] The United States' manganese ore imports during 1974-77 came from Gabon (36%), Brazil (33%), Australia (13%), and the Republic of South Africa (9%), with another 9% distributed among several minor exporters. Ferromanganese (a ferroalloy containing manganese as the special additive) was imported primarily from France (32%), the Republic of South Africa (30%), and Japan (14%).[547]**

Probable cumulative world demand from 1977 to the year 2000 will total 330 million tons. The world's annual demand will have increased from approximately 8.7 million tons in 1977 to a probable 12.6 million tons in 1985 and 19.4 million tons by the year 2000, a 2.9% average growth rate. Annual demand in the United States will likely expand from an anticipated 1.5 million tons in 1985 to 1.9 million tons by the year 2000. Between 1977 and the year 2000, the probable average annual demand growth rate will be 1.6%.[254]*

If production begins in 1985, it is possible for seabed nodule producers to supply 3% of the cumulative demand for manganese between 1985-2000, although it is difficult to judge the future because only one consortium (Deepsea Ventures)^c has actually indicated officially that it will commercially extract manganese from the nodules.[128] Thus, it is also hazardous to speculate about the impact of nodule-produced manganese on various world suppliers. Of the states that mine manganese, the Soviet Union, the Republic of South Africa, Gabon, and Brazil are the major producers. Together, the Soviet Union and the Republic of South Africa control 80% of the world's reserves. If Australia, Gabon, and Brazil are added in, over 97% of the world's total reserves can be accounted for.[254]

Gabon and Brazil have an important manganese export market that may be somewhat adversely affected by the nodule mining industry. Gabon will be the most sensitive to changes in manganese supplies because 15-20% of its export income derives from manganese. The problem, however, goes beyond the immediate market relationships of manganese. To illustrate, if Gabon's government feels manganese markets are threatened it may stop work on a rail link connecting inland mines with the coast; the result will be to retard the overall economic development of its interior regions. Brazil should be much less vulnerable to exchange losses, because only 2% of its total export income originates from manganese exports[399, 486] and because it has a broad commodities export base. A compensating factor for lost manganese export earnings for both states is that they have begun to produce significant quantities of petroleum from their continental shelf areas, which should offset some of the financial losses. This argument, however, may not be very satisfying to governments seeking to move forward in their overall economic development.

Given that the world has only an officially estimated 25-year reserve of manganese, why are not more first-generation seabed mining firms presently considering its extraction from nodules? The basic problem is a lack of major anticipated growth in demand and, as noted earlier, there is in fact a very large potential onland reserve available. Thus all but one of the seabed mining consortia presently see manganese only as a by-product. But if nodule processors desire, they can produce for silicomanganese or ferromanganese markets. Deepsea Ventures plans to enter the ferromanganese market which annually amounts to 4.5 to 6.4 million tons. Yet, until the steel market comes out of its current slump, demand for ferromanganese will remain stable, and any major addition to the ores available could cause a price decline.[135] Nodule miners could also produce pure manganese metal, but the world's small demand would soon be filled and the market saturated.[128] A single one-million-tpy plant could produce several times beyond current world demand. But as high-grade onland manganese reserves deplete and industrialization of the LDCs proceeds, second-generation seabed mining enterprises may find manganese extraction from nodules a more attractive prospect.

WHO WILL BENEFIT?

Market relationships of the "Big Four" indicate that the impact of deep seabed mining will be mixed. It has been suggested that the availability of large supplies of metals in the oceans could have a stabilizing effect on metal prices.[381] On the other hand, as pointed out by some mineral economists, a large output by seabed mining enterprises could create unstable market conditions, and a few states highly dependent on one export mineral commodity will probably be affected, at least initially. Yet many researchers friendly to the LDCs take a long-term view that most of these states will not suffer as a result of manganese nodule mining.[204] Seabed mining may cause short-term decreases in some metal prices, which may affect a state's foreign exchange earnings for one or two metals, but in the long run a lowering of other metal prices may more than offset such losses. A good example is India which may lose out to a degree in its manganese export income, but which will more than offset this loss due to cheaper nickel and copper prices,[482] both very important to it as a rapidly industrializing state. In fact, India's industrial growth may eventually demand most of its manganese production. Some economists suggest that the International Monetary Fund (IMF) could be used to help states weather the short-term crises resulting from lost export earnings. If Chile, Zambia, or Zaire were to have difficulty with copper export earnings or if Ghana, Morocco, or India were to need help as a result of losses in the manganese market, the IMF could be of significant help. The IMF could also assist the LDCs in diversifying their export base—an effort that should precede the start of seabed mining.[311] But should the IMF consider a given state's export earnings to be in jeopardy from seabed mining, it might be reluctant to extend credit to it.

The industrial states of Europe, North America, and Japan can benefit if metal prices are lowered and if precarious availability situations can be avoided.[9,494] Of the "Big Four," the United States is almost completely dependent on imports of manganese and cobalt, and nearly so for nickel. These conditions are likely to continue into the turn of the century. Copper imports are expected to more than double, reaching 35.7% of the United States' consumption by the year 2000,

Table 35. Projected United States Demand, Production, and Imports of
Metals in Seabed Nodules in the Year 2000

Metal	Primary demand	Production based on 20-year trend	Imports	Imports as percent of demand
	(thousand metric tons)			
Nickel	349	29.6	319	91.4
Copper	3,810	2,449.0	1,361	35.7
Cobalt	39	0.0	39	100.0
Manganese	1,932	0.0	1,932	100.0

Source: Calculated from [128].

although, once again, this situation may reflect price advantages of imports rather than any real domestic shortages (Table 35).

Mero feels that lowered prices will benefit both developed states and LDCs (a point many LDCs would debate) and that competition created for onland producers will be useful to the mining industry; he has admonished onland producers that firms steeped in traditionalism and given to inflexibility will be in trouble. On the other hand, ocean mining should enhance a renewed surge of efficiency and initiative among mineral producers. Although Mero does not necessarily predict cartelization, he has warned that "when ocean mining comes into full scale production it will upset the ... structure of much of the present world mining industry."[345] This statement probably strikes many in the mining industries as rather hyperbolic.

CARTELS

Some mineral industry analysts (perhaps overlooking likely antitrust actions) have stated that if future seabed mining costs were lowered to the point that land-based producers were put out of business, the world's few large seabed mining corporations might form cartels,[378, 431] which could play havoc with national economies. Others have turned this situation around and shown that seabed mineral producers may assure that onland producers cannot form OPEC-like cartels.[382] Many students of cartels, however, do not see them as a strong force in future metal markets. Past cartelization efforts in the lead, mercury, borax, asbestos, marble, molybdenum, nickel, phosphate, and silver industries have resulted only in internal animosities and dissolution.

Cartels can function efficiently if they (1) control a major portion of the commodity in relation to world trade and have consumers without stockpiles who are unable to find short-term alternative sources; (2) have members who can withstand revenue losses; (3) have a market with an inelastic demand; and (4) have a limited but inclusive membership with common goals. As Raymond has put it, "these are stringent conditions, which very few international materials markets meet."[443]

Which of these criteria could seabed mineral producers possibly meet? Amacher and Sweeney have pointed out that, because seabed mining enterprises will initially control only a small portion of the world market, short-run prospects for ocean-based cartels seem unlikely. But as they capture a larger share of the market, cartelization could become more probable.[9] As demonstrated, three of the "Big Four"—copper, nickel, and manganese—have considerable demand elasticity, with cobalt being somewhat inelastic in the short-term. But there are cobalt substitutes which could increase its demand elasticity. The "charter members" of the deep seabed mining club initially may have common goals of maintaining a stable metal supply and market, but as new producers enter the field, this consensus may melt away; and a seabed mining cartel would also be faced with a lengthy period during which national governments, as in the case of the United States, may tap stockpiles. Seabed mineral producers may also be faced with onland competitors and established cartels such as CIPEC (Counseil Intergouvernemental de Pays Exportateurs de Cuivre), although it has been a less than effective cartel and would probably pose little threat[d] (unless it drops its price, which could disrupt the nodule producers' cash flow while still in their pay-back phase).

If, rather than competing, ocean and onland producers collide (an unlikely prospect in most cases) cartelization could be of a significant concern. For example, firms like Kennecott Copper and Inco are leaders in seabed mining technology and land-based production.[9] Should a cartel succeed in pushing up the price of nickel, while onland production costs remain stable, companies in the United States with considerable potential nickel resources would likely enter the market and its self-sufficiency would increase. Also, if nickel prices were to rise significantly, new onland laterite producers would probably enter the market and established producers would expand. If, over a period of years, nickel were withheld from the market (a doubtful occurrence) to jack up prices, several of the economically disadvantaged nickel-producing areas, such as New Caledonia, could not withstand the revenue losses. To be effective, any nickel cartel would need to include Canada, and because the United States is Canada's primary market, it would not be in Canada's interest to economically damage its principal buyer.[128]

Although the possibility of developed state industries and governments forming seabed mining cartels may worry some government officials and people in industry, a more common concern is that an international agency dominated by the LDCs acting through a mandate of the Law of the Sea Conference will develop a cartel that may preclude free access to the oceans' mineral deposits.

SUMMARY AND CONCLUSIONS

Production keyed to one metal in nodules will affect the supply and price of the others, because to be economically viable the nodule producer will have to extract either three or four of the "Big Four"; Mero has driven this reality home, using copper as an example. He shows that, *hypothetically*, if nodules should form the main source of world copper supplies (produced at $0.20/lb. in 1975 prices), there would be major changes in the available quantities of other metals such as cobalt, nickel, and molybdenum. Assuming an exaggerated extraction rate of 400 million tons/year, nodules would supply 100% of the world's demand for copper. This output level would also result in the production of 2,230% of the free world's manganese consumption; 1,280% of its demand for nickel; 400% of its molybdenum requirements; 4,800% of its cobalt needs; and 12,000% of the demand for zirconium.[345] Needless to say, the copper contained in nodules is not now economically exploitable and such a large production of nodules is an unrealistic assumption that will not materialize in the foreseeable future. Although nodules eventually may be our primary source of copper, this will not take place until well into the next century. There are simply too many established onland capital investments for a rapid transition to seabed mining, especially if many of the land-based firms are subsidized by national governments. On the other hand, as pointed out by Pasho, national governments can easily "...subsidize their industries' efforts..." to mine the seabed, as is already the case in France and West Germany where between $5 and $15 million had been invested in each state by 1977.[420] But it may be that subsidies could be less important than unilateral political actions (quotas, for example) that could lead to a two-tiered price system—one for domestic firms and one for foreign producers.[e]

The involvement of national governments in the issues and economic problems of deep seabed mining is intricately related to jeopardized export markets of the LDCs and their desire to share in the bounty of the oceans as mining industry participants or as financial recipients of income generated by the technologically advanced states. It is with these issues and their successful resolution among the world's states that the future of the ocean lies.

NOTES

[a]Also Sax Cheshmah in Iran which might have been suspended or greatly curtailed in 1979.

[b]During mid-1979, world copper prices surged upward, in part due to labor strife in Peru. If this price trend continues, the world copper market in 1980 will be greatly improved for producers.

[c]Ocean Mgt., Inc., has also been reported to plan manganese extraction in its operations [135]:3-10.

[d]For a useful discussion of CIPEC, see [418].

[e]For a general discussion of these problems, as viewed by Canadians, see [116] and [151].

Chapter 13

Seabed Mining and the Conference on the Law of the Sea

Two philosophies of ocean space and resources are embodied in the current international debate within the Law of the Sea Conference—private territorial domain and free access. These antipodal views can be traced to the 17th century scholars Hugo Grotius, a Dutch jurist, and John Selden, a Britisher. Grotius, in his *De Juri Belli Ac Pacis*, spoke out against Portugal's claim to the oceans as its private property. He contended that the seas were a *mare liberum*, open to all users whether for commerce, military needs, or resource exploitation. Selden presented the case for dominion, i.e., if national self-interest and security needs demand it, a state can unilaterally restrict the right of passage and use of the seas. Since the 17th century, most major maritime powers have adhered to Grotius' position. Recently, however, this long-held concept has come under attack, with the views of John Selden emerging once again. International acquiescence to the extension of territorial seas and the establishment of offshore "economic zones" exemplify this trend. At the international level, the LDCs seek to constrain users of ocean space and resources, or the "common heritage of mankind." Some would argue that this quest is in the spirit of Grotius' view of open access, an option unavailable to most developing countries (LDCs) because of their lack of technology. Thus, those hoping to exploit the deep seabed for its potential mineral riches have been frustrated by those without the capability to exploit but who desire a share of the oceans' bounty. How this sharing should take place and how exploitation should proceed are the fundamental issues at stake. The controversy surrounding these questions is the primary focus of this chapter.

THE TRUMAN PROCLAMATION

On September 28, 1945, President Harry S. Truman issued a proclamation that has had important consequences for the questions of seabed ownership and the control of oceanic resources. The Truman Proclamation held that all resources of the contiguous continental shelf were under United States jurisdiction.[430] Truman's action helped set off events that are still in progress today. Soon after the proclamation was invoked, several states extended their territorial seas—Chile, Peru, and Ecuador—and claimed complete sovereignty over the ocean surface, fishing rights, and seabed for a distance of 200 mi. offshore.

THE NASCENT LAW OF THE SEA CONFERENCE

Fears by maritime states that numerous international straits might be closed encouraged the convening in Geneva of a Law of the Sea Conference (1958) which produced four conventions that helped codify existing maritime law. One of these four was the Convention on the Continental Shelf (which entered into force in 1964) that sanctioned coastal state jurisdiction over resources of adjacent submerged lands and defined the shelf as "... the seabed and subsoil of the submarine areas adjacent to the coast but outside the area of the territorial sea, to a depth of 200 meters or beyond that limit, to where the depth of the superjacent waters admits of the exploitation of the natural resources of the said areas."[100] This definition of sovereignty over seabed resources, in effect, allowed states to maintain a jurisdiction limited only by their technological capabilities, a highly significant allowance.

The decade of the 1960s saw rapid expansion of the petroleum industry into deep continental shelf waters, and several groups began to examine the possibilities of deep seabed mining for manganese nodules. Many observers of the marine environment began to ask whether unlimited access by a limited few (those with the technology) was politically wise or economically equitable.[160] Thus, in 1967, Arvid Pardo, Malta's United Nations ambassador, called for the establishment of some international mechanism for controlling seabed use and, most importantly, for reserving all nonterritorial seabed resources for the "common heritage" of all mankind.[24] Pardo assured that his proposal had substance by submitting to the United Nation's Secretary General an agenda item titled "declaration and treaty concerning the reservation exclusively for peaceful purposes of the seabed and of the ocean floor, underlying the seas beyond the limits of present national jurisdiction, and the use of their resources in the interests of mankind." The result was the establishment of the United Nations Ad Hoc Committee to Study the Peaceful Uses of the Sea-Bed and the Ocean Floor Beyond the Limits of National Jurisdiction, whose mandate was to examine the status of international seabed agreements then in force and to determine the best method for establishing a cooperative effort to explore, use, and conserve the seabed and its subsoil.[470] The Ad Hoc Committee met in 1968 and decided to request the establishment of a permanent committee, a request implemented by the United Nations late that same year.

The Committee membership was expanded and, as Glassner pointed out, began to function as a "... preparatory committee for a new Law of the Sea Conference."[205]

But in December 1969, before the Seabed Committee could finish its work, and despite a negative vote by the United States and 27 other United Nations members, the General Assembly passed a resolution for a moratorium on deep seabed mineral exploitation which could be initiated only under the administrative umbrella of an international regime. The resolution read:

> The General Assembly declares that, pending the establishment of the aforementioned international regime:
> (a) States and persons, physical or juridicial, are bound to refrain from all activities of exploitation of the resources of the area of the sea-bed and ocean floor, and the subsoil thereof, beyond the limits of national jurisdiction;
> (b) No claim to any part of that area or its resources shall be recognized.[195]

The United States' rationale for voting "no" on this resolution was that it felt the development of ocean technology should not be retarded.[383] The legal standing of this resolution has been frequently debated. Some LDCs claim it is legally binding upon states, but many United Nations members recognize it only as a moral commitment by the community of nations,[128] and like the United States, they have stated publicly that neither they nor their nationals are bound by it.

The efforts of the Seabed Committee finally reached the floor of the United Nations in 1970. Its report, titled the *Declaration of Principles Governing the Sea-Bed and the Ocean Floor and the Subsoil Thereof, Beyond the Limits of National Jurisdiction*, set down 15 "principles." Three of these are especially relevant to the problems of deep seabed mining: (1) All of the ocean seaward of coastal territorial jurisdiction was designated as mankind's common heritage and termed the "Area"; (2) the Area's resources were to be exploited for the benefit of all peoples; and (3) a treaty was to be established among all states that would provide for administering the Area and its resources and for equitably distributing the benefits generated, especially to assist the LDCs.

The United States, France, and the United Kingdom at this time also submitted alternative plans, all of which had similarities. The United States draft treaty suggested that an International Seabed Resource Authority (ISRA) be established. The proposal suggested the 200-m (656 ft.) isobath as the outer limit for coastal state territorial jurisdiction and called for the establishment of an International Trusteeship Area beyond the 200-m isobath to a point on the continental shelf where its slope reaches a certain degree of inclination. Beyond the Trusteeship Area would lie the deep seabed. Coastal states were to administer the Trusteeship Area, issuing permits for exploration and mining, supervising exploitation, and collecting royalty fees. Provisions for the LDCs' and landlocked states' economic participation were to be implemented through the transfer of 50–66.6% of the monies accruing from the Trusteeship Area to the ISRA, which would channel the funds to needy countries through various United Nations and regional development agencies. Composed of 4 organizational units—a Secretariat; an Assembly, containing all ISRA members; a 24-member Council; and a 5-to-9 member

Tribunal—the ISRA was to administer the deep seabed area. The Council was to be composed of 6 of the most highly developed industrial states and 18 elected states, at least 12 of which were to be LDCs; and at least 2 of the Council members had to be either landlocked or shelflocked states, i.e., those without access to the oceans' continental margin or the deep seabed.[160]

This initial proposal yielded much controversy and some compromise. Debate has continued within the framework of the *Third United Nations Conference on the Law of the Sea* (UNCLOS III) in Caracus, New York, and Geneva. Although the form of the United States proposal is still recognizable, the present-day negotiating text has evolved to contain numerous provisions that, as of mid-1979, are unacceptable to the United States, to other developed states, and to the LDCs. Divergent interests of the various Law of the Sea Conference participants have helped create an atmosphere of mistrust and misunderstanding that permeates the proceedings.

INTEREST GROUPS AND THE ISSUES

When the newly established United Nations Seabed Committee met in 1971, 1972, and early 1973 to develop an agenda for the first session of UNCLOS III (held in New York City in late 1973), various factions emerged. These factions represented two main blocs of states—the developed states and the Group of 77 (G-77), with the latter composed primarily of the LDCs. The G-77 represents common interest alliances that developed out of the first meeting of the United Nations Conference on Trade and Development (UNTAD) in 1964. Members, now numbering well over 100, seek to enhance G-77's political and economic clout by working closely together, seeking consensus on important issues before taking a position.[a]

Overlapping the developed states and G-77 blocs is a geographically disadvantaged bloc of states, i.e., those with only narrow coasts and those without access to the continental margin or to the deep seabed. Some 30 world states have no direct access to the sea, and many others are excluded from access to the deep sea (the shelflocked). Nepal, Botswana, and Paraguay are good examples of the landlocked states, whereas The Netherlands, Belgium, and Bulgaria represent the shelflocked, a condition whereby a state's access to the deep sea is blocked by other states' territorial waters.[b]

Thus, the participants' views on the priority of issues for discussion at the LOS session scheduled for Caracas, Venezuela, in 1974 were so divergent that they were unable to agree on a formal agenda or draft treaty for negotiation.[94] An especially important issue was whether the prospective ISRA should have its powers and functions defined prior to its establishment (the developed states' position) or to allow their evolution after the ISRA became operational (the LDCs' position). The LDCs also wanted to immediately determine who should exploit the International Seabed Area (many favored a monopoly by the Authority), whereas the developed states sought protection from a preemption of their mining industry's prior investments. Another important polarity of views relates to seabed mining production efficiency vs. distribution of benefits. According to Leipziger and Mudge, developed states want the international community to adopt a deep seabed mining regime that is economically efficient for production, but they are "... ignoring the distribution of benefits issue," whereas the LDCs are interested in the distribution of benefits and are "... willing to sacrifice efficient production."[311] To say that the developed states are "ignoring" the distribution of benefits is unfair, because—as clarified in subsequent discussion—they have done much to accommodate the financial interests of the Third World in ocean mining.

The LDCs' desire for a distribution of benefits goes beyond the issue of seabed mining. Many LDC members of the G-77 look upon the UNCLOS negotiations not so much as a financial windfall but as a "precedential" change (part of a "new international economic order") in the LDCs' and developed countries' economic-political relationships.[470] In short, the LOS debates have become symbolic confrontations that discourage consideration of ocean affairs as a set of technical problems.[79] Keeping in mind that a significant part of the slow progress of the LOS Conference is attributable to an ideological struggle between the haves and have-nots, what are the substantive issues facing those attempting to hammer out an acceptable international agreement and an effective administrative organ that can assure a viable economic-political environment for ocean resources' exploitation?

UNCLOS TODAY

The Third Law of the Sea Conference has a much broader focus than seabed mining; also included among its concerns are oceanic research, marine pollution, navigation rights, and fisheries problems. The Conference currently has seven informal negotiating groups. Three of these groups focus on issues directly related to seabed mining: Group 1 is concerned with resource exploitation policies for the International Seabed Area and the creation of an ISRA (now commonly referred to as the ISA); Group 2 deals with financial arrangements for the proposed ISA; Group 3 focuses upon the composition, powers, and functions of the ISA.

What Kind of International Seabed Authority?

From a broad perspective, developed states would like to see an Authority with limited and specifically defined functions, whereas the LDCs want an Authority with comprehensive and flexible management functions. As was noted earlier, the currently proposed ISA structure is similar to the United States' 1970 proposal. The Council has been expanded to include 36 members; Tribunal membership is now 21. The Secretariat remains an integral part of the organization, as do numerous special commissions. One important change is the addition of a body called the "Enterprise," composed of a 15-member governing board and a director general whose functions would be to develop seabed resources on behalf of the Authority.

The Seat of Power: The Assemby vs. the Council. It is not the ISA's structure that poses the major difficulty to the establishment of a LOS treaty. Rather, the difficulty can be traced to the determination of the functions and scope of authority to be held by its administrative components and the competition for the seat of power within it. Various negotiating texts produced at the several (eight as of mid-1979) marathon meetings of UNCLOS III (1973-79) have seen frequent changes in the seat of power, again reflecting a polarity of views between developed states and the LDCs. The developed countries want the main power base lodged in the Council, because its membership structure and voting procedures could allow more opportunity for controlling events. If the Council's voting decisions were based on a three-fourths-of-a-quorum rule, the developed states could avoid passage of unfavorable decrees. The LDCs, on the other hand, would prefer a two-thirds voting rule which could avoid another United Nations Security Council situation, whereby the will of the majority, those in the Assembly, may be vetoed.[79] The LDCs want the Council to function only as an executive body, with the Assembly authorized to consider general policy for the ISA; they want decisions based on one country-one vote with a two-thirds majority needed for passage,[219] a procedure that will effectively neutralize the voting power of the developed states.

The Enterprise. In 1975, within the then current negotiating format (the Informal Single Negotiating Text), the LDCs succeeded in introducing a very significant component into not only the proposed structural organization but also into the proposed functional operation of the ISA. This component—the "Enterprise"— was to become the operating arm of the Authority, which could mine the seabed in its own right and/or regulate the mining activities of others. The developed states were at first cool to the Enterprise concept, but in late 1976, Secretary of State Henry Kissinger said that the United States would finance the Enterprise, a step that not only enhanced its credibility, but seemed to solve what had been a significant dilemma, how to fund it.[470] It is unfortunate that Kissinger did not make his offer sooner and did not provide a concrete plan for implementing it. Because he did not, considerable momentum toward a possible treaty may have been lost.[140] Elliot Richardson, the Carter Administration's Ambassador-at-large and head of delegation to UNCLOS III, has suggested that the Enterprise should be funded through the World Bank or some other functioning international financial agency, with the United States only guaranteeing that funds would be available, these funds would be gradually reduced as the Enterprise became self-supporting.[470]

If an Authority is established, the majority of the Enterprise's membership will probably be comprised of LDCs. Consequently, the United States and other developed countries may have difficulty obtaining a hearing for their views. There is also likely to be a divergence of views among the LDCs themselves, especially concerning mineral production levels. LDCs without much mineral resource production will probably want the Enterprise to allow a maximum output, whereas those states producing cobalt, nickel, copper, and manganese (the main nodule metals) may want to restrict output.[315] The United States recently proposed the idea of tying total seabed mineral output to the overall growth in the world's nickel market. If this idea of a production ceiling were to prove acceptable, it would help eliminate debate on how rapidly the ocean mining industry should be allowed to develop.[302] According to some analysts, a major problem could develop for seabed mining firms if the Authority is given the right to unilaterally set seabed production quotas for individual operations, rather than coming to an agreement through mutual consent. Restrictive policies for nickel, for example, could cost United States consumers anywhere from $2.8 million to $77 million/year at anticipated 1985 levels of consumption, and if all of the "Big Four" metals are taken together, the United States each year could potentially pay almost $310 million more than necessary. The rest of the world would experience a similar pattern of unnecessary costs.[271] This whole issue, however, is highly controversial and needs much additional study.

Alternative Approaches to Exploitation Arrangements

The basic dichotomy of the several suggested methods of providing access to the oceans' minerals is whether the Authority should be the exclusive manager and exploiter of deep seabed minerals or whether exploitation should be undertaken by private corporations or state agencies.

Registry. Under this system a mining firm would merely record with the Authority its claim and intent to mine a given site. This method is unacceptable to the LDCs, because they view it as merely a continuation of the old commercial imperialism of the developed states, and feel it would do nothing to further their aspirations for a new international economic order.

Licensing. One of the primary methods suggested and debated for administering mining under the ISA is licensing. But who will decide, and by what criteria and under what conditions, who is eligible to be licensed? Developed states fear that if the Authority determines the rules, capricious decisions could result, with qualified applicants being rejected.[216] Independent operators would be licensed for a specific period and for a specific payment to the Authority. For several years the licensing idea was castigated by the LDCs[300] as a "sellout" to the developed states, because the income generated would be minimal. The LDCs may have a point. According to Shyam, the technologically advanced states favor this approach because it would provide both significant profits for mine operators and help lower world metal-market prices.[482] The latter condition would be an unwelcome situation for states that sell and export copper, nickel, cobalt, and manganese. On the other hand, as noted in the preceding chapter, mineral-poor LDCs could benefit from the lower prices (although most have only a small total consumption); and the license fees would help cover the Authority's operational costs, regardless of whether the mining companies made a profit. Other disadvantages of licensing, in the view of the LDCs, are that the Authority may not have as tight a control over mining operations as they desire it to have, and it probably could not determine where the minerals would be processed and marketed.[300]

Service Contract. More recently, the licensing concept has been discussed in the context of a contract. Under this system the Enterprise would set production goals, hire the mining and processing firms, and could do its own marketing. This approach might require that the Enterprise have a much greater financial capability than would be necessary under a licensing program. La Que feels that one of the major problems that the Authority might face with this system is an inability to attract contractors willing to work under the constraints set by the Enterprise.[300] Mining firms will be especially discouraged under this system if the ISA, in an effort to protect LDC producers, establishes output levels below the point of maximum efficiency.

Joint Venture. Development of seabed minerals under this arrangement could be organized so that the Enterprise would supply the capital and the private mining firm or consortium would provide the technological expertise and mining equipment; more likely, however, the contracting party would also have to supply most of the capital, a situation unlikely to attract private corporations.[300]

Parallel Development. The idea of parallel development seems to be one of the most appealing and often discussed exploitation arrangements, whereby both private groups and the Authority would pursue mining operations. The main problem with this system is that the Enterprise would still be short on technology, competent personnel, and scientifically measured and mapped manganese nodule deposits. Therefore, negotiators have been working with the possibilities of tying the right of private mining firms to extract nodules to their providing the Authority with an area that is likely to contain a viable mine site. In 1975, the United States proposed this possibility within what it called the "banking system," designed to persuade the G-77 to agree to guaranteed access for private mining firms. Under this system the prospective mining applicant would submit data on two potentially exploitable areas. The Authority would choose one of these; the other could be mined by the applicant, providing stipulations set down in the LOS treaty were met. The Soviet Union submitted a somewhat similar plan while the G-77 was considering the United States proposal. The Soviets suggested that selected portions (one-half) of the minable seabed should be set aside for the Authority. Both the United States and Soviet proposals were rejected by the G-77 because it felt opportunities for the LDCs to directly participate in seabed mining were not really provided for.[470]

Direct Exploitation. The Enterprise could be given a mandate to undertake mining for the Authority, without others being allowed to mine. This approach seems unlikely, however, because financial arrangements would be a stumbling block. Many banking institutions would probably be hesitant to lend (without political guarantees) the large amounts of capital required to finance an economically hazardous venture that may be directed by inexperienced administrators. Another major hurdle would be the Authority's lack of technical competence to mine or process the nodules, unless special arrangements were made for the transfer or purchase of technology for its use.[154]

Technology Transfer and Revenue Sharing

Because most of the LDCs have neither the technological capability nor the financial resources to mine the deep seabed, one of the important and bitter issues debated within the LOS has been technology transfer. The LDCs want the developed states to provide them with the skills and equipment to mine the deep seabed. But industry feels it would be grossly unfair for mining firms to bear the expense of technology development only to have it handed over to the LDCs or the Enterprise.

Although many taxpayers would question its equitableness, the United States or other developed state governments could pay the mining firms and development laboratories for the technology delivered to the LDCs or Enterprise.[295, 485, c] The United States government has also offered to make loans to the ISA which would equal 20% of the total costs for the purchase of technology and mining equipment.[302] Several other approaches to the technology-transfer problem could be used. Mining firms might be allowed to exploit seabed minerals only if they make available to the LDCs or Enterprise a certain amount of technology,[470] although this situation would mean that a large amount of power would have to rest in the Authority. One drawback to this system would be that mining firms after transferring a technical innovation to the LDCs could set about developing a superior device or technique that could make the item transferred obsolete and economically uncompetitive. Another variation of handling the problem would be to have the Authority guarantee patent protection in return for the full disclosure of technology systems transferred, making those possessing the technology less reluctant to divulge their information; alternatively, the Authority could grant a short exclusivity period, after which the LDCs could use the technology upon payment of a royalty.[485]

An integral part of mine site access and technology transfer problems is the determination of equitable systems of revenue sharing from production in the Area. The negotiating text of the 1978 LOS meetings provided for three basic methods of financing the Authority and of sharing the ocean wealth with the LDCs and geographically disadvantaged states. These are licensing fees, production charges on mineral output, and charges (taxes) on net income.[382] Johnson and Logue, writing in the mid-1970s, noted that if a 50% royalty from the total revenues generated from seabed mining went to the Authority, as has been suggested by the LDCs, it would be debatable whether a seabed mining industry could exist. They believed that any royalty tax probably would contribute to premature closure of mining operations, because it "... artificially increases the marginal cost of production ...." Johnson and Logue suggested in 1975 that a better solution for generating revenue while encouraging seabed mineral production would be a lump-sum tax of perhaps $50,000/year for each mine site.[271] A good example of the LDC's efforts to establish not only a royalty tax but other levies as well surfaced during the September 1978 meeting of UNCLOS III in New York City. Singapore introduced a proposal which would have seen the mining companies pay a $500,000 fee for each mine site application and also a royalty and/or profit sharing fee of 40% during the initial 6 years of commercial production. The fee would increase to 70% of profits during the 7th through 12th years and would become 80% in the 13th through 20th years. The 80% fee would not be instituted until the company had recovered twice its original capital investment.[473] Needless to say, with concepts of income distribution such as this, the LDCs would surely laugh at the suggestion of Johnson and Logue of a flat $50,000 levy for each mine site/year.

The 200-mile Exclusive Economic Zone and the Continental Margin Beyond

A fundamental difficulty in initial negotiations at the LOS was the problem of where to set the shoreward limit of the Area. Coastal states were very apprehensive about this issue.[268] By the end of the 1974 Caracas meeting of the UNCLOS III, numerous states favored the establishment of an Exclusive Economic Zone (EEZ) or "Patrimonial Sea" to extend 200 nautical mi. out from a state's shore. All economic resources of the waters and seabed within the EEZ were to be reserved for the exclusive use of the coastal state. Although the United States and Soviet Union at first resisted this approach, the two superpower governments have come to accept the concept, having really the most to gain of any of the world states. In 1976, United States declared a 200-mi. fishing zone. Soon thereafter, the Soviet Union, Norway,[392] and India,[257] in 1977, and France and Australia, in 1978, declared either a comprehensive EEZ or fishing-rights zones.[356] These may have helped set the stage for what is becoming, as Darman so aptly defines it, a *territorial tide*[122]—80 states have now claimed some type of the 200-mi. fishing or resource zone (or terrirotiral sea).[207] If the 200-mi. zone is applied to coastal United States, Soviet Union, Brazil, India, New Zealand, Australia, Japan, Canada, and Norway, a major portion of the world's oceans is encompassed (Fig. 86). The move toward a general acceptance of the EEZ arrangement only makes the problems of access to the oceans' resources an even greater problem for the landlocked states.[140] By accepting the 200-mi. EEZ idea, the international community has handed over to a select few the richest oceanic fisheries, a major portion of marine petroleum resources, and important hard mineral deposits. One danger of the 200-mi. EEZ may be what has been termed a "creeping jurisdiction," whereby the EEZ becomes a 200-mi. territorial sea.[140]

By 1977 some statesmen were having second thoughts about the sagacity of the EEZ concept. But in July of that year UNCLOS III participants included still another giveaway plan in their Informal Composite Negotiating Text (ICNT), which officially was "... only a procedural device to facilitate negotiation ..." but which functioned "... as a draft treaty."[205] The proposal was that all continental margin areas lying seaward of a coastal state's 200-nautical-mi. EEZ be designated as under the jurisdiction of the coastal state but with the obligation to "share" the income from any minerals produced. But, as Nigrelli notes, there may not be much to share, because those LDCs that are net importers of minerals produced in the designated zone may be exempted from contributing to the fund.[378] Oxman contends that any state importing minerals in such quantity is unlikely to be among the truly poor LDCs.[416] The ICNT plan provided that states would not begin to pay for 5 years after the start of production so that they could recover their investments. In the sixth year the state would begin remitting a specific percent of the annual gross production value, with the percentage increasing up to the tenth year when the rate would be frozen.[378]

Nigrelli estimated in early 1978 that—given its potential area of continental margin available for exploitation beyond the 200-mi. EEZ and oil prices of $11–13/bbl—the United States might expect to receive gross revenues of $61 billion to

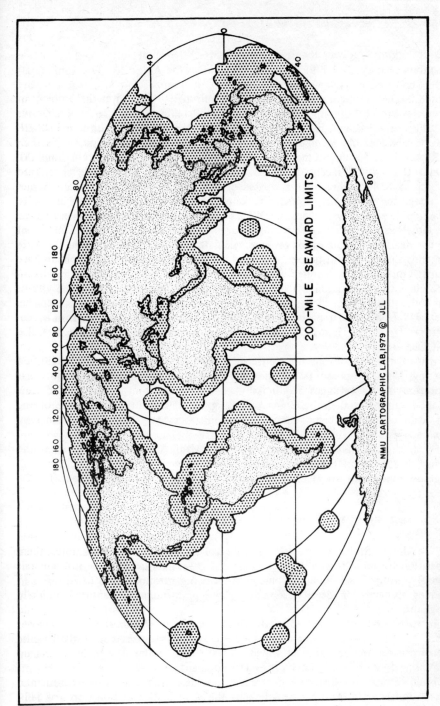

Fig. 86. The 200-mi. seaward limits.

$4,641 billion "... over the last 15 years of a 20-year depletion period," while contributing only $1.1 billion to $58.9 billion to the Authority. This sum equals a "... contribution/value-of-production ratio of about 1.8%, in contrast to the United States' foreign aid/GNP ratio of about 0.3%." World states' overall income beyond the 200-mi. EEZ might be $440 billion to $18,367 billion, with the Authority receiving $7.2 billion to $333 billion. Similarly, natural gas should provide considerable income to coastal states. The United States should accrue $12.9 to $566.8 billion; the Authority would receive $0.233 to $10.2 billion. World states would receive $19.6 to $786.4 billion, and they would contribute $0.353 to $14.1 billion to the coffers of the Authority. The ranges of revenue calculated by Nigrelli are so great that one might question whether they are really meaningful, especially so because OPEC's chaotic petroleum price increases after the late December 1979 meeting in Caracas to a range of $24.00 to $30.00/bbl require nearly a tripling of his revenue calculations.^{d} Furthermore, Nigrelli himself contends that onland resources of oil and gas, as well as those located out to 200-nautical mi. are large enough that most major producing states will not likely proceed with exploitation of petroleum resources in the Area before the 1990s, and consequently even the small portion of revenues that may be generated under this 1977 UNCLOS III proposal will not be available until after the turn of the century.[378] As of mid-1979, this proposal was still under consideration by UNCLOS III.

It is of note that during the course of a 1978 (March 28–May 19) UNCLOS III meeting, Nepal proposed a "common heritage" fund to redistribute to the LDCs a portion of the revenues acquired by coastal states from production of nonliving resources in their EEZ.[192] According to Glassner, the Nepalese proposal has little chance of being adopted.[207] Similar revenue-sharing proposals have been suggested previously. If a revenue-sharing plan were established and applied to United States petroleum production within the EEZ, total annual payments to the international community could be considerable—perhaps as much as $6.23 billion annually in 1970 dollars.[206, 271]

UNITED STATES POLICY AND OCEAN MINERAL RESOURCES

The United States Congress has been involved with developing marine-related legislation for more than a decade. Numerous governmental agencies, independent advisory groups, and industrial and academic organizations have had a hand in advising Congress on ocean resources policies or formulating ocean resource-use programs.

In 1966, Congress legislated the *Marine Resources and Engineering and Development Act* (PL 89-454) which established objectives and policies for United States use of ocean resources. The act provided for the constitution of a National Council on Marine Resources and Engineering Development, chaired by the Vice President, to evaluate and coordinate various federal marine activities at both domestic and international levels. Another provision of the act included the creation of a temporary 15-member advisory group, the Commission on Marine Sciences, Engineering, and Resources, composed of members from the federal and state governments, and independent marine-science research institutions and laboratories; four members of Congress provided liaison with federal lawmakers. The function of the Commission was to evaluate the contemporary marine science situation and to make an advisory report on current and future needs for a viable national oceanographic program. The Commission's report, *Our Nation and the Sea*, was published in early 1969; its recommendations prompted the formation within the federal government of NOAA in 1970 and the establishment of a 25-member body of marine specialists titled the National Advisory Committee on the Oceans and Atmosphere. Although the latter became defunct in 1971, it was replaced by a policy-making body titled the Interagency Law of the Sea Task Force, chaired by the legal advisor of the State Department and composed of members from the Departments of State, Justice, Interior, Commerce, and Transportation, as well as representatives from the National Security Council and the National Science Foundation.[147]

The government has helped with financing of major deep ocean research programs. DOI established the Ocean Mining Administration (OMA) in February 1975 to coordinate its seabed-mining-related functions, and DOC established the Office of Marine Minerals (OOMM) in September of the same year to perform functions similar to its counterpart in the DOI. As this brief discussion tends to illustrate, the federal government has not neglected oceanic affairs, even though some critics might suggest that it did, given the current LOS impasse. It is rather evident that it is vitally interested in encouraging United States mining companies to mine the deep seabed.

Unilateral Legislation for Deep Seabed Mining

As the work of various mining interests intensified in the late 1960s, particularly following the passage of the 1969 United Nations moratorium on deep seabed mining, heavy pressure was exerted on the United States Congress to pass unilateral legislation for a beginning of mining activities in the sea. Groups such as the American Mining Congress especially desired legislation guaranteeing industry that it would not suffer financial losses as a result of actions taken at the international level, such as placing all mining rights under the control of the Authority.[e] The first federal legislation designed to guarantee mining industry investments was introduced in the House of Representatives in November 1971 and in the Senate in March 1972. This legislation failed to pass. In each succeeding Congress, similar bills have been introduced to provide not only investment guarantees but also recognize the legality of mining claims staked out by mining companies, until such time as an international seabed regime is established. Various arguments against this legislation have been presented by the State Department and private organizations at Congressional hearings. The Izzaak Walton League of America, Sport Fisheries Institute, Sierra Club, Environmental Defense Fund, Center for Law and Social Policy, and the World Federalists are representative of the many groups that have testified before congressional committees. Some of the arguments heard at the myriad hearings include: (1) too little provision is being made to assure protection of the marine environment;[f] (2) investment guarantees extending up to 40 years are too long; (3) the federal government should not be in the insurance business; (4) proposed lease sizes (one bill set this at 15,400 mi.^{2}/mine site and 154,000 mi.^{2}/firm) are too large; (5) fees for licenses are too small (one bill set these at as low as $5,000); (6) the federal government should not be locked into a set system of block sizes, tenure, fees, royalties, and work requirements; and (7) unilateral action by the United States might jeopardize the LOS Conference.[399] Proponents of the unilateral interim legislation have argued that: (1) deep seabed areas are relatively sterile of biota and that preliminary studies have shown no major environmental effects resulting from mining;[399] (2) the magnitude of investments and the long lifetime of given mine sites and mining equipment would require a long-term investment guarantee program, although some in industry would probably be willing to accept a shorter period;[184] (3) mine sites of approximately 4,000 mi.^{2} would be barely adequate for a 1-million-ton/year operation, and most establishments are expected to be three times this size; (4) insurance programs contained in the legislation are, conceptually, no different from those provided for United States firms located in foreign countries through the Overseas Private Investment Corporation (OPIC); (5) the interim legislation in no way would compromise the freedoms and activities of others in the deep seabed and it would not jeopardize the successful conclusion of a LOS treaty, either provisional or final[152]—any regulations established would concern only the activities of the mining firms and would not sanction actual ownership of the seabed mine sites; and (6) the threat of unilateral action should help prod UNCLOS negotiators into compromise.[545]

Prior to 1978, all bills to establish unilateral deep seabed mining legislation were proposed by members of Congress without the support of the State Department. On July 26, 1978, the House of Representatives with support from President Carter's office, reached a benchmark in its efforts to establish unilateral deep seabed mining legislation by passing (on a vote of 312 to 80) *The Deep Seabed Hard Minerals Act* (H.R. 3350). This bill was introduced in the House in 1977, but a similar bill in the Senate (S. 2053) was allowed to die on the calendar (October 15, 1978) when held from a floor vote. Once again, mining companies were faced with the dilemma of continuing to pour research and development funds into the rat hole of international and domestic political indecision.

Many in government considered the 1978 House legislation an improvement over previous bills because it deleted investment guarantees related to losses incurred, should an unfavorable LOS treaty be adopted by the world community; the bill also disallowed any mining until January 1, 1980 (the Senate bill disallowed mining until 1981). Under the House bill, NOAA was to be in charge of mining operations and licensing. Three-fourths of 1% of the gross revenues of a given mining operation was to be collected for distribution to the world community.[g]

In the first session of the 96th Congress, the perennial deep sea mining bills were introduced once again. On February 26, 1979, Senator Spark M. Matsunaga (Hawaii) introduced S. 493[98] and on March 27, 1979, Congressman Bob Wilson (California) placed H.R. 3268 before the House of Representatives.[99] Will these bills finally give American companies a "green light" to proceed with deep seabed mining? Perhaps. It is certain, however, that the longer concrete action is delayed at the international level the more likely the passage of domestic legislation to fill the void. It is also possible the world will see an increasing number of deep seabed mining claims of the kind made by Deepsea Ventures in 1974.

Deepsea Ventures' Claim

The LDCs and developed states were startled when Deepsea Ventures made a formal claim to a mine site in the northeastern Equatorial Pacific. Deepsea Ventures makes no claim to ownership of the seabed, but only to its use. This 1974 claim has been the focus of much debate, because it brought to a head the issues of unilateralism. At question is whether a private mining firm, apart from the United States government, can declare an exclusive right to use the seabed. Traditionally, individuals and private corporations do not have an international personality, although this concept seems to be changing. Although not truly comparable to a seabed mining claim, there are instances when United States firms have laid claim to resources outside the country, as in the case of some guano islands and on the island of Spitzbergen (where coal is mined), that the federal government has intervened with legislative or executive assistance.[187]

Deepsea Ventures filed its claim in formal letters, maps, and affidavits with the U.S. State Department and with foreign governments. It has also advertised its claim to all interested parties by publishing statements in newspapers.[187] S. J. Burton, writing in the *Stanford Law Review*, argues that an analysis of precedents in international law shows deep seabed mining by private groups to be legal, but exclusive mining claims like that made by Deepsea Ventures are not. "Such claims purport to establish exclusivity as to all the world, whereas high seas principles limit jurisdiction of each state to the activities of its flag vessels and nationals on the high seas, ensure that all parts of the high seas are open to all nations, and accordingly forbid appropriation of high seas areas."[76]

WILL THERE BE A LAW OF THE SEA TREATY, OR WHAT ALTERNATIVES?

In a detailed analysis of the stated and estimated positions of the more than 150 members negotiating at the LOS Conference, Friedheim and Durch reached the pessimistic conclusion that the Conference is destined for stalemate as long as its negotiators' referent, conceptual framework, differs so radically. They contend that the basic concepts focusing on the organization and functions of a seabed authority must be closer to consensus before meaningful negotiations can proceed.[191] A start in this direction might be achieved if the negotiators could move from the purely technical level to the political level, whereby the top leaders of the various states become involved.[162]

Some students of the Law of the Sea Conference have argued that the United States should take great care not to give too many concessions on seabed mining issues, despite the possibility it may not obtain a Law of the Sea treaty. Too many concessions may only further slow the negotiations, as the G-77 continue to attempt to squeeze more advantages from the developed states. Johnston feels that unless the provisions of the treaty are carefully worked out there will be misallocations of resources with subsequent efforts by states to rectify these which will create grave conflicts within the International Seabed Authority and the family of states.[273] Some have suggested that the United States' negotiating position would be enhanced if an effort were made to develop a "mini-treaty" whose membership would be composed of the mining states but open to all states. This arrangement would provide a mechanism for protecting investment interests and establish a form of leverage to get the G-77 to compromise on the contents of the treaty. The "mini-treaty" would establish a licensing enterprise that could (1) control or regulate environmental aspects of mining; (2) provide loans to Third World states who were signatories; (3) transfer technology to the LDCs through their joint participation in mining ventures; and (4) generate taxes that would be given to the poorest-of-the-poor states.[122, 289]

If UNCLOS III fails to establish a treaty, the effort should not be viewed as a total loss, because world statesmen will have gained a better notion of the problems involved in managing ocean space and ocean resources.[289, 290] It has also shown that "a strategy for the oceans cannot be thought of as belonging to the USA or to the USSR, to the northern hemisphere or to the southern. It must be a strategy for all mankind."[573]

SUMMARY AND CONCLUSIONS

The Third Law of the Sea Conference negotiations have now spanned nearly a decade, with no end in sight, although some participants emerged from the 8th session (summer 1979) with renewed confidence that a treaty may soon be a reality. The LDCs see the UNCLOS III proceedings as a means for establishing a new international economic order, whereas the developed states view them as a mechanism for establishing an efficient and dependable regime for the utilization of oceanic resources. How the LDCs and the developed countries will reconcile their differences cannot now be determined, but that they must do so should be apparent to the least perceptive observer of world affairs. Guarantees for an equitable sharing of the oceans' mineral wealth among the world's rich, poor, and disadvantaged states are imperative, if not from a moralistic stance then from a pragmatic recognition that, ultimately, world peace may be at stake. The LDCs must come to recognize that there is no such thing as a "free lunch," but the developed states, in turn, must confront the reality that the most patient of those in breadlines may eventually demand butter for their crusts.

THE PROSPECT

Some view the world's oceans as a domain belonging to no one, *res nullius*, to be exploited upon demand without constraints. Others view them as belonging to everyone, *res communis*, a common heritage to be used but husbanded. For the moment, the communal view seems dominant. Let us hope it continues, for as industrial man pushes his quest into the outer continental margins and into the deep seabed, the delicate balance of one of earth's most precious resources will come under increasing stress.

Petroleum producers continue to build offshore platforms and city-like work sites, supported by a sophisticated technology, and they have dispatched their technicians to probe the outermost reaches of the continents, the continental slope and rise. These activities can only intensify, as an energy-hungry world becomes ever more mechanized and overpopulated. Hard mineral producers in the nearshore extract a wide array of resources, including coal for steam-powered electrical generating plants and sand and gravel for construction industries. Their deep-sea counterparts seek out the new "black gold" of manganese nodules, which have been the focus of extraordinary capital investments and intensive research. As onland mineral supplies deplete, these potato-like concretions may become a mainstay of the world's mineral sustenance.

A recognition that mankind is fast becoming capable of laying claim to all of earth's last resource frontier, the oceans, has created a tense international atmosphere between those with and those without the ability to tap these riches. The dawning of this recognition came with the Truman Proclamation that helped set in motion a wave of international reaction that is now approaching the shores of all world states. No state can turn its back upon this wave and expect to withstand its force, for it is an engulfing wave, one which may sweep the beaches clean. It is also a wave that could lift mankind to a higher and mutual vantage point, whereby all the world's peoples can see a brighter future.

NOTES

[a]K. Schoonover presents a good discussion of the structure and operational procedures of the G-77 in [470].

[b]M. Glassner makes an excellent presentation of the problems of landlocked and shelf-locked states in [205].

[c]To obtain a good overall view of the technology transfer problem, examine the entire issue of [334]; especially useful for marine mining technology transfer problems, as viewed by many of the developed states, is a presentation by U.S. Congressman J. Breaux [55]:19–23. He sees the developed states' acquiescence to the LDCs' demands for seabed mining technology transfer as a step toward insistence for technology transfer in other economic sectors.

[d]Nigrelli's data are based on the assumption that all oil fields are contiguous 700-million-bbl. fields, with each providing $140 million for the ISA during the last 15 years of a 20-year depletion period [378].

[e]This pressure continues today (late 1979). The American Mining Congress has taken a very strong stand on this issue. On Oct. 8, 1978, it called upon the U.S. government to stiffen its negotiating position at the LOS Conference, and encouraged Congress to pass legislation allowing domestic firms to begin mining the deep seabed. It expressed the view that if the present situation continues much longer, deep seabed mining research teams may begin to disband and planned development programs may be abandoned [125].

[f]S. Whitney has investigated in detail the provisions the U.S. government should incorporate into legislation regulating any interim arrangements for seabed mining [579].

[g]For a helpful discussion of the House bill see [249], and for a brief summary of the mining industry's disappointment over the death of S. 2053 and its hopes for future legislation see [571].

References

[1] Aagaard, P. M. "Extreme Environmental Factors in Offshore Structure Design," *Pet. Eng. Internat.*, 46 (Nov. 15, 1974): 30 ff.
[2] Abrahami, A., "India Makes Mark in Marine Pipelining," *Offshore*, 38 (July 1978): 48, 54.
[3] "Accidents Hit Production from Four North Sea Fields," *Offshore Services*, 9 (Jan. 1976): 11.
[4] Adams, M. V. et al., *Mineral Resource Management of the Outer Continental Shelf*, USGS Circ. 720 (Wash., D.C.: GPO, 1975): 2, 12-13, 21.
[5] Ahern, W. R., Jr., *Oil and the Outer Coastal Shelf: The Georges Bank Case* (Cambridge, Mass.: Ballinger, 1973).
[6] "Aker Chosen for Backfill Operation," *Offshore*, 38 (Oct. 1978): 104.
[7] Albers, J. P., "Seabed Mineral Resources: A Survey," *Sci. and Public Aff.: Bul. of the Atomic Scientists*, 29 (Oct. 1973), 34-35, 37-38.
[8] Amacher, R. C. and R. J. Sweeney, eds., *The Law of the Sea: U.S. Interests and Alternatives* (Wash., D.C.: Amer. Enterprise Inst. for Public Pol. Res., 1976): 54.
[9] Amacher, R. C. and R. J. Sweeney, "International Commodity Cartels and the Threat of New Entry: Implications of Ocean Mineral Resources," *Kyklos*, 29 Fasc. 2 (1976): 292-309.
[10] Amman, H., H. Backer and E. Blissenbach, "Metalliferous Muds of the Marine Environment," paper presented at the Offshore Tech. Conf., Dallas, Tex., Vol. 1 No. 1759 (1973): 345-58, as cited in [43].
[11] Amos, A. F. and O. A. Roels, "Environmental Aspects of Manganese Nodule Mining," *Marine Pol.*, 1 (Apr. 1977): 160, 162.
[12] Anderson, A., "The Rape of the Seabed," *Saturday Rev. World*, 1 (Nov. 6, 1973): 21.
[13] "Andrus Planning Two Georges Bank Lease Sales," *O&G J.*, 77 (Apr. 30, 1979): 140.
[14] Angerstein, J., "Jack-ups—A Future in the North Sea: An Interview with P. M. Lovie," *North. Offshore*, 3 (Apr. 1974): 33, 37-38.
[15] "Aragonite: White Gold in the Bahamas," *Carib*, 1 (1978).

[16] Archer, A. A., "Progress and Prospects of Marine Mining," *Min. Eng.*, **25** (Dec. 1973): 31.
[17] Archer, A. A., "Prospects for the Exploitation of Manganese Nodules: The Main Technical, Economic and Legal Problems," in [203] : 21–38.
[18] Archer, A. A., "Sand and Gravel Demands on the North Sea," in E. D. Goldberg, ed., *North Sea Science* (Cambridge, Mass.: MIT Press, 1973): 343.
[19] Arndt, R. H., "The Shell Dredging Industry of the Gulf Coast Region," Chap. 2, in [52] : 13–48.
[20] Arrhenius, G. et al., "Counterions in Marine Manganates," in *Proceedings: CNRS International Colloquim on Genesis of Manganese Nodules* (Paris: Centre Nat. de la Recherche Scientifique, 1979): 1–7.
[21] "Artificial Seaweed Prevents Pipeline Erosion," *New Scientist*, **75** (Sept. 22, 1977): 732.
[22] "As Industry Sets Sites on Labrador, Experts Eye Coast's Unique Problems," *Offshore*, **37** (Nov. 1977): 149.
[23] "Asia Searching Harder for Offshore Crude," *Bus. Wk.* (Aug. 4, 1973): 32.
[24] Auburn, F. M., "The International Seabed Area," *The Internat. and Comparative Law Qrt.*, 4th Series, **20** (Apr. 1971): 176.
[25] BLM (Bur. Land Mgt.), *Draft Environmental Impact Statement: Proposed 1979 Outer Continental Shelf Oil and Gas Lease Sale Offshore the Mid-Atlantic States*, Vol. 2 of 3, OCS Sale No. 49 (Wash., D.C.: GPO, 1978): 404–28.
[26] "BP's Shetland Well Center of Attraction Again," *Offshore*, **38** (Oct. 1978): 160.
[27] Bachman, W. A., "Embargo Lessons Largely Unheaded," *O&G J.*, **76** (Oct. 9, 1978): 31.
[28] Bailey, R., "Drafting Britain's First Energy Policy," *Nat. Westminster Bank Qrt. Rev.* (Nov. 1976): 21.
[29] Bakke, D. R., "U.S.S.R. Production Increases in 1978," *Offshore*, **38** (June 20, 1978): 175–76, 180.
[30] Baldwin, P. L. and M. F. Baldwin, *Onshore Planning for Offshore Oil: Lessons from Scotland* (Wash., D.C.: The Cons. Foundation, 1975): 11–17, 32, 48, 51–60, 79–94, 96–97, 108–12.
[31] "Bar 347 Will Lay 36-in. Pipe in 1000-ft. Water," *Ocean Ind.*, **10** (Sept. 1975): 229–30.
[32] Baram, M., D. Rice and W. Lee, *Marine Mining of the Continental Shelf: An Assessment of the Legal, Technical and Environmental Feasibilities* (Cambridge, Mass.: Ballinger, 1978): 31, 36, 71, 111, 117, 159–80, 203, 205–13, 221–47.
[33] Barry, L., Minister, Mines and Energy, "Natural Resources off Canada's Coasts," notes for a panel discussion at the Canadian Bar Assoc. Annual Meeting in Quebec City, Aug. 26, 1975, p. 2. (mimeo.)
[34] Bartley, F. R., *The Tidelands Controversy: A Legal and Historical Analysis* (Austin: Univ. of Tex. Press, 1953).
[35] Beaudell, M., "UK North Sea: Forecasts of Production Build-up," *Pet. Econ.*, **44** (June 1977): 218.
[36] Beck, F. W. M. and K. M. Wiig, *The Economics of Offshore Oil and Gas Supplies* (Lexington, Mass.: D. C. Heath, 1977).
[37] Beeston, K., Press Officer, Nat. Coal Board, London, U.K., pers. com., Aug. 29, 1978 and Mar. 1, 1979.
[38] Bell, W. E., "The Equipment Requirements for Oil and Gas in the North Sea," *North. Offshore*, **3** (Apr. 1974): 119, 124.
[39] "Benn Flexes His North Sea Muscles," *Economist*, **264** (July 2, 1977): 85.
[40] Berkovitch, I., "How the Difficulties of Subsea Treasure Recovery Are Tackled," *Engineer*, **240** (May 29, 1975): 43.
[41] Berquist, A. J. R. and B. N. Mason, "Subsea Production System Steps-out to the North Sea," *North. Offshore*, **3** (May 1974): 41, 45.
[42] Berryhill, H. L., Jr., *The Worldwide Search for Petroleum Offshore—A Status Report for the Quarter Century, 1947–72*, USGS Circ. 694 (Wash., D.C.: GPO, 1974): 1–2, 8, 10, 14–15.
[43] Bignell, R. D., "Genesis of the Red Sea: Metalliferous Sediments," *Marine Min.*, **1**, No. 3 (1978): 212, 216–20; after H. Backer, "Rezente Hydrothermalsedimentare Lagerstattenvidung," *Erzmetall*, **26** (1973): 544–55.

[44] Blaikley, D. R., "Environmental Protection in North Sea Exploration and Production Operations," *Marine Pol.*, **1** (Apr. 1977): 144-46, 148-50.
[45] Blissenbach, E., "Technology Assessment of Deep Sea Nodule Mining," in [581] : 82-85.
[46] Blumberg, R., "Weather Key to Economics in Offshore Operations," *Pet. Eng. Internat.*, **46** (Nov. 15, 1974): 46 ff.
[47] Bonatti, E., "Metallogenesis at Oceanic Spreading Centers," *Annual Rev. of Earth and Planetary Sci.*, **3** (1975): 401-31.
[48] Bonatti, E., "Origin of Metal Deposits in the Oceanic Lithosphere," *Sci. Amer.*, **238** (Feb. 1978): 54, 57, 59.
[49] Booth, D., "Limpet Resistance Secures All-In-one Platform to Sea Bed," *Engineer*, **240** (Jan. 16, 1975): 43; "North Sea Oil: More Than a Drop in the Ocean for British Industry," *Engineer*, **239** (July 18, 1974): 49;* "One Way to Develop North Sea Oilfields," *Engineer*, **241** (Aug. 14, 1975): 34.**
[50] Borgese, E. M., *The Ocean Regime: A Suggested Statute for the Peaceful Uses of the High Seas and the Sea-Bed Beyond the Limits of National Jurisdiction*, Occasional Paper Series, Vol. 1, No. 5 (Santa Barbara, Calif.: Ctr. for the Study of Democratic Institutions, 1968): 2.
[51] Borgese, E. M. and N. Ginsburg, eds., *Ocean Yearbook I* (Chicago: Univ. of Chicago Press, 1978): 143-44.
[52] Bouma, A. H., ed., *Shell Dredging and Its Influence on Gulf Coast Environments* (Houston, Tex.: Gulf Pub., 1976).
[53] Bowen, R. and A. Gunatilka, "Copper Ores and Plate Tectonics," Chap. 2 in *Copper: Its Geology and Economics* (N.Y.: John Wiley, 1977): 21-60, 326.
[54] Bratbak, B. and T. Moen, Norw. Pet. Directorate, Stavanger, Nor., pers. com., Mar. 2 and 21; May 8; June 29, 1979.
[55] Breaux, J., "Technology Transfer: A Case Study of the Inequity of the New International Economic Order," *Marine Tech. Soc. J.*, **13** (June-July 1979): 19-23.
[56] "The British Offshore Search—for Foreign Business," *Offshore*, **38** (Feb. 1978): 37-05.
[57] Broecker, W. S., *Chemical Oceanography* (N.Y.: Harcourt Brace Jovanovich, 1974): 98-103, 105.
[58] Brondel, G., "Offshore Oil and Gas," in [581] : 102, 104.
[59] Brown, C. L. and R. Clark, "Observations on Dredging and Dissolved Oxygen in a Tidal Waterway, *Water Resources Res.*, **4** (Dec. 1968): 1381-84.
[60] Brown, R. J., "Examining New Pipe Burial Methods," *Offshore*, **38** (Apr. 1978): 70 ff.
[61] Brown, S. and L. L. Fabian, "Diplomats at Sea," *For. Affairs*, **52** (Jan. 1974): 302.
[62] Buckman, D., "New Licenses Will Spur Irish Oil and Gas Search," *Ocean Ind.*, **10** (Aug. 1975): 48-50.
[63] "Bullet System Profiles Sediments," *Offshore*, **37** (Dec. 1977): 154, 156.
[64] Bunich, P., "The Economics of Developing the Resources of the World's Oceans," *Problems of Econ.*, **18** (Aug. 1975): 79-81.
[65] Burk, C. A., "Global Tectonics and World Resources," *Amer. Assoc. of Pet. Geol. Bul.*, **56** (Feb. 1972): 200.
[66] Burk, C. A. and C. L. Drake, "Continental Margins in Perspective," in [67] : 956-1000, 1003-09.
[67] Burk, C. A. and C. L. Drake, eds., *The Geology of Continental Margins* (N.Y.: Springer-Verlag, 1974).
[68] Burke, K. C. and J. T. Wilson, "Hot Spots on the Earth's Surface," in *Readings from Scientific American: Continents Adrift and Continents Aground* (San Francisco: W. H. Freeman, 1976): 61.
[69] Burnett, B., "Worldwide Drilling and Production," *Offshore*, **38** (June 20, 1978): 66; and reprint.
[70] Burnett, B., "Drilling Rigs Slated for Chinese Waters," *Offshore*, **38** (Sept. 1978): 11.
[71] Burnett, B., "Survey Tallies Increases in Capital Expenditures," *Offshore*, **38** (Dec. 1978): 43-45.
[72] Burnett, B., "Worldwide Drilling and Production," *Offshore*, **38** (June 20, 1978): 54, 66, 70.

[73] Burnett, W. C., "Geochemistry and Origin of Phosphorite Deposits from off Peru and Chile," *Bul. of the Geol. Soc. of Amer.*, **88** (June 1977): 813.
[74] Burns, R. G. and B. A. Brown, "Nucleation and Mineralogic Controls on the Composition of Manganese Nodules," in [241*]: 51.
[75] Burns, R. G. and V. M. Burns, "Mineralogy," Chap. 7 in [202]: 185, 247.
[76] Burton, S. J., "Freedom of the Seas: International Law Applicable to Deep Seabed Mining Claims," *Stanford Law Rev.*, **29** (July 1977): 1180.
[77] Burzminski, M. J. and Q. Fernando, "Extraction of Copper and Nickel from Deep Sea Ferromanganese Nodules," *Analytical Chem.*, **50** (July 1978): 1177.
[78] Bynum, D., Jr. and P. M. Lovie, "How Jack-ups Fit in North Sea Boom," *Pet. Eng. Internat.*, **46** (Oct. 1974): 94.
[79] Byrum, H. C., Jr., "International Seabed Authority: The Impossible Dream?" *Case Western Reserve J. of Internat. Law*, **10** (Summer 1978): 630-31, 658.
[80] Calvert, S. E. and N. B. Price, "Shallow Water, Continental Margin and Lacustrine Nodules: Distribution and Geochemistry," Chap. 3 in [202]: 85.
[81] "Cameron Predicts Dry Tree Production System as the Future Seabed Trend," *Offshore*, **38** (June 5, 1978): 50-53 ff.
[82] "Campeche Wild Well Spawns Fallout in Norway," *O&G J.*, **77** (July 16, 1979): 41.
[83] "Canadian Firm Converts Submarine into Data-collecting Unit for All Seasons," *Offshore*, **37** (Sept. 1977): 190.
[84] Carmichael, J., "Companies Cooperate in Program to Develop Continental Shelf," *Offshore*, **35** (June 1975): 31, as adapted from a Nat. Pet. Council chart, "The Seagap Program Will Be Exploring the 200 Mile and Beyond Areas of Offshore Around the World."
[85] Carmichael, J., "Geology of the Indian Ocean Is among Least Known of World's Oceans," *Offshore*, **35** (Apr. 1975): 126.
[86] "Cerromatoso Nickel Project," *Colombia Today*, **14**, No. 2 (1979): 2.
[87] Chapman, K., "North Sea Hydrocarbons too Precious to Burn," *Geog. Mag.*, **50** (May 1978): 492 ff.
[88] Cheng, C-Y., *China's Petroleum Industry: Output, Growth and Export Potential* (N.Y.: Praeger Pub., 1976): 12-13.
[89] Childers, M. A., "Deep Water Mooring-Part II: The Ultradeep Water Spread Mooring System," *Pet. Eng. Internat.*, **46** (Oct. 1974): 108, 110.
[90] "China Orders Revolving Floating Crane from IHI," *Offshore*, **39** (Feb. 1979): 13.
[91] Christian, E., "Minerals from the Sea," in [581]: 88-92.
[92] Church, T. H., ed., *Marine Chemistry in the Coastal Environment*, ACS Symposium Series 18 (Wash., D.C.: Am. Chem. Soc., 1975): 557-71.
[93] Coleman, R. G. and W. P. Irwin, "Ophiolites and Ancient Continental Margins," in [67]: 925.
[94] Comm. on Int. and Insular Affairs, U.S. Sen., *The Law of the Sea Crises: An Intensifying Polarization*, a staff rept. on the UN Seabed Comm. for the OCS and Marine Mineral Develop., 92nd Cong., 2d sess. (Wash., D.C.: GPO, 1972): 26.
[95] Comm. on Oceanography, Nat. Res. Council, Nat. Acad. of Sci. and the Comm. on Ocean Eng., Nat. Acad. of Eng., *The International Decade of Ocean Exploration: An Oceanic Quest* (Wash., D.C.: Nat. Acad. of Sci., 1969): 30.
[96] "Computerized Safety for Offshore Rigs," *Pet. Eng. Internat.*, **46** (May 1974): 96.
[97] "Condeep Platform Readied for Beryl Field," *Pet. Eng. Internat.*, **47** (June 1975): 20.
[98] *Cong. Rec.*, **125**, No. 21 (Feb. 26, 1979): S1800.
[99] *Cong. Rec.*, **125**, No. 38 (Mar. 27, 1979): H1737.
[100] "Convention on the Continental Shelf," Article 1, adopted Apr. 26, 1958 (U.N. Doc. A/Conf. 13/L. 53).
[101] Conwell, C. N., Min. Eng., Dept. of Nat. Res., State of Alaska, Fairbanks, pers. com., Aug. 17, 1978.
[102] Conwell, C. N., *Progress and Prospects in Marine Mining in Alaska*, Info. Circ. 22 (Fairbanks: State of Alaska Dept. of Nat. Res., 1976): 1-22.
[103] Cooper, B. and T. F. Gaskel, *North Sea Oil—The Great Gamble* (Indianapolis, Ind.: Bobbs-Merrill, 1966): 94-104.

[104] Corliss, J. B., "The Origin of Metal-Bearing Submarine Hydrothermal Solutions," *J. of Geophys. Res.*, 76 (Nov. 20, 1971): 8128-38.
[105] Cornell, N. W., "Manganese Nodule Mining and Economic Rent," *Nat. Res. J.*, 14 (Oct. 1974): 524, 530.
[106] Cozens, A., "Ocean Mining: The Treasure Hunt Beings," *Offshore*, 38 (Feb. 1978): 162.
[107] Cozens, A., "Status of Offshore Africa: Trying to Drill through Man-made Obstructions," *Offshore*, 37 (June 20, 1977): 187, 191, 193-94.
[108] Crawford, D., "Baltimore Canyon: Exploring the Industry's New Frontier," *Offshore*, 38 (Apr. 1978): 47.
[109] Crawford, D., "Development of East Coast Basins Still Entangled in Environmental Red Tape," *Offshore*, 37 (June 20, 1977): 135.
[110] Crommelin, M., "Offshore Oil and Gas Rights: A Comparative Study," *Nat. Res. J.*, 14 (Oct. 1974): 492-95, 499.
[111] Cronan, D. S., "Deep-Sea Nodules Distribution and Geochemistry," Chap. 2 in [202]: 12-13.
[112] Cronan, D. S., "Metalliferous Sediments and Manganese Nodules on the Ocean Floor," *Chem. and Ind.*, No. 14 (July 16, 1977): 577-78.
[113] Cruickshank, M. J., "Mineral Resources Potential of Continental Margins," in [67]: 965-1000.
[114] Cruickshank, M. J., Bu. Min., Wash., D.C., pers. interview, Jan. 12, 1979.
[115] Cruickshank, M. J., and H. D. Hess, "Marine Sand and Gravel," *Oceanus*, 19 (Fall 1975): 34-35.
[116] Cundiff, W. E., *Nodule Shock? Seabed Mining and the Future of the Canadian Nickel Industry*, Occasional Paper No. 1 (Montreal: Inst. for Res. on Public Pol., 1978).
[117] Cutting Red Tape Is Real Path to Diligent OCS Development," *O&G J.*, 76 (July 3, 1978): 21.
[118] DOI (Dept. Int.), *First Annual Report of the Secretary of the Interior under the Mining and Minerals Policy Act of 1970* (Wash., D.C.: Dept. Int., 1972): 62-65.
[119] DOI, *USGS Survey Yearbook, Fiscal Year 1977* (Wash., D.C.: GPO, 1978): 137-39, 140-41;* DOI, *USGS Survey Yearbook, Fiscal Year 1978* (Wash., D.C.: GPO, 1979): 203.**
[120] Dam, K. W., *Oil Resources: Who Gets What How?* (Chicago: Univ. of Chicago Press, 1976): 103-23.
[121] Danenberger, E. P., Cons. Div., USGS, Reston, Va., pers. com., Jan. 25, 1979.
[122] Darman, R. C., "The Law of the Sea: Rethinking U.S. Interests," *For. Affairs*, 56 (Jan. 1978): 378, 393-94.
[123] "Davis Strait: Not Even Threatening Icebergs Will Deter the Search for Hydrocarbons," *Offshore*, 37 (Oct. 1977): 48, 50, 53-55.
[124] De La Mahotiere, S., "Vital Role of North Sea Oil in the U.K.'s Energy Requirements," *Port of London*, 51, No. 590 (1975): 2-4.
[125] "Declaration of Policy of the American Mining Congress," *Min. Cong. J.*, 64 (Nov. 1978): 79-91.
[126] "Deep Ocean Floor Nodule Mining—First Generation Techniques Are Here," *Min. Eng.*, 27 (Apr. 1975): 47-50, 52.
[127] *Deep Seabed Hard Minerals*, "Hearings," on Mar. 28, 1973, House of Rep. Subcomm. on Oceanography of the Comm. on Merchant Marine and Fisheries, 93rd Cong., 1st sess., on H.R. 7732 (Wash., D.C.: GPO, 1973): 86-87.
[128] *Deep Seabed Minerals: Resources, Diplomacy, and Strategic Interest*, prepared for the Subcomm. on Internat. Org. of the Comm. on Internat. Relations, House of Rep., 95th Cong., 2d sess., by the Sci. Pol. Res., For. Affairs and Nat. Defense, and Econ. Div., Cong. Res. Service, L.C. (Wash., D.C.: GPO, 1978): 13-18, 43, 45-51, 59, 64-65, 76, 82-84, 99-100, 112-13, 115-17, 119, 121.
[129] *Deep Seabed Mining*, "Hearings," Subcomm. on Oceanography of the Comm. on Merchant Marine and Fisheries, House of Rep., 95th Cong., 1st sess., on H.R. 3350 and H.R. 4582, H.R. 4922, H.R. 5624, H.R. 6846, and H.R. 6784 (Wash., D.C.: GPO, 1977).

[130] Degens, E. T. and D. A. Ross, eds., *Hot Brines and Recent Heavy Metal Deposits in the Red Sea: A Geochemical and Geographical Account* (N.Y.: Springer-Verlag, 1969): 1009.
[131] Denisov, S. V., "Perspectives in Reconnaissance for Coastal Marine Placers on Southern and Western Continental Shores of Okhotsk Sea," *Internat. Geol. Rev.*, 13 (Mar. 1971): 301-04.
[132] Dept. Int., *An Assessment of Mineral Resources in Alaska*, prepared for the Comm. on Int. and Insular Affairs, U.S. Sen., 93rd Cong., 2d sess. (Wash., D.C.: GPO, 1974).
[133] *Description of Manganese Nodule Processing Activities for Environmental Studies, Vol. I, Processing Systems Summary*, prepared for the Dept. Com. by Dames & Moore and EIC Corp. (Rockville, Md.: NOAA, 1977): 1-2–1-3, 2-1–2-5.
[134] *Description of Manganese Nodule Processing Activities for Environmental Studies, Vol. II, Transportation and Waste Disposal Systems*, prepared for the Dept. Com. by Dames & Moore and EIC Corp. (Rockville, Md.: NOAA, 1977): 3-1, 3-3–3-4, 3-6.
[135] *Description of Manganese Nodule Processing Activities for Environmental Studies, Vol. III, Processing Systems Technical Analysis*, prepared for the Dept. Com. by Dames & Moore and EIC Corp. (Rockville, Md.: NOAA, 1977): 3-6–3-7, 3-9–3-12.
[136] "Despite Scanty Results, Wildcatters Continue to Probe Mediterranean Sea," *Offshore*, 37 (June 20, 1977): 171.
[137] "Detecting Ocean Resources," *Nuclear Eng. Internat.*, 18 (June 1973): 502.
[138] "Development of the Oil and Gas Resources of the United Kingdom 1978," presented to Parliament by HMSO (London: Apr. 1978): 21, 29-36.
[139] "Diamond Core Drilling for Seabed Samples," *Marine Min. Eng.*, 25 (Nov. 1973): 46.
[140] Dickey, M. L., "Should the Law of the Sea Conference Be Saved?" *Internat. Law*, 12 (Winter 1978): 7, 10, 14.
[141] Dickinson, W. R., "Subduction and Oil Migration," *Geology*, 2 (Sept. 1974): 421-24.
[142] Dienes, L. and T. Shabad, *The Soviet Energy System: Resource Use and Policies* (Wash., D.C.: V. H. Winston & Sons, 1979): 46-47, 51, 70, 254.
[143] Dighe, B. Y., "India Finds the Road to Self-reliant Production—Thanks to Bombay High," *Offshore*, 37 (June 20, 1977): 230, 243.
[144] Dikshit, O. and J. T. Henry, Sr., "Mineral Resources of the Ocean," *Deccan Geographer*, 12 (July-Dec. 1974): 79.
[145] Dohrs, F. E. and L. L. Sommers, "North Sea Oil: Changing Offshore and Onshore Development," Chap. 15 in *World Regional Geography: The Problem Approach* (St. Paul, Minn.: West Pub., 1976): 206.
[146] Dolan, E., Jr., "Shell Roads in Texas," *Geog. Rev.*, 55 (Apr. 1965): 223-40.
[147] Doumani, G. A., *Ocean Wealth: Policy and Potential* (Rochelle Park, N.J.: Hayden Book Co., 1973): 57, 59, 80, 84.
[148] Dowley, C., Public Relations Mgr., Consolidated Diamond Mines of South West Africa, Orangemund, S.W. Africa/Namibia, pers. com., Dec. 7, 1978.
[149] Dravo-Van Houten Assoc., "Demand for Offshore Loading Facilities Reaches an Industrial Record High," *Offshore*, 37 (Sept. 1977): 88, 90.
[150] "Drive Is on for Worldwide Exploration," *Offshore*, 38 (Oct. 1978): 41.
[151] Drolet, J-P., *Deep Seabed Mining: A Canadian Perspective in Relation to the Nickel Industry* (Montreal: Canadian Inst. of Min. and Metallurgy, 1978).
[152] Dubs, M. A., Dir. of Ocean Res., Kennecott Copper Corp. (but speaking as a private citizen) in testimony in [127]: 66-70.
[153] Dubs, M. A., Dir. of Ocean Res., Kennecott Copper Corp., N.Y., N.Y., pers. com., Aug. 28, 1979.
[154] Dubs, M. A., Dir. of Ocean Res., Kennecott Copper Corp., testimony in [129]: 56-57.
[155] Duncan, L., "UK: The Taxation of Oil Profits from the UK Sector of the North Sea," *Intertax* (Neth., Feb. 1978): 70-78.
[156] "E. Coast Test Shows Noncommercial Gas," *O&G J.*, 76 (Dec. 4, 1978): 29.
[157] Earney, F. C. F., "Mining and the Environment," *Ecumene*, 5 (May 1973): 19, 21.
[158] Earney, F. C. F., "Mining, Planning, and the Urban Environment," *CRC Critical Reviews in Environmental Control*, 7 (Apr. 1977): 2-10, 63, 65.

[159] Earney, F. C. F., "New Ores for Old Furnaces: Pelletized Iron," *Annals of the Assoc. of Amer. Geographers*, **59** (Sept. 1969): 532.
[160] Earney, F. C. F., "Ocean Space and Seabed Mining," *J. of Geography*, **74** (Dec. 1975): 541-43.
[161] Econ. and Social Commission for Asia and the Pacific, *Offshore Hydrocarbon and Heavy Minerals Investigation in East Asia*, Technical Bul., **11** (Tokyo: ESCAP, 1977).
[162] "The Economic Aspects of the Law of the Sea," *Morgan Guarantee Survey* (Feb. 1976): 9, 13.
[163] Ehrlich, H. L., "Bacteriology of Manganese Nodules, I: Bacterial Action on Manganese in Nodule Enrichment," *Applied Microbiology*, **11** (Jan. 1963): 19.
[164] Ehrlich, H. L., "The Role of Microbes in Manganese Nodule Genesis and Degredation," in [241*]: 63-70.
[165] "Ekofisk Blow-out Threatens North Sea Oil Development," *New Scientist*, **74** (Apr. 28, 1977): 179.
[166] "Ekofish Bravo: Oil under Pressure," *Economist*, **263** (Apr. 30, 1977): 95.
[167] "Electro-oceanology Expands Boundaries of Marine Knowledge," *Offshore*, **37** (Aug. 1977): 108.
[168] Ely, N. and R. F. Pietrowski, Jr., "Boundaries of Seabed Jurisdiction off the Pacific Coast of Asia," *Nat. Res. Lawyer*, **8**, No. 4 (1976): 624-25.
[169] Emery, K. O., "The Continental Shelves," *Sci. Amer.*, **221** (Sept. 1969): 108-09.
[170] Emery, K. O. and B. J. Skinner, "Mineral Deposits of the Deep-Ocean Floor," *Marine Min.*, **1**, Nos. 1/2 (1977): 4-5, 7, 12-13, 15, 21, 26, 34-43, 45.
[171] Enright, R. J., "Surging North Sea Oil, Gas Flow Fast Cutting Imports," *O&G J.*, **76** (Aug. 28, 1978): 63-64, 66-68.
[172] "Esso Declines to Apply for U.K. Offshore Licenses," *O&G J.*, **76** (Nov. 27, 1978): 39.
[173] "Esso Starts Developing New Malaysian Field, Expands Hunt," *O&G J.*, **76** (July 24, 1978): 27.
[174] "Establishing Reliability in the Offshore Leasing Program," *Offshore*, **37** (July 1977): 41.
[175] Estrup, C., "Minerals from the Sea," in [581]: 91.
[176] "Exploring the North Sea Spectrum," *Offshore*, **38** (Oct. 1978): 74.
[177] Farris, J., Office of Public Info., Tex. Dept. of Water Res., Austin, pers. com., Oct. 25, 1978.
[178] *Federal Register*, **35** (Feb. 12, 1970): 3301.
[179] Fernie, J., "The Development of North Sea Oil and Gas Resources," *Scot. Geog. Mag.*, **93** (Apr. 1977): 21, 24-25.
[180] "First Mining Test," *Ocean World*, **1** (May-June 1978): 16.
[181] "Fiscal Regime Blamed for Drop in North Sea Drilling," *O&G J.*, **77** (Jan. 29, 1979): 85.
[182] Flipse, J. E., "Ocean Mining—Its Promises and Its Problems," *Ocean Ind.*, **10** (Aug. 1975): 133-36.
[183] Flipse, J. E., "Ocean Mining Stiffled by Lack of U.S. and U.N. Action," *Sea Tech.* (Jan. 1974): 33.
[184] Flipse, J. E., then Pres. of Deepsea Ventures, Inc., testimony in [127]: 86-87.
[185] Flood, L. B., "Record Offshore Oil and Gas Spending Indicated for 1975-80," *Ocean Ind.*, **10** (Sept. 1975): 255-57.
[186] Ford, E., "New Platform Uses Seabed Suction," *Pet. Eng. Internat.*, **47** (May 1975): 12.
[187] Frank, R. B. and B. W. Jenett, "Murky Waters: Private Claims to Deep Ocean Minerals," *Law & Pol. in Internat. Bus.*, **7** (Fall 1975): 1254-55, 1258.
[188] Frazer, J. Z., Geol. Res. Div., Scripps Inst. Ocean., Univ. of Calif., La Jolla, pers. com., July 30, 1979.
[189] Frazer, J. Z., "Manganese Nodule Reserves: An Updated Estimate," *Marine Min.*, **1**, Nos. 1/2 (1977): 103, 105, 107-08, 110, 113, 119-21.
[190] Frazer, J. Z. and M. B. Fisk, *Sediment Data Bank Users Handbook* (La Jolla: Scripps Inst. Ocean., Univ. of Calif., 1978).
[191] Friedheim, R. L. and W. J. Durch, "The International Seabed Resources Agency Negotiations and the New International Economic Order," *Internat. Org.*, **31** (Spring 1977): 378-79.

[192] "Further Sea Law Conference Session Unavoidable," says Mr. Zuleta in discussing hard-core issues, Apr. 28, 1978 interview with Zuleta, *UN Mo. Chronicle*, 15 (May 1978): 31-33.
[193] Garrand, L. J., "Offshore Phosphorite—World Occurrences, Origin, Recovery, and Economics," paper presented at the Ninth Underwater Min. Inst., San Diego, Calif., Oct. 19-20, 1978.
[194] Garrand, L. J., Pres., Garrand Corp., Salt Lake City, Utah, pers. com., Feb. 22, 1979.
[195] "General Assembly Resolution 2574D-XXIV," (Dec. 15, 1969).
[196] "Geneva Convention on the High Seas," Article 24, adopted Apr. 26, 1958 (U.N. Doc. A/Conf. 13/L. 53).
[197] Geyer, R. A., ed., *Submersibles and Their Use in Oceanography and Ocean Engineering* (Amsterdam: Elsevier Scientific Pub., 1977): 5.
[198] Ghosh, S. K., "Rivalry in the South China Sea," *China Rept.* (Mar.-Apr. 1977): 3-8.
[199] Gill, R., "Singapore Awaits Offshore Orders," *Far E. Econ. Rev.*, 96 (June 24, 1977): 80.
[200] Gillman, K., *Oil and Gas in Coastal Lands and Waters: A Report by the Council on Environmental Quality* (Wash., D.C.: GPO, 1977): 82, 131-32.
[201] Gittinger, L. B., Jr., "Sulfur," in S. J. Lefond, ed., *Industrial Minerals and Rocks*, 4th ed. (N.Y.: Amer. Inst. of Min., Metallurgical, and Pet. Engs., 1975): 1103-25.
[202] Glasby, G. P., ed., *Marine Manganese Deposits* (Amsterdam: Elsevier Scientific Pub., 1977).
[203] Glasby, G. P. and H. R. Katz., eds., *Papers presented at the I.D.O.E. Workshop, Suva, Fiji, Sept. 1-6, 1975*, Technical Bul. No. 2 CCOP/SOPAC (U.N. Econ. and Social Commission, 1976): 21-38.
[204] Glassner, M. I., "The Illusory Treasure of Davy Jones' Locker," *San Diego Law Rev.*, 13 (1976): 533-51.
[205] Glassner, M. I., "The Law of the Sea," *Focus*, Amer. Geog. Soc., 28 (Mar.-Apr. 1978): 14-15, 19-24.
[206] Glassner, M. I., New Haven, Conn., telephone interview, July 16, 1979.
[207] Glassner, M. I., Southern Conn. State Col., New Haven, pers. com., Aug. 9, 1979.
[208] Goblot, G. R., "Diverting Icebergs from Labrador Oil Rigs," *Canadian Geog. J.*, 94 (June/July 1978): 53-57.
[209] "Going to Stavanger," *Offshore*, 38 (Apr. 1978): 202.
[210] Goldberg, E. D. and G. Arrhenius, "Chemistry of Pacific Pelagic Sediments," *Geochimica Acta*, 13 (1958): 153-212.
[211] "The Great Submarine Snatch," *Time*, 105 (Mar. 31, 1975): 25.
[212] Greenslate, J., "Microorganisms in the Construction of Manganese Nodules," *Nature*, 249 (May 10, 1974): 181-83.
[213] Grigalunas, T. A., *Offshore Petroleum and New England: A Study of the Regional Economic Consequences of Potential Offshore Oil and Gas Development* (Kingston: NOAA Sea Grant, Univ. of R.I., 1975): 4, 6, 14-15, 88.
[214] "Guarantees: Necessary to Develop Seabed Mining . . . or 'Bail Out' That Could Harm U.S. Relations Abroad?" *Congressional Qrt. Weekly Rept.*, 36 (Jan. 21, 1978): 124.
[215] Gutierrez, R. M., "Ocean Exploitation: Deep Seabed Mining," a research essay submitted to the Norman Patterson School of Internat. Affairs, Carleton Univ., Ottawa, Ont., Canada, Oct. 1978, especially pp. 96-111.
[216] Haight, G. W., "Law of the Sea Conference—Why Paralysis?" *J. of Maritime Law and Commerce*, 8 (Apr. 1977): 288-89.
[217] Hammond, A. L., "Manganese Nodules II: Prospects for Deep Sea Mining," *Science*, 183 (Feb. 15, 1974): 644-46.
[218] Hanemann, A. R., Jr., Asst. Dir. Public Relations, J. Ray McDermott & Co., pers. com., Aug. 6, 1979.
[219] Hardy, M., "The Implications of Alternative Solutions for Regulating the Exploitation of Seabed Minerals," *Internat. Org.*, 31 (Spring 1977): 335.
[220] Hardy, R. W., *China's Oil Future: A Case of Modest Expectations* (Boulder: Colo.: Westview Press, 1978): 19-25.

[221] Harris, K. L., Div. of Nonferrous Metals, Bu. Min., Wash., D.C., pers. com., Oct. 20, 1978.
[222] Harris, P. M., Inst. of Geol. Sciences, London, U.K., pers. com., Nov. 7, 1978.
[223] Harris, W. M., B. F. McFarlane, S. K. Piper, and D. J. Arrit, *Outer Continental Shelf Statistics 1953 through 1978* (Wash., D.C.: Dept. Int., 1979): 12, 16-17, 28-29, 54, 70, 98-100.
[224] Harrison, S. H., *China, Oil, and Asia: Conflict Ahead?* (N.Y.: Columbia Univ. Press, 1977): 43-44, 77, 96-99, 118, 123, 168, 197, 211-12, 256.
[225] Hawrylshyn, G., "Industry Shares Enthusiasm over the Production Prospects in South America," *Offshore*, 37 (Sept. 1977): 128.
[226] Head, I. L., "The Canadian Offshore Minerals Reference: The Application of International Law to a Federal Constitution," *Univ. of Toronto Law J.*, 18, No. 2 (1968): 131-57.
[227] Heath, G. R. and T. Dymond, "Genesis and Transformation of Metalliferous Sediments from the East Pacific Rise, Bauer Deep, and Central Basin, Northwest Nazca Plate," *Geol. Soc. of Amer. Bul.*, 88 (May 1977): 723-33.
[228] Hedberg, H. D., "Continental Margins from Viewpoint of the Petroleum Geologist," *Amer. Assoc. of Pet. Geol. Bul.*, 54 (Jan. 1970): 22, 38-39.
[229] Hedberg, H. D., "Ocean Boundaries and Petroleum Resources," *Science*, 191 (Mar. 12, 1976): 1013.
[230] Hemphill, D. P., "Repair and Construction of Offshore Pipelines in Water Depths to 3,000 Feet," *J. of Pet. Tech.*, 27 (Apr. 1975): 415.
[231] *Heritage of the Sea . . .: Our Case on Offshore Mineral Rights*, published under the authority of A. B. Peckford, Minister of Mines & Energy, Govt. of Newf. and Lab. (St. John's, Newf.: Saga Communications, 1977): 9-10.
[232] Heywood, C. H., "Shut-in System Allows Safe Production during Tropical Storms," *Pet. Eng. Internat.*, 46 (Aug. 1974): 10.
[233] "Highlands Center for Deep Sea Diving," *British Rec.*, No. 12 (July 1976).
[234] Hill, K. M., "Technology Assessment and Seawater Resource Engineering," in [581]: 98-101.
[235] Hogg, A., Aberdeen Col. of Educ., Aberdeen, Scotland, pers. com., June 26, 1979.
[236] Holbik, K., "US—Reliance on Foreign Mineral Resources," *Intereconomics*, No. 12 (1971): 377.
[237] Holden, C., "OCS Oil: Mammoth Lease Plan Encountering Heavy Opposition," *Science*, 186 (Nov. 15, 1974): 611.
[238] Holden, W. M., "Miners under the Sea—Right Now," *Oceans*, 8 (Jan. 1975): 55-57.
[239] Holser, A. F., *Manganese Nodule Resources and Mine Site Availability* (Wash., D.C.: Ocean Min. Adm., Dept. Int., 1976): 11.
[240] Holter, J. P. and H. Roseng, Norges Bank, Oslo, Norway, pers. com., May 21, 1979.
[241] Horn, D. R., ed., *Papers from a Conference on Ferromanganese Deposits on the Ocean Floor* (Wash., D.C.: Office of the Int. Decade Oc. Expl., NSF, 1972): 9, 15, 31-49; and Horn, D. R., B. M. Horn and M. N. Delach, "Distribution of Ferromanganese Deposits in the World Ocean": 9, 15.*
[242] Horn, D. R., B. M. Horn and M. N. Delach, *Factors Which Control the Distribution of Ferromanganese Nodules and Proposed Research Vessels Track North Pacific*, Phase II, Ferromanganese Program, Technical Rept., No. 8, NSF-GX 33616 (Wash., D.C.: Int. Decade Oc. Expl., NSF, 1973): 2.
[243] Horn, D. R., B. M. Horn and M. N. Delach, *Ferromanganese Deposits of the North Pacific Ocean*, Technical Rept. No. 1, NSF-GX 33616 (Wash., D.C.: Int. Decade Oc. Expl., NSF, 1972): 6.
[244] Horn, D. R., B. M. Horn and M. N. Delach, *Ocean Manganese Nodules: Metal Values and Mining Sites*, Technical Rept. No. 4, NSG-GX 33616 (Wash., D.C.: Int. Decade Oc. Expl., NSF, 1973): 32, 47.
[245] Horne, D., Mineral Statistics and Econ. Unit, Inst. of Geol. Sciences, London, U.K., pers. com., Jan. 19, Apr. 5, and June 27, 1979.
[246] Horsfield, B. and P. B. Stone, *The Great Ocean Business* (London: Hodder and Stoughton, 1972): 259, 261, 263-65.

[247] "Hotel or Hospital, Workshop or Firehouse–The Sedco/Phillips SS Keeps it Afloat," *Offshore*, **38** (Apr. 1978): 125.
[248] Housden, A. J. D., "Nine Subsea Wells for Brazil's Garoupa Field," *Pet. Eng. Internat.*, **49** (June 1977): 52, 54.
[249] "House Sanctions Seabed Mining, 312–80," *Ocean Reporter* (Aug. 1978): 1-2, 6.
[250] "How Atlantic Oil Search May Be Affected by Oil Prices," *Ocean Ind.*, **10** (Aug. 1975): 167-68.
[251] "How Barite Is Recovered from an Offshore Consolidated Deposit," *Min. Eng.*, **27** (Apr. 1975): 46.
[252] Howard, P. F., "Exploration for Phosphorite in Australia," *Econ. Geol.*, **67** (Dec. 1972): 1181.
[253] Howell, B., "Safeguarding Our Oceans in Metal Mining," *Sea Frontiers*, **24** (Sept.-Oct. 1978): 268.
[254] Huff, De G. L., "Manganese," *Mineral Facts and Problems*, 1975 ed., Bu. Min. Bul. 667 (Wash., D.C.: GPO, 1976): 654, 659; Huff, De G. L. and T. S. Jones, "Manganese," *Mineral Commodity Profiles* (Wash., D.C.: Dept. Int., 1979): 17.*
[255] Hutcheson, A. M. and A. Hogg, eds., *Scotland and Oil* (Edinburgh: Oliver & Boyd, 1975): 20, 30, 46, 67, 80, 116-17.
[256] "India Invites Offshore Exploration," *Pet. Eng. Internat.*, **45** (June 1973): 90.
[257] "India Rules on Protective Steps," *Offshore*, **38** (May 1978): 313.
[258] "India's Offshore Pace Quickens," *Pet. Eng. Internat.*, **46** (Aug. 1974): 70.
[259] "Industry Eyes New Markets for Submersibles," *Offshore*, **37** (Dec. 1977): 11.
[260] Ingo, W., "Economics of Undersea Resources," *Intereconomics*, Nos. 7/8 (July/Aug. 1977): 181.
[261] "Investment in the North Sea," *Norwegian Commercial Banks Financial Rev.*, **49** (Mar. 1975): 2-3.
[262] Ippolito, J. T., "Newfoundland and the Continental Shelf: From Coal to Oil to Gas," *Columbia J. of Transnational Law*, **15** (1976): 161.
[263] "Irish Sea Gas Field Destined for Development," *O&G J.*, **76** (July 24, 1978): 22.
[264] Irving, R. R., "How North Sea Pipelines Bring Oil and Gas Ashore," *Iron Age*, **220** (Nov. 14, 1977): 35-40.
[265] Ives, G., "Ice Platform Concept Proven for Arctic Offshore Drilling," *Pet. Eng. Internat.*, **46** (Jan. 1974): 14.
[266] "JIM Looks Good," *Ocean Ind.*, **10** (Aug. 1975): 105-06.
[267] "Jack Up Use High, Floaters Firming Up," *O&G J.*, **76** (Sept. 1978): 107.
[268] Janis, M. W. and D. C. F. Daniel, "The U.S.S.R.: Ocean Use and Ocean Law," *Maritime Studies and Mgt.*, **2**, No. 2 (1974): 71-87.
[269] *The Jaramac* (Summer 1975): 24-25.
[270] Jenkins, P., "Norway No Model for U.K. Oil Policy," *North. Offshore*, **3** (May 1974): 48, 51-52.
[271] Johnson, D. B. and D. E. Logue, "U.S. Economic Interests in Law of the Sea Issues," in [8]: 54-55, 60.
[272] Johnson, L. P. "High Cost of Offshore Oil," *Chilton's Oil and Gas Energy*, **1** (Jan. 1975): 20-21.
[273] Johnston, J. L., "Geneva Update," in [8]: 192-94.
[274] Joiner, T. J., State Geologist, Geol. Survey of Ala., University, Ala., pers. com., Dec. 4, 1978.
[275] Kalyniuk, G. W., Support Mgr., Esso Resources, Canada, Edmonton, Ala., pers. com., Mar. 13, 1979.
[276] Kash, D. E. et al., *Energy under the Oceans: A Technology Assessment of Outer Continental Shelf Oil and Gas Operations* (Norman: Univ. of Okla. Press, 1973).
[277] Kaufman, A., "The Economics of Ocean Mining," *Marine Tech. Soc. J.*, **4** (July-Aug. 1970): 58-59, 62, 64.
[278] Kaufman, R., "The Selection and Sizing of Tracts Comprising a Manganese Nodule Ore Body," OTC 2059, Offshore Tech. Conf. Preprints, **11** (1977), as cited in [189]: 113.

[279] Kennedy, C., "All This and Oil Too: Rich Rewards of North Sea Technology," *Director*, **30** (Aug. 1977): 50.
[280] Kennedy, H. T., "Malaysian Project to Flow by 1983," *Offshore*, **38** (Sept. 1978): 158 ff.
[281] Kennedy, H. T., "Malaysia's Fields Expected on Stream Soon," *Offshore*, **38** (Apr. 1978): 106, 111.
[282] Kennedy, J. L., "Southeast Asia," *O&G J.*, **73** (Mar. 3, 1975): 69-71, 74, 77.
[283] Kennett, J. P. and N. D. Watkins, "Deep-sea Erosion and Manganese Nodule Development in the Southeast Indian Ocean," *Science*, **188** (June 6, 1975): 1011-13.
[284] Kenward, M., "Getting Oil from Troubled Waters," *Holland Shipbuilding*, **24** (June 1975): 76.
[285] Keto, D. B., *Law and Offshore Oil Development: The North Sea Experience* (N.Y.: Praeger, 1978): 9, 34, 41-42, 46, 58, 84-89, 96-97, 101-07.
[286] Kilpatrick, J. E., "Legal Problems in Offshore Resource Development, Pt. II: The Role of North Carolina in Regulating Offshore Petroleum Development," *CRC Critical Reviews in Environmental Control*, **6**, No. 2 (1976): 202.
[287] Kiyozuka, N. et al., "Japan Hopeful First Oil/Gas Field Will Prove to Become Major Producer," *Offshore*, **37** (Sept. 1977): 170.
[288] Klinger, F. L., Div. of Ferrous Metals, Bu. Min., Wash., D.C., pers. com., Oct. 27, 1978.
[289] Knight, H. G., "Alternatives to a Law of the Sea Treaty," in [8] : 133-47.
[290] Knight, H. G., *Consequences of Non-Agreement at the Third U.N. Law of the Sea Conference: A Report for the Working Group on Technical Issues in the Law of the Sea*, Studies in Transnational Legal Policy No. 11 (Wash., D.C.: Amer. Soc. of Internat. Law, 1976): 2.
[291] Kovach, A. J., "An Assessment of the Merits of Newfoundland's Claim to Offshore Mineral Resources," *Chitty's Law J.*, **23** (Jan. 1975): 18-23.
[292] Krasov, D. D., "Limnological Hypothesis of the Origin of Hot Brines in the Red Sea," *Nature*, **221** (Mar. 1, 1969): 850-51.
[293] Krueger, R. B., "International and National Regulation of Pollution from Offshore Oil Production," Chap. 24 in D. W. Hood, ed., *Impingement of Man on the Oceans* (N.Y.: John Wiley, 1971): 627.
[294] Krueger, R. B., "The Background of the Doctrine of the Continental Shelf Lands Act," *Nat. Res. J.*, **10** (July 1970): 454-55.
[295] Krutein, M., "Ocean Mining: Problems and Today's Results," paper presented at the 9th Underwater Min. Inst., San Diego, Calif., Oct. 19-20, 1978, pp. 13-14, 19. (mimeo.)
[296] Kulm, L. D. et al., "Evidence for Possible Placer Accumulations on the Southern Oregon Continental Shelf," *Ore-Bin*, **30** (May 1968): 88-90.
[297] Kuraoka, T., "Development of Under-Sea Coal Field at Miike Coal Mine," paper presented at the 4th Internat. Conf. on Coal Res., Oct. 8-11, 1978, Vancouver, B.C., Can.
[298] Laedal, S., Dir. of Public Affairs, Phillips Pet. Co., Stavanger, Norway, pers. com., Apr. 25, 1979.
[299] Lafferty, C. F., "The British Diving Industry," *Offshore*, **38** (Feb. 1978): 37-24, 37-26.
[300] La Que, F. L., "Different Approaches to International Regulation of Exploitation of Deep Ocean Ferromanganese Nodules," *San Diego Law Rev.*, **15** (Apr. 1978): 478-79, 483.
[301] "Last Gulf Sale Dubbed Lifeless," *Offshore*, **38** (Dec. 1979): 161.
[302] "Law of the Seas III Draws Mixed Reactions," *Offshore*, **38** (Feb. 1978): 175.
[303] LeBlanc, L., "China Sets New Sights," *Offshore*, **38** (Feb. 1978): 92-93.
[304] LeBlanc, L., "Chinese Officials Ponder Next Move," *Offshore*, **39** (Jan. 1979): 50, 53.
[305] LeBlanc, L., "Mediterranean Picture Brightens as Mobile Rigs Drill Deeper Waters," *Offshore*, **37** (Apr. 1977): 62-66.
[306] LeBlanc, L., "Nations Scramble for Unclaimed Seabed," *Offshore*, **37** (Mar. 1977): 43.
[307] LeBlanc, L., "North Sea Blowout Tamed Quickly," *Offshore*, **37** (Jan. 5, 1977): 33.
[308] Lee, G. C., "Deep Thoughts on Conventional Concepts," *Offshore*, **38** (Apr. 1978): 98.

[309] Lee, H., "Decision to Lease Outer Continental Shelf Lands," *Coastal Zone Mgt. J.*, **2**, No. 1 (1975): 38.
[310] Lee, K. L. and J. A. Focht, Jr., "Liquefaction Potential at Ekofisk Tank in North Sea," *Amer. Soc. of Civil Engs., Proc.*, **101**, GT 1, No. 11504 (Jan. 1975): 1-18.
[311] Leipziger, D. M. and J. L. Mudge, *Seabed Mineral Resources and the Economic Interests of Developing Countries* (Cambridge, Mass.: Ballinger, 1976): 130, 150-51, 160-61, 164-70, 183.
[312] Lewis, J. F., S. J. Green and W. J. McDonald, "Tunnel-Chamber Production System Proposed for Arctic Offshore Fields," *O&G J.*, **75** (Jan. 3, 1977): 71-76.
[313] Loftas, T., "Seabed Miners Ready to Take the Plunge," *New Scientist*, **76** (Oct. 13, 1977): 70.
[314] Logan, R. M., *Canada, the United States and the Third Law of the Sea Conference* (Montreal: Canadian-Amer. Comm. sponsored by C. D. Howe Res. Inst. and Nat. Planning Assoc., 1974): 54-57, 59-60.
[315] Logue, D. E., R. J. Sweeney and B. N. Petrou, "The Economics of Alternative Deep Seabed Regimes," *Marine Tech. Soc. J.*, **9** (Apr.-May 1975): 11.
[316] Longworth, H. C., "The North Sea Oil Rush Is On: But Britain and Norway's European Neighbors Shouldn't Count on Energy Bonanza," *European Community*, (Apr. 1975): 3.
[317] Lowe, A. V., "International Law and Federal Offshore Lands Disputes," *Marine Pol.*, **1** (Oct. 1977): 311-17.
[318] Lowery, E. L., "Rig Instrumentation Systems: A New Type of Safety Insurance," *Pet. Eng. Inter.*, **47** (Jan. 1975): 10-11.
[319] Lund, S., "Statfjord-Norway Oil Line Would Be North Sea's Deepest," *Pipeline & Gas J.*, **204** (Apr. 1977): 38.
[320] Mabon, J. D., "Increased Oil Production, More Exploration Due in U.K.," *World Oil*, **184** (June 1977): 92, 96.
[321] McCaslin, J. C., ed., *International Petroleum Encyclopedia* (Tulsa, Okla.: Pet. Pub. Co., 1978): 130-33.
[322] McCusker, T. G. and P. J. Tarkoy, "The Use of Tunneling to Develop Arctic Oil and Gas Reserves," *Tunneling Technology Newsletter*, No. 15 (Sept. 1976): 1-5.
[323] MacKay, D. I. and G. A. Mackay, *The Political Economy of North Sea Oil* (Boulder, Colo.: Westview Press, 1975): 44-45, 62, 69.
[324] McKelvey, V. E. et al., "Subsea Physiographic Provinces and Their Mineral Potential," in *Mineral Resources and Problems Related to Their Development*, USGS Circ. 619 (Wash., D.C.: GPO, 1969): 2, as modified from L. R. Heselton, Jr., "The Continental Shelf," Contract 106, Naval Studies Inst., Ctr. for Naval Analyses Res., 1969.
[325] McKelvey, V. E. et al., *World Subsea Mineral Resources*, Misc. Geol. Investigations–Map 1–632, Sheet 2 of 4 (Wash., D.C.: USGS, 1970).
[326] McLin, J., "Resources and Authority in the North-East Atlantic, Part I: The Evolving Politics and Law of the Sea in Northern Europe," *Amer. Field Staff Repts.*, West Europe Series, Vol. VIII, No. 5 (1973): 8.
[327] Maddox, P., "Construction Boom Assures Healthy Submersible Market," *Offshore*, **37** (Aug. 1977): 67-75.
[328] Maddox, P., "Subsea Completions Expected to Reach 140 by 1978," *Offshore*, **37** (Aug. 1977): 61-66.
[329] Mancke, R. B., *Mexican Oil and Natural Gas: Political, Strategic, and Economic Implications* (N.Y.: Praeger Pub., 1979): 65-66.
[330] Manheim, F. T., *Mineral Resources off the Northeastern Coast of the United States*, USGS Circ. 669 (Wash., D.C.: USGS, 1972): 8, 18.
[331] "Marathon Outlines Capital Spending Plan," *O&G J.*, **76** (Sept. 25, 1978): 63.
[332] "Marine Construction Market Continues to Thrive," *Offshore*, **38** (Nov. 1978): 57-64.
[333] *Marine Science Affairs: Annual Report of the President to the Congress on Marine Resources and Engineering Developments*, p. 67, as cited in [542]: 28.
[334] *Marine Technology Society Journal*, **13** (June-July 1979).

[335] Martin, C., "Newfoundland's Case on Offshore Minerals: A Brief Outline," *Ottawa Law Rev.*, 7 (Winter 1975): 45-48, 59.
[336] Mason, B. N., "Arctic Subsea Completions," *Pet. Eng. Internat.*, 47 (Jan. 1975): 44, 48, 50.
[337] Matthews, N. A., "Nickel," *Mineral Commodity Profiles* (Wash., D.C.: Dept. Int., 1979): 6, 15.
[338] Menard, H. W., "Time, Chance, and the Origin of Manganese Nodules," *Amer. Scientist*, 64 (Sept.-Oct. 1976): 529.
[339] Menard, H W. and J. Z. Frazer, "Manganese Nodules on the Sea Floor: Inverse Correlation between Grade and Abundance," *Science*, 199 (Mar. 3, 1978): 969-71.
[340] Mero, J. L., "Economic Aspects of Nodule Mining," Chap. 11 in [202] : 336.
[341] Mero, J. L., "Future Promise of Mining in the Ocean," *Canadian Min. and Metallurgical Bul.*, 65 (Apr. 1972): 21-27.
[342] Mero, J. L., *The Mineral Resources of the Sea* (Amsterdam: Elsevier Pub., 1964): 70, 73, 75-76, 97.
[343] Mero, J. L., "Ocean Mining: An Historical Perspective," *Marine Min.*, 1, No. 3 (1978): 245, 247, 253.
[344] Mero, J. L., "Potential Economic Value of Ocean-Floor Manganese Nodule Deposits," in [241*] : 199.
[345] Mero, J. L., "Will Ocean Mining Revolutionize World Industry?" in [581] : 77-80.
[346] Mes, M. J., "What You Need to Know about Ocean Bottoms," *Ocean Eng.*, 46 (Nov. 15, 1974): 13-14.
[347] "Mexico," map, in *Offshore*, 39 (May 1979): 127.
[348] Meyerhoff, A. A. and J-O. Willums, "China's Potential Still a Guessing Game," *Offshore*, 39 (Jan. 1979): 54.
[349] Mills, H., "Eastern Canada's Offshore Resources and Boundaries: A Study in Political Geography," *J. of Canadian Studies*, 6 (Aug. 1971): 36-50.
[350] Moen, T., Norwegian Pet. Directorate, Stavanger, Norway, pers. com., Mar. 21; May 28, 1979.
[351] Moncrieff, A. G., "Offshore Mining of Manganese Nodules," *Marine Observer*, 46 (Apr. 1976): 72, 74-75.
[352] Moncrieff, A. G. and K. B. Smale-Adams, "The Economics of First Generation Manganese Nodule Operations," paper presented at the Amer. Min. Cong., Oct. 10, 1974.
[353] Monger, N. R., Supt. of Collieries, Broken Hill Proprietary Co., Belmont, N.S.W., Australia, pers. com., Feb. 2, 1979.
[354] Moore, G. W. and E. A. Silver, *Gold Distribution on the Sea Floor off the Klamath Mountains, California*, USGS Circ. 605 (Wash., D.C.: GPO, 1968): 9.
[355] Moore, W. D. and L. Auldridge, "Slump Ends in S.E. Asia, Action Back on Uptrend," *O&G J.*, 77 (Jan. 8, 1979): 48-52, 54.
[356] "More 200-mile Zones," *Ocean Reporter* (Feb. 1978): 7.
[357] "More Wildcats Urged in U.K. North Sea," *O&G J.*, 76 (Nov. 27, 1978): 38-40.
[358] Morse, R. G., "Second Decade for Norwegian Oil: Second Thoughts?" *Scandinavian Rev.*, 65 (Mar. 1977): 29-31.
[359] Mullins, P. J., "Steel Technology Sets Course for North Sea," *Iron Age*, 211 (June 7, 1973): 48.
[360] Murakami, T., Min. Engineer, Mitsui Mining Co., Tokyo, Japan, pers. com., Mar. 16, 1979.
[361] Murray, J. and A. Renard, "Report on Deep-sea Deposits," Vol. 5, in C. W. Thompson, ed., *Report on the Scientific Results of the Voyage of H.M.S. "Challenger"* (London: Eyre and Spottiswoods, 1891): 1-525, as cited in [342] : 147-48.
[362] Murray, J. W. and P. G. Brewer, "Mechanisms of Removal of Manganese Iron and Other Trace Metals from Sea Water," Chap. 10 in [202] : 297-98.
[363] Mutch, D., "BNOC Control Worries UK Operators," *Offshore*, 38 (June 20, 1978): 98, 108.
[364] Mutch, D., "U.K. Setting Drilling Milestone," *Offshore*, 38 (Feb. 1978): 112.

[365] Nakajima, S., Vice Pres., Taiheiyo Engineering, Inc., Sydney Liaison Office, Sydney, Australia, pers. com., Mar. 23, 1979.
[366] Nat. Commission on Materials Pol., *Material Needs and the Environment Today and Tomorrow: Final Report of the National Commission for Materials Policy, June, 1973* (Wash., D.C.: GPO, 1973): 3-3.
[367] Nat. Energy Info. Ctr., Energy Info. Adm., Dept. of Energy, *Mo. Energy Rev.* (Feb. 1978): 60.
[368] Nat. Res. Council Rept., "Mining in the Outer Continental Shelf" (Wash., D.C.: Nat. Acad. of Sci., 1975), as cited in [11]: 162.
[369] Nealon, J. and J. Oshinsky, Statistics Dept., Amer. Pet. Inst., Wash., D.C., pers. com., Mar. 2, 1979; Dec. 7, 1979. The Institute developed the world data, excluding those for the U.S., from *Offshore*.
[370] Nelson, T. S., Asst. Exec. Consultant, Harding A/S, Rosendal, Norway, pers. com., May 30, 1978.
[371] Neudecker, S., "Coral Mining in Sri Lanka," *Sea Frontiers*, 22 (July-Aug. 1976): 215, 220.
[372] "New Design: Floating Concrete Production Platform," *Ocean Ind.*, 10 (Aug. 1975): 55.
[373] "New OCS Law Adds More Red Tape," *O&G J.*, 76 (Sept. 25, 1978): 68-69.
[374] "New Sight in the Ocean: Offshore Petroleum Operations Will Benefit," *Compressed Air Mag.*, 78 (July 1973): 16-18.
[375] "New Tools Created for Soil Testing," *Offshore*, 38 (Oct. 1978): 138.
[376] "Newfoundland: Last Harrumph," *Time Canada*, 22 (Sept. 29, 1975): 13.
[377] "Newsletter," *Offshore*, 38 (Apr. 1978): 5.
[378] Nigrelli, V. J., "Ocean Mineral Revenue Sharing," *Ocean Develop. and Internat. Law*, 5 (1978): 154, 156-57, 159, 161, 164-65, 168.
[379] Nisbet, J. M., Gen. Mgr., Profunda Ltd., Hong Kong, pers. com., Aug. 8, 1979.
[380] Noakes, L. C., "Review of Provenance for Mineral Sands and Tin in Southeast Asia" in [161]: 166.
[381] "Nontechnical Barriers Continue to Hold Back Deep Ocean Mining," *Chemical & Eng. News*, 55 (Dec. 5, 1977): 24-25, 27.
[382] Nordquist, M., "Deep Seabed Mining: Who Should Pay?" *Marine Tech. Soc. J.*, 10 (June 1976): 14, 24.
[383] Nordquist, M., "International Legal Aspects Concerning Exploitation of Manganese Nodules," in D. R. Horn, ed., *Ferromanganese Deposits on the Ocean Floor* (Wash., D.C.: GPO, 1978): 82-84.
[384] *Norges Bank Econ. Bul.*, Tables 35 and 36, 50 (Mar. 1979), Oslo, Norway.
[385] "North Sea: A Worksite for Manned Submersibles," *Pet. Eng. Internat.*, 46 (Jan. 1974): 11.
[386] "North Sea Oil Finance," *Banker* (London), 127 (May 1978): 73-77.
[387] "North Sea Oil Needs Assurances of Stable Conditions," *Offshore*, 37 (Feb. 1977): 41.
[388] "North Sea Platform Operating Costs Top Estimates," *O&G J.*, 76 (Nov. 13, 1978): 112.
[389] "North Sea Repercussions from Bravo Blowout," *Pet. Econ.* (London), 44 (June 1977): 241.
[390] Northeast Scotland Development Authority, *Offshore Oil Directory 1979*, 1st ed. (Aberdeen: The Authority, 1979).
[391] "Norway, UK Officials Take Steps in Resolving Statfjord Problems," *Offshore*, 39 (Apr. 1979): 138 ff.
[392] "Norwegian Exploitation Focuses North of 62nd Parallel," *Offshore*, 37 (Feb. 1977): 68, 70.
[393] "Not Yet a Northern Kuwait," *Economist*, 263 (Apr. 30, 1977): 241.
[394] Nyhart, J. D. et al., *A Cost Model of Deep Ocean Mining and Associated Regulatory Issues*, MIT Sea Grant Program Rept. No. MITSG 78-4 (Cambridge, Mass.: MIT Sea Grant Info. Ctr., 1978): ES12.
[395] "OCS Amendments," *Ocean World*, 1 (Sept.-Oct. 1978): 16.

[396] "OCS Leasing Fight to Be Replayed," *Congressional Qrt. Weekly Rept.*, 35 (July 9, 1977): 1410.
[397] OECD, *World Energy Outlook*, Rept. of the OECD Secy. Gen. (Paris: Org. for Econ. Cooperation and Develop., 1977), as cited in [285] : 46.
[398] OTA, U.S. Cong., *Coastal Effects of Offshore Energy Systems* (Wash., D.C.: GPO, 1976): 152, 156-58, 161, 164-66.
[399] *Ocean Manganese Nodules*, 2d ed., prepared for the Comm. on Int. and Insular Aff., Sen., 94th Cong., 2d sess., by the Cong. Res. Service (Wash., D.C.: GPO, 1976): 6, 15-18, 27-29, 31, 33, 51, 53, 57, 76-77.
[400] Odell, P. R., "Oil and Gas Exploration and Exploitation in the North Sea," in [51] : 143-44.
[401] Odell, P. R. and K. E. Rosing, "North Sea Oil Province: A Simulation Model of Development," *Energy Pol.*, 2 (Dec. 1974): 326.
[402] Oden, H. A. Acting Chief, Cons. Div., USGS, Reston, Va., pers. com., May 2, 1979.
[403] Office of Internat. Energy Affairs, "United Kingdom," in *The Relationship of Oil Companies and Foreign Governments* (Wash., D.C.: Fed. En. Adm., 1975): 47-48, 184.
[404] Office of Pol. Analysis and Office of Energy Develop., *Initial Report on Oil and Gas Resources, Reserves, and Productive Capacities* (Wash., D.C.: Fed. En. Adm., 1975): 1.
[405] "Offshore Argentina Looks Promising," *Offshore*, 38 (June 20, 1978): 142, 144.
[406] "Offshore Reserves Top PRC Priority," *Offshore*, 38 (June 20, 1978): 188.
[407] "Offshore Service Vessels, Helicopters and Submarines Play an Important Role," *Marine Eng./Log*, 82 (Apr. 1977): 80-81.
[408] "Offshore Service Vessels: Special Report," *Marine Eng./Log*, 82 (Apr. 1977): 80.
[409] Olander, J. A., "Applications of Chemistry in Deep Ocean Mining," Chap. 34 in [92] : 557-71.
[410] Oliver, L., "Tenneco Tests Diverless System for 2000-ft. Waters," *Pet. Eng. Internat.*, 49 (May 1977): 48 ff.
[411] "Onshore Effects of Offshore Oil," *Exxon USA*, 15, No. 2 (1976): 14.
[412] "Open Ocean Petroleum Storage," *Pet. Eng. Internat.*, 49 (Aug. 1977): 14.
[413] Outer Continental Shelf Advisory Comm., Exec. Secy., *Virginia and the Outer Continental Shelf Problems, Possibilities and Posture* (Richmond: Council on the Environment, Office of the Governor, Commonwealth of Va., 1974): 6.
[414] Owen, R. M., "An Assessment of the Environmental Impact of Mining on the Continental Shelf," *Marine Min.*, 1, Nos. 1/2 (1977): 98-99.
[415] Owen, R. M. and J. R. Moore, "Sediment Dispersal Patterns as Clues to Placer-Like Platinum Accumulation in and Near Chagvah Day, Alaska," paper presented at the Eighth Annual Offshore Conf., Houston, Tex., May 3-6, 1976.
[416] Oxman, B. H., "The Third United Nations Conference on the Law of the Sea: The 1977 New York Session," *Amer. J. of Internat. Law*, 72 (Jan. 1978): 81.
[417] Pallister, A. E., "Beaufort Sea Environment and Development," *Pet. Eng. Internat.*, 48 (July 1976): 14.
[418] Park, C. F., Jr., *Earthbound: Minerals, Energy, and Man's Future* (San Francisco: Freeman & Co., 1975): 100-03.
[419] Park, C-H., "The South China Sea Disputes: Who Owns the Islands and the Natural Resources?" *Ocean Develop. and Internat. Law*, 5, No. 1 (1978): 27-59.
[420] Pasho, D. W., "Review of the Development of Deep Seabed Manganese Nodules," *North. Miner*, 63 (Apr. 14, 1977): B6, B9.
[421] Pastoors, W. C. B., "Pipeline Trenching in Rock," *World Dredging and Marine Construction*, 11 (Oct. 1975): 18, 20.
[422] Pearson, J. S., *Ocean Floor Mining* (Park Ridge, N.J.: Noyes Data, 1975).
[423] Peckford, A. B., Minister, Dept. of Mines and Energy, Govt. of Newf. and Lab., "Predictions for 1979," Dec. 22, 1978, p. 2. (mimeo.)
[424] Pepper, J. F., "Potential Mineral Resources of the Continental Shelf of the Western Hemisphere," in *An Introduction to the Geology and Mineral Resources of the*

Continental Shelves of the Americas, USGS Bul. 1067 (Wash., D.C.: GPO, 1958): 43–65.
[425] "Permanent Production Facilities at Ekofisk Nearly Complete," *North. Offshore*, 3 (Apr. 1974): 103.
[426] "Petrobas Racks up Two New Oil Fields off Brazil," *O&G J.*, 77 (Mar. 12, 1979): 40.
[427] "Petroleum Subs Work in the North Sea," *Work Boat*, 32 (Aug. 1975): 32.
[428] Pietrowski, R. F., Jr., "Hard Minerals on the Deep Ocean Floor: Implications for American Law and Policy," *William and Mary Law Rev.*, 19 (1977): 74.
[429] Piper, D. Z., "Rare Earth Elements in Manganese Nodules from the Pacific Ocean," in [241*]: 123–50.
[430] "Policy of the United States with Respect to the Natural Resources of the Subsoil and Seabed of the Continental Shelf," Proclamation No. 2667, Sept. 28, 1945.
[431] Pontecorvo, G. and M. Wilkinson, "From Cornucopia to Scarcity: The Current Status of Ocean Resource Use," *Ocean Develop. and Internat. Law*, 5 (1978): 392.
[432] Power, L. D. et al., "Tests Recognize Guyed Tower Potential," *Offshore*, 38 (May 1978): 221.
[433] Pratt, R. M. and P. F. McFarlin, "Manganese Pavements on the Blake Plateau," *Science*, 151 (Mar. 4, 1966): 1080–82.
[434] "Private Firms Slate Tests off Viet Nam," *O&G J.*, 77 (Jan. 8, 1979): 28–29.
[435] "Progress in Offshore Development and Production," *Pet. Rev.*, 31 (Apr. 1977): 15–16.
[436] "Prospecting and Exploration Techniques for Ocean Resource Development," *Min. Eng.*, 27 (Apr. 1975): 36, 38, 43.
[437] Prow, W., "Mn–Resource or Rancor?" paper prepared for the annual meeting of the Canadian Geographers Assoc., London, Ont., May 23–27, 1978, pp. 47–48. (mimeo.)
[438] Psuty, N., Dir. Marine Sci. Ctr., Rutgers Univ., testimony before the Office Tech. Assess., Trenton, N.J., Feb. 14, 1975.
[439] Raab, W., "Physical and Chemical Features of Pacific Deep Sea Manganese Nodules and Their Implications to the Genesis of Nodules," in [241*]: 31–49.
[440] Rahmer, B. A., "Offshore Prospects and Problems in the U.S.S.R.," *Pet. Econ.*, 44 (May 1977): 191–92.
[441] Rasmussen, J. W., "China to Initiate Offshore Oil Exploration," *Ocean Ind.*, 10 (Aug. 1975): 142.
[442] Raymond, N., "Sea Law: The Unpleasant Options," *Ocean World*, 1 (Jan. 1978): 28.
[443] Raymond, R. C., "Seabed Minerals and the U.S. Economy: A Second Look," *Marine Tech. Soc. J.*, 10 (June 1976): 14.
[444] Read, E. C. K., Asst. Vice Pres. and Dir. of Public Relations, Freeport Minerals Co., N.Y., N.Y., pers. com., Jan. 8 and 26, 1979.
[445] Read, E. C. K., Asst. Vice Pres. and Dir. of Public Relations, Freeport Minerals Co., N.Y., N.Y.; calculated by the author from data provided by pers. com., Dec. 13, 1978.
[446] Reed, A. H., Branch of Cement and Aggregates, Div. of Nonmetallic Minerals, Bu. Min., Wash., D.C., pers. com., Oct. 31, 1978.
[447] Reimnitz, E. and G. Plafker, *Marine Gold Placers along the Gulf of Alaska Margin*, USGS Bul. 1415 (Wash., D.C.: GPO, 1976): 12–13.
[448] Reynolds, J. M., "Mining Wealth from the Sea Floor," *Geog. Mag.*, 51 (Oct. 1978): 23.
[449] Richardson, R. M. and S. C. Solomon, "Intraplate Stress as an Indicator of Plate Tectonic Driving Forces," *J. of Geophys. Res.*, 81 (1976): 1847–56.
[450] Riley, J. G., "How Imperial Built First Arctic Island," *Pet. Eng. Internat.*, 46 (Jan. 1974): 25–28.
[451] Riley, J. G., "The Construction of Artificial Islands in the Beaufort Sea," *J. of Pet. Tech.*, 28 (Apr. 1976): 366.
[452] Rintoul, B., "Cook Inlet Sale Abounds with Support," *Offshore*, 37 (Dec. 1977): 119 ff.
[453] Rintoul, B., "Ocean Minerals Outlines Mining Plans," *Offshore*, 88 (Oct. 1978): 132.
[454] Rintoul, B., "West Coast Highlights," *Offshore*, 37 (June 20, 1977): 127.

[455] Robertson, R., "Countries Take Control of Oil and Gas," *Offshore*, 38 (June 20, 1978): 175.
[456] Robinson, B., "Satellite Could Speed Data," *Engineer*, 238 (Jan. 24, 1974): 11.
[457] Roels, O. A. et al., *The Environmental Impact of Deep Sea Mining: Progress Report*, Technical Rept. ERL 290-0011 (Boulder, Colo.: NOAA, 1973).
[458] Rona, P. A., "Criteria for Recognition of Hydrothermal Mineral Deposits in Oceanic Crust," *Econ. Geol.*, 73 (Mar.-Apr. 1978): 135-60.
[459] Rona, P. A., "Pattern of Hydrothermal Mineral Deposition: Mid-Atlantic Crest at Latitude 26°N.," *Marine Geol.*, 21 (Aug. 1976): M59-M66.
[460] Rona, P. A., "Plate Tectonics and Mineral Resources," *Sci. Amer.*, 229 (July 1973): 86, 90, 92.
[461] Rona, P. A., "Plate Tectonics: Energy and Mineral Resources: Basic Research Leading to Payoff," *EOS*, Amer. Geophys. Union, 58 (Aug. 1977): 636.
[462] Rona, P. A. and L. D. Neuman, "Energy and Mineral Resources of the Pacific Region in Light of Plate Tectonics," *Ocean Mgt.*, 3 (1976): 57.
[463] Ross, J. M., "Sea Spider: An Ocean-floor Oil Wellhead Monitoring and Maintenance System," *Marine Tech. Soc. J.*, 10 (Oct. 1976): 26.
[464] Rothstein, A. J. and R. Kaufman, "The Approaching Maturity of Deep Ocean Mining— The Pace Quickens," *Min. Eng.*, 26 (Apr. 1974): 33.
[465] Royal Ministry of Industry and Handicraft, *Legislation Concerning the Norwegian Continental Shelf with Unofficial English Translation*, 5th ed. (Oslo: Staten Oljedirektorat, 1977): 99-164, 189-273.
[466] "Russia to Double Caspian Oil Flow by 1985," *O&G J.*, 73 (May 26, 1975): 140.
[467] Sacerdoti, G., "Indonesian Oil Hopes Revive," *Far. E. Econ. Rev.*, 96 (Apr. 29, 1977): 42-43.
[468] Sano, S., Chief, Overseas Geology Branch, Geol. Survey of Japan, Tokyo, pers. com., Mar. 30, 1979.
[469] Scholle, P. A., ed., *Geological Studies on the COST No. B-2 Well, U.S. Mid-Atlantic Outer Continental Shelf Area*, USGS Circ. 750 (Arlington, Va.: USGS, 1977): 1, 5.
[470] Schoonover, K. W., "The History of Negotiations Concerning the System of Exploitation of the International Seabed," *J. of Internat. Law and Politics*, 9 (1977): 485-86, 488, 497, 503-08, 513.
[471] Sci. Pol. Res. Div., Cong. Res. Service, *Energy from the Ocean*, prepared for the Subcomm. on Advanced Energy Technologies and Energy Cons., Res., Develop., and Demonstration of the Comm. on Sci. and Tech., House of Rep., 95th Cong., 2d sess. (Wash., D.C.: GPO, 1978): 3, 23.
[472] Scott, R. W., "The Gulf of Suez: Where the Action Is in Egypt," *World Oil*, 182 (Jan. 1976): 75.
[473] "Sea Law Conference: Back to Geneva," *Ocean World*, 1 (Sept.-Oct. 1978): 18.
[474] "Seafloor Unit to Boost Campos Production," *Offshore*, 38 (May 1978): 188.
[475] "Seasat Mission Formally Ends," NASA News Release, Nov. 15, 1978, text provided by D. McCormack, Public Aff. Officer, Office of Space and Terrestrial Applications, NASA, Wash., D.C., pers. com., Jan. 29, 1979.
[476] Secy. of the Int., *Mining and Minerals Policy, 1977: Annual Report under the Mining and Minerals Policy Act of 1970* (Wash., D.C.: GPO, 1977): 67-69, 120, 122, 124-26, 130, 135, 140.
[477] Secy. of the Int., *Mining and Minerals Policy: 1976 Bicentennial Edition—Annual Report of the Secretary of the Interior under the Mining and Minerals Policy Act of 1970* (Wash., D.C.: GPO, 1976): 33.
[478] Senftleben, W., "Political Geography of the South China Sea," *Philippines Geog. J.*, 20 (Oct.-Dec. 1976): 163-75.
[479] Shapley, D., "Offshore Oil: Supreme Court Ruling Intensifies Debate," *Science*, 188 (Apr. 11, 1975): 135.
[480] Sharp, J. M., "How GURC Establishes a Control Profile of Offshore Environment," *Pet. Eng. Internat.*, 46 (Nov. 15, 1974): 20 ff.
[481] "Shetlands Well Proves Noncommercial," *O&G J.*, 76 (Sept. 4, 1978): 54.

[482] Shyam, M., "International Seabed Regime: An Empirical Analysis of State Preferences," *Cooperation & Conflict*, 11, No. 2 (1976): 116, 123-24.
[483] Silas, C. J., "Ekofisk Field Development Will Have Two Long Offshore Pipelines," *Pipeline & Gas J.*, 201 (Jan. 1974): 62.
[484] Sillitoe, R. H., "A Plate Tectonic Model for the Origin of Porphyry Copper Deposits," *Econ. Geol.*, 67 (Mar.-Apr. 1972): 184.
[485] Silverstein, D., "Proprietary Protection for Deepsea Mining Technology Transfer: New Approach to Seabeds Controversy," *J. of the Patent Office Soc.*, 60 (Mar. 1978): 143, 145-46, 169.
[486] Sisselman, R., "Ocean Miners Take Soundings on Legal Problems, Development Alternatives," *Eng. & Min. J.*, 176 (Apr. 1975): 79.
[487] Sitje, O., Deputy Dir. Gen., Royal Ministry of Pet. and Energy of Norway, Oslo, personal interview, Sept. 29, 1979.
[488] Skipper, K., "Offshore Petroleum Developments in Northwest Europe—An Update," *Geoscience Canada*, 4, No. 1 (1977): 31.
[489] Slack, J. R. and R. A. Smith, *An Oilspill Risk Analysis for the South Atlantic Outer Continental Shelf Lease Area*, USGS Open File Rept. 76-653 (Reston, Va.: USGS, 1976).
[490] Smith, H. D., "The Environment and the Offshore Oil Industry in Scottish Waters—A Review," Part 2, *Maritime Studies and Mgt.*, 3, No. 4 (1976): 208, 210-13, 217.
[491] Smith, H. D., A. Hogg and A. M. Hutcheson, "Scotland and Offshore Oil: The Developing Impact," *Scot. Geog. Mag.*, 92 (Sept. 1976): 80, 83, 86-87, 89-90.
[492] Smith, R. A., J. R. Slack and R. K. Davis, *An Oilspill Risk Analysis for the Mid-Atlantic Outer Continental Shelf Lease Area*, USGS Open File Rept. 76-451 (Reston, Va.: USGS, 1976): 1, 11-12, 16, 21.
[493] Smith, R. A., J. R. Slack and R. K. Davis, *An Oilspill Risk Analysis for the North Atlantic Outer Continental Shelf Lease Area*, USGS Open File Rept. 76-620 (Reston, Va.: USGS, 1976).
[494] Smith, W. J., "International Control of Deep Sea Mineral Resources," *Naval War College Rev.*, 24 (June 1972): 85.
[495] Snider, W. D. et al., "Management of Mid-Atlantic Offshore Development Risks," *Marine Tech.*, 14 (Oct. 1977): 336-37.
[496] "Solid Waste Disposal Offshore," *Pet. Eng. Internat.*, 45 (Jan. 1973): 88.
[497] Solvang, O., "Financing Required to Develop Norway's North Sea Sector," *Ocean Ind.*, 10 (Aug. 1975): 62.
[498] Sorem, R. K. and R. H. Fewkes, "Internal Characteristics," Chap. 6 in [202]: 181.
[499] Sorokin, Y. I., "Role of Biological Factors in the Sedimentation of Iron, Manganese, and Cobalt and in the Formation of Nodules," *Oceanology*, 12 (1972): 1-11, as cited in [80]: 85.
[500] "South Atlantic Sale Draws 99 Bids," *Offshore*, 38 (May 1978): 289-90.
[501] "Southeast Asia Recovers from Slump," *Offshore*, 38 (June 20, 1978): 165.
[502] "Spain's Supply May Not Meet Demand," *Offshore*, 38 (June 20, 1978): 202.
[503] Spencer, M. P., "Blood Bubble Detection Prevents Decompression Sickness," *Naval Res. Reviews*, 28 (June 1975): 18.
[504] Standish-White, D. W., "Diamonds in the Surf," *Compressed Air Mag.*, 77 (July 1972): 8-11.
[505] Stasaitis, L. J., Operations Mgr., Marcona Ocean Industries Div., Marcona Sales, Inc., Ft. Lauderdale, Fl., pers. com., Dec. 12 and 21, 1978.
[506] Statham, A., Vice Pres., Inco, Toronto, Ont., Can., telephone interview, Aug. 21, 1979.
[507] "Steady Pounding of North Sea Ages Platforms," *Ocean Ind.*, 10 (Aug. 1975): 70.
[508] Stephen-Hassard, Q. D. et al., *The Feasibility and Potential Impact of Manganese Nodule Processing in Hawaii* (Honolulu: Dept. of Planning and Econ. Develop., 1978): 2-6–2-7, 3-2–3-6, 3-9, 4-11–4-21, 5-1–5-2, 5-7.
[509] Sterling, G. H. and G. E. Strohbeck, "Failure of the South Pass 70 Platform B in Hurricane Camille," *J. of Pet. Tech.*, 27 (Mar. 1975): 263-68.
[510] Steven, R., "Stormy Politics Deter U.K. Drilling," *Offshore*, 39 (Feb. 1979): 61.

[511] Stevens, J. F., "Mining the Alaskan Seas," *Ocean Ind.*, 5 (Nov. 1970): 47–49.
[512] Stivers, G. S., "Subsea Trees Eyed for Deep North Sea Waters," *Pet. Eng. Internat.*, 46 (Oct. 1974): 60.
[513] Stowasser, W. F., Div. of Nonmetallic Minerals, Bu. Min., Wash., D.C., pers. com., Feb. 16, 1979.
[514] "Strongest, Lightest Mini-Sub Operates in North Sea," *British Rec.*, No. 12 (Sept. 1977): 6.
[515] Stuart, D., "North Sea Blowouts," *Best's Rev.: Property/Liability Insurance*, 78 (Dec. 1977): 64.
[516] "Sub-bottom Profiling System for Shallow Water," *Pet. Eng. Internat*, 46 (Sept. 1976): 20.
[517] Sujitno, S., "Some Notes of Offshore Exploration for Tin in Indonesia, 1966–1976," in [161]: 169–82.
[518] "Sullom Voe Project Has Long Way to Go," *Offshore*, 39 (Feb. 1979): 84.
[519] *Sulphur: Ally of Agriculture and Industry* (New Orleans, La.: Freeport Sulphur Co., 1978): 7, 10–15.
[520] Sumpter, R., "Acreage in Gulf Draws Spirited Bidding," *O&G J.*, 76 (Dec. 25, 1978): 76.
[521] Sumpter, R., "Why U.S. Independents Aren't Rushing Offshore," *O&G J.*, 77 (Mar. 5, 1979): 67–70.
[522] *Supreme Court Reporter*, 95, p. 1155.
[523] Surveys and Mapping Branch, Dept. of Energy, Mines and Res., *Surveying Offshore Canada Lands for Mineral Resources Development*, 2d ed. (Ottawa: Minister of Oil and Services, 1975): 3.
[524] Swan, G. S., "The Newfoundland Offshore Claims: Interface of Constitutional Federation and International Law," *McGill Law J.*, 22 (Winter 1976): 541–73.
[525] "Tax Policy Revisions Proposed in U.K.," *Offshore*, 38 (Oct. 1978): 60.
[526] Taylor, D. M., "North Sea Report Overview," *Ocean Ind.*, 8 (Feb. 1973): 61.
[527] "Teng's Triumphant Tour," *Time*, 113 (Feb. 12, 1979): 16.
[528] "Texaco Logs First Discovery off East Coast of U.S.," *O&G J.*, 76 (Aug. 21, 1978): 32–33.
[529] "Three North Sea Sectors Enjoying High Point in Offshore Development," *Offshore*, 38 (Oct. 1978): 64–65.
[530] Tinsley, C. R., "Activities and Economics of Existing Nodule Mining Consortia," in *Oceans' 78: The Ocean Challenge* (Wash., D.C.: Marine Tech. Soc. and Inst. of Electrical and Electronics Eng., 1978): 602–05.
[531] Tinsley, C. R., "Economics of Deep Ocean Resources: A Question of Manganese or Non-Manganese," *Min. Eng.*, 27 (Apr. 1975): 33.
[532] Tinsley, C. R., Second Vice Pres., Continental Bank, Chicago, Ill., pers. com., Aug. 26, 1979.
[533] Tinsley, C. R., "The Future Markets for Nodule Metals," *Proc.*, AIME/MMIJ Conf., Denver, Colo., Sept. 1976.
[534] "Transfer Capsule: A Giant Step in Effecting Diver Safety Methods," *Offshore*, 37 (Nov. 1977): 136–38.
[535] "Two Baltimore Canyon Tests to Be Abandoned," *O&G J.*, 76 (Dec. 18, 1978): 34.
[536] "Two Specialized Offshore Devices Developed by Norwegian Firms," *Marine Eng./Log*, 65 (Apr. 1977): 100–01.
[537] "U.K. Cuts Offshore Reserve Estimate," *O&G J.*, 76 (Dec. 11, 1978): 46.
[538] "U.K. Government Consults Firms on PRT," *O&G J.*, 77 (Mar. 5, 1979): 88.
[539] "U.K. Norway Join Forces for Emergencies," *Offshore*, 38 (Oct. 1978): 58–60.
[540] "U.K. to Hit Report Opposing N. Sea Gas Network," *O&G J.*, 76 (July 24, 1978): 30–31.
[541] "UN A/Conf. 62/25 (May 22, 1974)," as reported in [311]: 161.
[542] U.N. Comm. on the Peaceful Uses of the Sea-bed and the Ocean Floor beyond the Limits of Nat. Jurisdiction, *Economic Significance, in Terms of Seabed Mineral*

Resources, of the Various Limits Proposed for National Jurisdiction: Report of the Secretary General (A/AC. 138/87), June 4, 1973, pp. 29-30, 32.
[543] UNCTAD, *The Effects of Possible Exploitation of the Seabed on the Earnings of Developing Countries from Copper Exports* (TD/B/484), May 28, 1974, p. 7.
[544] *U.S. Reports*, **420**, pp. 515-28.
[545] "U.S. to U.N.' Stop Pushing Us Around," *Forbes*, **122** (Sept. 18, 1978): 47.
[546] "U.S. Urged to Spur Atlantic Hunt via More Leases," *O&G J.*, **76** (July 3, 1978): 49.
[547] USBM (Bu. Min.), *Mineral Commodity Summaries, 1978* (Wash., D.C.: Dept. Int.): 42-43, 47, 113;* *Mineral Commodity Summaries, 1979* (Wash., D.C.: Dept. Int.): 38-39, 42-43, 94-95, 104-05.**
[548] "The Ubiquitous SPM," *Surveyor*, **9** (Aug. 1975): 4-5.
[549] *The Uncontrolled Blowout: The Ekofisk Field (The Bravo Platform) 22 April, 1977*, Storting Rept. No. 65, 1977-78, Oslo, Norway.
[550] *United Kingdom Prevention of Oil Pollution Act of 1971*, section c(1), as quoted in [285]: 98-99.
[551] "United Kingdom: Report on Offshore Oil and Gas Policy," *Internat. Legal Materials*, **14** (Mar. 1975): 460-61.
[552] "Update on Offshore Mining: The Unheralded Mineral Producer," *Min. Eng.*, **27** (Apr. 1975): 42-44.
[553] Urabe, T. and T. Sato, "Kuroko Deposits of the Kosaka Mine, Northeast Honshu, Japan—Products of Submarine Hot Springs on Miocene Sea Floor," *Econ. Geol.*, **73** (Mar.-Apr. 1978): 161-79.
[554] Van Meurs, A. P. H., *Petroleum Economics and Offshore Mining Legislation: Geological Evaluation* (Amsterdam: Elsevier Pub., 1971): 10-20.
[555] Vernon, J. W. and L. D. Furse, "Working Small Submersibles," *Oceanology: Internat. Offshore Tech.*, **7** (Apr. 1972): 60-62.
[556] Vestdal, J., "Iceland Relies on Sea for New Cement Industry," *Pit and Quarry*, **53** (Feb. 1961): 82-91.
[557] Vielvoye, R., "Conservatives Move to Split and Trim BNOC," *O&G J.*, **77** (July 30, 1979): 118.
[558] Vielvoye, R., "Phillips Oil Find Revives Irish Hopes," *O&G J.*, **76** (Oct. 9, 1978): 32.
[559] Vielvoye, R., "Problems Mount at Sullom Voe," *O&G J.*, **77** (Mar. 19, 1979): 59.
[560] Vielvoye, R., "Royalty Relief Awaits Test," *O&G J.*, **76** (Dec. 18, 1978): 35.
[561] Vinogradova, N. G., "Some Problems of the Study of Deep Sea Bottom Fauna," *J. of the Oceanographic Soc. of Japan*, 20th Anniversary Volume (1962): 725-46.
[562] Wakefield, S. A., "Atlantic Outer Continental Shelf," *Amer. Gas Assoc. Mo.*, **55** (Sept. 1973): 11.
[563] Wallenberg, B., "The Ekofisk Accident: Its Environmental and Political Implications," *OECD Observer*, No. 88 (Sept. 1977): 9-11.
[564] Wang, K. P., *Mineral Industry of the People's Republic of China* (Pittsburgh, Pa.: Bu. Min., 1978): 6, reprinted from "Mining Annual Review 1978," *Min. J.* (June 1978): 420 ff.
[565] Wargo, J. G., "Cobalt—A Supply Crisis," *Min. Cong. J.*, **65** (Apr. 1979): 38-39.
[566] Watson, T. N., "Scour in the North Sea," *J. of Pet. Tech.*, **26** (Mar. 1974): 290-91.
[567] Watt, D. C., "Britain and North Sea Oil: Policies Past and Present," *Pol. Qrt.*, **47** (Oct.-Dec., 1976): 378-79, 382-83.
[568] Wearne, C. E. C., Mgr. of Operations, King Island Scheelite Pty., Grassy, Tas., Aust., pers. com., Apr. 5, 1979.
[569] Weeks, L. G., "Petroleum Resources Potential of Continental Margins," in [65]: 953-64.
[570] "Wellhead Remote Control Unit Takes Subsea Technology Another Step Forward," *Offshore*, **38** (June 1978): 68-70.
[571] Welling, C., "1979 Forecast Ocean Mining," *Offshore*, **38** (Dec. 1978): 70-71.
[572] Welling, C. G., testimony in *Mining of the Deep Seabed*, Joint Hearings before the Subcomm. on Public Lands and Res. of the Comm. on Energy and Natural Res. and the

Comm. on Commerce, Sci., and Transportation, Sen., 95th Cong., 1st sess., on S. 2053 (Wash., D.C.: GPO, 1978): 130, 139–42.
[573] Wenk, E., Jr., "Oceans and the Predicament of Humankind," in [581] : 17.
[574] "We're a Lot Deeper Than Just a Diving Companion," *Offshore*, 38 (May 1978): 172.
[575] Werenskiold, K., "The Condeep Project Group: Dedicated to Dealing with Maritime Problems, Plans," *Offshore*, 37 (Nov. 1977): 170.
[576] *West's Federal Practice Digest*, 2d., "Navigable Waters," 60, p. 631.
[577] Whebell, C. F. J., Dept. of Geography, Univ. of Western Ont., Canada.
[578] White, I. L., *North Sea Oil and Gas: Implications for Future United States Development* (Norman: Univ. of Okla. Press, 1973): 12, 15 16, 26–27, 31–42.
[579] Whitney, S. C., "Environmental Regulations of United States Deep Seabed Mining," *William and Mary Law Rev.*, 19 (Fall 1977): 77-97.
[580] Williams, L. W., Acting Dir., Bureau of Mineral Res., Geology, and Geophysics, Canberra, Aust., pers. com., Oct. 27, 1978.
[581] Wilmot, P. D. and A. Slingerland, eds., *Technology Assessment and the Oceans: Proceedings of the International Conference on Technology Assessment, Monaco, 26-30 October, 1975* (Boulder, Colo.: Westview Press, 1977).
[582] Wilson, H. M., "Interior Pressing Action in Santa Barbara Channel," *O&G J.*, 76 (June 26, 1978): 47-51.
[583] Wold, K. G., "Economic Situation," *Norges Bank Econ. Bul.*, 50 (Mar. 1979): 8, 12.
[584] Wold, K. G., "North Sea Oil and its Impact on Norway's Future Energy," *Norges Bank Econ. Bul.*, 45 (Dec. 1974): 173, 177-80.
[585] "'Workhorse' Awaits Finishing Touches," *Offshore*, 37 (Dec. 1977): 11.
[586] "The World Offshore," *Offshore*, 38 (Feb. 1978): 102-03.
[587] "The World Offshore," *Offshore*, 37 (Oct. 1977): 82.
[588] "Worldwide Production," *O&G J.*, 76 (Dec. 25, 1978): 98-148.
[589] Wright, R. L., *Ocean Mining: An Economic Evaluation*, professional staff study, Ocean Min. Adm., Dept. Int., May 1976, in [129] : 95-96.
[590] Wu, P., Dir., Dept. of Mines, Ministry of Econ. Affairs, Taipei, Taiwan, pers. com., May 28, 1979.
[591] Young, G., "Everything's Coming Up Diamonds," *Popular Mechanics*, 133 (Mar. 1970): 118-19, 211, 216.
[592] Zumberge, J. H., "Mineral Resources and Geopolitics in Antarctica," *Amer. Scientist*, 67 (Jan.-Feb. 1979): 73-74.

Index

ARMS bell and life support system, 118; see also submersible vehicles
Abu Dhabi, 54-55
Abyssal plains, 176
Adriatic Sea, 54
Aegean Sea, 54
Afernod (French Association for Nodule Exploration), 194, 196
Africa, 54, 98
Aggregates dredging, 24
Ahern, W. R., 108
Alabama, 20
Alaska, 19, 27, 29, 43, 60; Beaufort Basin and Sea, 41, 43, 60, 98, 159-61, (map) 44; Cook Inlet, 111; Goodnews Bay, 27; Gulf of Alaska, 27; Harrison Bay, 160; Kodiak Basin, 98; Platinum Bay, 27; Prudhoe Bay, 43, 160
Aluminum, 182, 230
Amacher, R. C., 240
American Mining Congress, 255
Amos, A. F., 22
Andrus, C., 100, 107
Angola, 53, 58, 235; Loanga oil field
Antarctica, petroleum potential, 41
Atimony, 172
Arab oil embargo, 63, 69, 101-02, 104
Arabian Gulf, 55

Aragonite, 20-22
Arctic Ocean, 56, 160, 175, 189
Argentina, petroleum potential, 58-60
Artificial sea-ice islands, 160
Asbestos, 239
Asia, 45-53, 172
Atlantic City, 109
Atlantic Ocean, 177, 182-83, 189
Australia, 51-52, 252; Bass Strait oil field, 52; Arafura Sea, 52; cobalt, 234; Cockatoo Island iron ore, 12; Exmouth Plateau, 52; manganese, 237; Rankin Trend, 52; scheelite mining, 15
Ayatollah Khomeini, 55

BRGM, 171
Bahamas, 20-22; Bimini Islands, 21; Ocean Cay, 21
Baja California, 26
Baldwin, M. F., 108
Baldwin, P. L., 108
Baltic Sea, 57
Baltimore Canyon, 104-05, 108-09, 146, 158; map, 107
Bangka Island, Indonesia, 19
Barite, 182; Castle Island, 19
Basalts, 185-86

283

284 Index

Battelle Institute, 116-17, 197; Northwest Laboratories, 204
Bay of Biscay, 54
Belgium, 24, 194, 233, 246
Belle Island, iron ore, 10, 12
Benim, 54
Benn, A., 73
Benthic organism destruction, 219-20
Berryhill, H., 37, 108
Bethlehem Steel Corp., 197
Birnessite, 180
Blake Plateau, 175, 177, 196, 211, 221
Blissenbach, E., 193, 230
Bonn Agreement, 153
Borax, 239
Borgny Dolphin, 47
Botswana, 246
Bottom sampling, 115-17
Brazil, 252; manganese, 237; oil fields and reserves, 58, 128; map, 59
Brewer, P. G., 188
Bristol Channel, 24
British National Oil Corp., 75, 134; *see also* United Kingdom
Broecker, W. C., 186
Broken Hill Proprietary Co., Ltd., 13
Brown & Root, Inc., 135
Bulgaria, 246
Burma, 51
Burns, V. M., 180
Burton, S. J., 257

CLB Group, 194, 206
CNEXO, 206
Calcareous algae and shells, 20
Calcite, 182
Calcium, 182
California, 25-26, 29-30, 92-93; *Coastal Act of 1976*, 30-31; Santa Barbara Channel, 100, 145; State Land Commission, 30; Tanner Banks, 97
Californium-252, 204
Cambodia, 51
Cameroon, petroleum production, 53
Canada, 33, 43, 98, 194, 235, 240, 252; British Columbia, 41; Cape Breton Island coal, 13; Labrador, 159; Mackenzie River Delta, 160; nickel resources, 229-30; Nova Scotia, 25, 41-43; oil concession by Vietnam, 49; provinces vs. Ottawa in petroleum resources dispute, 44-45; Supreme Court, 44-45, 60
Cape May, New Jersey, 109
Cartels, 239-40
Carter, J., 96, 256
Celtic Sea, 65, 67

Center for Law and Social Policy, 255
Central America, 177, 219
Centre National pour l'Exploitation des Oceans, 194
Chalcopyrite, 167
Charles Anthony Diamond Investment Co., 23
Chicago Bridge & Iron Co., 140
Chile, 58, 231, 233, 238, 244
Chinese Petroleum Corp., 50
Chromium, 188, 229, 234
Church, F., 197
Clarion and Clipperton Fracture Zones, 183
Coal, 12-15
Coastal Zone Management Act of 1972, 29, 31
Cobalt, 182-83, 185, 187-88, 191-95, 200, 204, 213-15, 222, 227, 233-36, 238, 249; statistics, 234-35, 248
Colombia, 58, 230
Columbium, 229
Condeep platform, 140
Congo, 53
Connecticut, 28, 43
Conprod platform, 140
Consolidated Diamond Mines, 23
Continental Offshore Stratigraphic Test (COST) Group, 105
Continuous line bucket mining system, 194, 206-08, 210-11, 220
Convention on the Continential Shelf, 244
Convergence Zones, 165
Copper, 167-68, 170, 172, 182-83, 185, 187-88, 191, 195, 200, 204, 213-15, 222, 227, 231-33, 238, 240, 248-49; statistics, 233
Copper sulphate, 222
Coral production, 22
Corsica, 22
Counseil Intergouvernemental de Pays Exportateurs de Cuivre, 240
Crawford Marine Specialists, Inc., 171
Crommelin, M., 94
Cronan, D. S., 170
Cuba, 177
Cuprion processing of manganese nodules, 195
Cyprus, 167

Dames & Moore, 192, 211, 214, 216
Dansville, Rhode Island, 111
Darman, R. C., 252
Davis Strait, 41, 159
DeBeers, 23

Declaration of Principles Governing the Sea-Bed and the Ocean Floor and the Subsoil Thereof, Beyond the Limits of National Jurisdiction, 245
Deep Sea Driller, 155
Deep Ocean Mining Environmental Study (DOMES II), 219
Deep seabed mining consortia, 194-97; ownership statistics, 195; R&D investment statistics, 198
Deepsea Miner, 221
Deepsea Ventures, Inc., 194, 196-97, 200, 211, 214, 237-38, 257
Delaware, 107
Denmark, petroleum resources, 24, 41, 64-65, 67-68, 70
Destin Dome, 97
Diamonds, 22
Diapirs, 34, 107
Directional drilling, 126
Divergence zones, 165-71
Diving, 119-22; Donald Duck speech, 120; JIM, 120; nitrogen narcosis
Dolan, E., 20
Dominican Republic, 230
Dominion Coal Co. of Canada, 13
Douglas-Home, A., 72
Dredging, aragonite, 20; barite, 19; calcareous shells, 20; diamonds, 22-23; iron sands, 18; *see also* hard minerals
Dubs, M. A., 205
Drilling rigs and platforms, jackups, 122; semisubmersibles, 122; drill ships, 122-23
Dubai, 140

EIC Corp., 192, 211, 214-16
East China Sea, 49-50, 60; map, 46
East Pacific Rise, 172
Eastern Gulf petroleum lease sale, 97
Economic Commission of Europe, 153
Ecuador, 244
Egypt, petroleum resources, 54-55
Ehrlich, H. L., 187
Eire, petroleum exploration, 65, 67, 88
Ekofisk, 65, 67, 79, 86, 127, 134, 137; blowout, 151-53; producing complex, 70; cartogram of complex, 71
Ekofisk-Emden pipeline, 138
Emery, K. O., 166-67, 171
Emirate of Ras al-Khaimah, 55
Environmental Defense Fund, 255
Environmental impact statements, 146
Equatorial Pacific, 200, 218, 221, 257
European Economic Community, 63, 88, 153
European Space Research Organization, 156

Falkland Islands, 58, 60
Falconbridge Nickel, 235
Feldspars, 182
Ferric hydroxides, 167
Ferric oxides, 167
Fiji, 20
Finland, 12, 233
Foraminifera and nodule formation, 187
France, 22, 153, 194, 237, 252; Bay of Biscay and petroleum exploration, 54
Frasch mining, 31
Frazer, J. Z., 179, 200
Freeport Sulphur Co., 21
Frictional drag on platforms, 158

Gabon, petroleum production, 53; manganese, 237
Galveston Bay, Texas, 20
Garrand, L. J., 26
Geevor tin mine, 15
Geneva Convention on the High Seas, 143
Geological exploration on the U.S.' eastern seaboard, 105
Geomarine Corp., 171
Georges Bank, 42-43, 98, 104-05, 108-09
Geothite, 180
Ghana, 54, 238
Glassner, M., 245, 254
Glomar Explorer, 197
Gold, 168
Goodnews Bay Mining Co., 27
Grand Isle sulphur mine, 16
Granites, 185-86
Greece, 54
Greenland, 41, 67, 158
Grigalunas, T. A., 108-10
Grotius, H., 243
Group of 77, 246-47, 250, 258
Gulf of Finland, 12; Gabes, 54; Mexico, 92-93, 97-98, 108, 126, 156; St. Lawrence, 44; Siam, 51; Suez, 54; Tonkin, 47, 49, 60
Gulf Universities Research Consortium, 161

H.M.S. *Challenger*, 175
Hard minerals dredging regulations, OCS, 28-29; state territorial waters, 29-31
Harris regional forecasting model, 109-10
Harrison, S. S., 48-49
Hawaiian Islands, 183, 196, 218-24; Hawaii, 177; Hilo, 222; Keaau, 222-23; Maui, 22
Hogg, A., 83
Hollings, E. F., 100
Hong Kong, 24, 52
Hot spots, 166
Hottentot Bay, 22

286 Index

Hughes, H., 197
Hurricane Camille, 158
Hutcheson, A. M., 83
Hydrometallurgy, 213-14

Ice floes and icebergs, 159
Iceland, 20, 67, 153
Ikeshima coal mine, 13
Illite, 182
India, 33, 52, 238, 252; Bengal Basin, 52; Bombay High oil field, 52-53; Kerala, 53; Kutch Basin, 52; Mahanadi Basin, 53; Oil and Natural Gas Commission, 53
Indian Ocean, 52-53; Ridge, 167
Indonesia, 20, 49, 51
Informal Composite Negotiating Text, 252
Interagency Law of the Sea Task Force, 255
International Monetary Fund, 238
International Nickel Co. of Canada (Inco), 229-30, 240
International Seabed Area, 246-47
International Seabed Resource Authority, 245, 249-51
Ionian Sea, 54
Iran, 55, 60
Ireland, 65, 153
Irish Sea, petroleum exploration, 65, 67
Iron ore, 10, 12, 234
Ivory Coast, 54
Izzak Walton League of America, 255

Japan, 24, 31, 47, 52, 57, 194, 237-38, 252; Aga, 51; Ariake Bay, 13; Hokkaido, 13; Honshu, 18; Kyushu, 12; Tokyo Bay, 18
John Darling coal mine, 13
Johnson, D. B., 251
Johnston, J. L., 258
Jussaro Island, Finland, 12; *see also* iron ore

Kaufman, A., 26
Kaufman, R., 192, 199
Kennecott Consortium, 194-95
Kennecott Copper Corp., 194-95, 198, 205, 214, 240
Kennecott Exploration Corp., 187
Keto, D. B., 72, 88
Khazzan Dubai I, II, III oil storage units, 140-41
Kinsale Head gas field, 65
Kissinger, H., 50, 248
Kleppe, T. S., 108
Krasov, D. D., 171
Krutein, M., 209
Kuroko metallic deposits, 172
Kushiro coal mine, 13

Lamont Institute, 58
Lamont-Doherty Geological Observatory, 186
Landfill islands, Immerk, 160-61; map, 160
La Que, F. L., 250
Latin America, 58-59
Lead, 168, 172, 183, 188, 234, 239
Le Blanc, L., 47
Leipziger, D. M., 247
Levant tin mine, 15
Lewes, Delaware, 109
Libya, 54
Limestone, 10
Line Islands, 177
Lockheed Corp., 196
Lockheed Missiles and Space Co., 26
Logue, D. E., 251
London Reefs, 49
Louisiana, 37, 91-93, 156; Creole oil field, 92
Louisiana-Texas petroleum lease sale, 98
Luxembourg, 233

Mabon, J. D., 75
Madagascar, 52, 179
Magmatic segregation, 166-67
Magnesium, 182
Maine, 43
Malta, 244
Manganese metal, 183, 185-88, 191, 204, 213-15, 227-28, 236-38, 248-49; statistics, 236
Manganese nodule mining, site characteristics, 180, 182, 199; environmental relationships, 217-24; exploitability, 199-204; gathering systems, 205-09; global estimator approach, 200; grid estimator approach, 200; lifting systems, 210-11; mining costs, 211-12, 223, statistics, 212; potential revenues and profits, 192, 215-18; processing costs, 213, 223, statistics, 215-16
Manganese nodule, coring programs, 179-80; distribution, 175-80, 185-89; nuclei, 185; onland processing, 218, pavements, 175-76; transportation, 217-19, cost statistics, 223
Manganite, 187
Manganous manganese, 167
Malaysia, 121
Marble, 239
Marcona Conveyor, 22
Marcona Ocean Industries, 21
Marcos, F., 49
Marine Associates, 51
Marine Diamond Corp., 23

Maryland, 107
Massachusetts, 108
Massachusetts Institute of Technology, 192
Matsunaga, S. M., 256
Mediterranean Sea, 58, 172
Mercury, 10, 239
Mero, J., 12, 25-26, 188, 191-92, 199-200, 204, 208, 213, 219, 239-40
Metallgesellschaft AG, 194
Metalliferous muds, 168
Mexico, 231; Bay of Campeche, 58-59
Meyerhoff, A. A., 45
Middle East oil fields, 37, 54-56, 74; Fateh, 140; July, 54; Morgan, 54; Ramadan, 54; Safinaya, 54; production statistics, 55
Mid-oceanic ridge, 167
Mitsui Mining Co., Ltd., 13
Moller, A. P., 70, 72
Molybdenum, 182, 195, 214, 229, 239
Moncrieff, A. G., 192-93
Montmorillonite, 182
Moore, J. R., 27
Moores, F., 45
Morocco, 22, 54, 238
Mudge, J. L., 247
Murray, J. W., 188

Nanshan Archipelago, 48
Naples, 22
Natural hazards and the working environment, storms, frictional drag, currents and scour, 155-59
Naval Research Laboratory, 197
Nepal, 246, 254
Netherlands, 24, 64-65, 153, 194, 199, 246; natural gas, 67; see also North Sea oil and gas fields
Neto, M., 58
Neuman, L. D., 172
New Caledonia, nickel, 230; cobalt, 233, 240
New England, 98, 109-10
New Jersey, 107
New York, 107; Harbor, 25; Long Island, 25, 43
New Zealand, 252; Maui gas/condensate field, 52; Taranaki Bight, 52
Newfoundland, offshore claims, 44-45; iceberg hazards, 159
Newport News Shipbuilding and Drydock Co., 194
Nickel, 182-83, 185, 187-88, 191, 195, 200, 204, 213-15, 222, 227-31, 234, 236, 238, 240, 248; statistics, 228, 231-32
Nigeria, 53, 58
Nigrelli, V. J., 236, 252, 254

North Atlantic, 177
North Equatorial Current, 221
North Korea, 50
North Sea oil and gas fields, 65, 67, 69, 73-74, 134-39, 151-58; discoveries, map, 66; diving accidents, 120; Ekofisk blowout, 151-53; exploration, map, 68; gas flaring, 134-35, 140; pipelines, map, 134; weather conditions, 155-57; individual fields, Auk, Argyll, Beatrice, Dunlin, Fulmar, Maureen, Montrose, Murchison, Ninian, Piper, South Cormorant, Thistle,˜73; Beryl, 156-57; Brent, 73, 84, 119, 140; Buchan, 69; Cormorant, 69, 135; East Frigg, Heimdal, Odin, Sleipner, 65; Forties, 69, 134; Frigg, 65, 138; Heather, 69; Hutton, 86; Magnus, 65, 73; North Cormorant, 75; see also Ekofisk
North Pacific Ocean, 197
Northeast Channel, territorial dispute—Canada and U.S., 42-43, 60
Norway, 33, 64, 67, 86, 153-55, 252; concessions limitations, 72; Den Norsk Credit Bank, 76; environmental policies, 153-54; investment, statistics, 77; leasing policies, 80; Norsk Hydro, 79; ownership structure of oil industry, 78-79; 62nd parallel, 79-80, 88; Storting, 79; taxes and fixed capital formation, statistics, 78; use policies, 79-80
Norwegian oil port/service centers, Bergen, 68, 87, 138; Karmoy, 138; Mongstad, 87; Sotra, 68, 138; Stavanger, 87, 138, 157; Tananger, 135
Norwegian Trench, 135, 138, 140
Nuclear probes, 116-17, 204

Obduction, 172
Ocean Management, Inc., 194, 196, 198, 220
Ocean Minerals Co., 26, 194, 196-98
Ocean Mining Associates, 194, 196
Odell, P., 63, 73
Offshore hard minerals production, 9-32; leasing regulations, 29; production, map, 11; tunneling, 10, 13-16; statistics, 12
Offshore hazards, human and industrial wastes, 144; blowouts, 150-53; oil spills, 144-54; U.S. oil spills, statistics, 145
Offshore petroleum companies, private, Arco, 100; Amerada, 75; Amoco, 49; Armco, 58; British Petroleum, 65, 67, 134; Chevron, 75, 100, 107; Cities Service, 49; Conoco, 50, 65; Deminex, 67, 72, 75; Dansk Undergrunds Consortium, 70; Duabi Petroleum, 140; Elf, 67; Esso,

67, 75; Exxon, 54, 97, 107; Gelsenberg and Veba, 72; Getty, 97; Gulf, 50, 75, 107, 158; Hamilton, 65; Hess, 75; Houston Oil and Minerals, 107; ICI and Santa Fe, 75; Imperial Oil Ltd. of Canada, 160; Kerr-McGee, 92; Marathon, 65; Mitchell Energy & Development, 98; Mobil, 48, 67, 75, 97, 107, 156; Murphy-Odeco, 75; Occidental Group, 75, 134; Pan Canadian Petroleum, 97; Petronord, 79; Phillips Petroleum Group, 65, 70, 79, 138; Piper Group, 74; SAGA Petroleum, 79; Shell, 48, 74, 97, 107, 158; Shell/Esso, 65, 75, 134; Sun, 97; Tenneco, 97, 107, 194, 197; Texaco, 65, 75, 105, 107; Texas Eastern, 75; Total, 134, 139; Trasco Exploration, 97; Union North Sea Group, 75; Union Oil, 160

Offshore petroleum companies, state owned, British National Oil Corp. (BNOC), 75; Pertamina (Indonesia), 51; Petrobalt (Soviet Union, Poland, East Germany), 57; Petrobas (Brazil), 58; Petroleos Mexicanos, 59; Petronas (Malaysia), 51; Petro-Vietnam, 49; Statoil (Norway), 79

Offshore petroleum, storage systems, 139--40; seismic exploration, 100; survival systems, 154-55; tunneling and potential production, 161; well fires and explosions, 151; production statistics, 39-40; world's distribution of giant fields, map, 38

Oil spills, 144-54; Monte Carlo simulation of, 146; Santa Barbara, 119; U.S. mid-Atlantic, maps, 147, 149-50; U.S. statistics, 145; *see also* Ekofisk and Offshore hazards

Okhotsk Sea, 57

Onshore impacts of offshore petroleum, Scotland, 81-86; oil industry associated facilities, map, 82; Norway, 86-87

Ophiolites, 167, 172

Orangemund, Namibia, 22

Organization for Economic Cooperation and Development, 63, 153

Organization of Petroleum Exporting Countries (OPEC), 1, 4, 37, 47, 52, 55, 60, 63, 88, 100, 111, 254

Orogenic belts, 35

Owen, R. M., 27, 161

Oxman, B. H., 252

Pacific Ocean, 179, 182-83, 189, 196
Palawan Island, 49; Trough, 49
Panama Canal, 218
Paracels (Hsi-sha), 48

Paraguay, 246
Pardo, A., 244
Pasho, D. W., 241
Pennsylvania, 109
Pelagic carbonate detritus, 177
People's Republic of China, 33, 45-51, 60; Gulf of Chihle, 47; Liaotung Bay, 47; Luichow Peninsula, 47; Po-Hai Gulf, 47; Vice Premier Teng Hsiao-p'ing, 47
Peroxide of manganese, 175
Persian Gulf, 54-55, 57
Peru, petroleum resources, 58; copper resources, 233, 244
Petroleum exploitation policies, Norway and United Kingdom, 80-81
Philippines, 18, 27, 172
Phosphates, 239
Phosphorites, 25, 175
Phuket Island, Thailand, 19
Pipelines, bottom towing, 135; North Sea, 132-35, 137-39; origin and terminal map, 134; maintenance, 138-39; scour problems, 138; United States, 133
Platinum, 234
Plate convergence, petroleum formation, 35
Port Sulphur, Louisiana, 16
Potassium-argon radioisotope dating, 185
Preusaag AG, 171
Prince Edward Island, 44
Production platforms, column stabilized submersibles, 126-28; fixed steel, 124-26; guyed tower, 127-28
Project Jennifer, 197
Pumping, 16
Pyrite, 167
Pyrometallurgy, 213-14
Pyrrhotite, 167

Qatar, 55
Quartz, 182
Quebec, 44

Raab, W., 187
Radio-carbon dating, 171
Raymond, R. C., 240
Red Sea, 54; divergence zone deeps and metalliferous sediments, 168-72
Reed Bank, 49
Rensselaer Polytechnic Institute, 187
Republic of South Africa, 177, 237
Richardson, E., 248
Rockall, United Kingdom, 65
Roels, D. A., 220-21
Romania, 54
Rona, P. A., 167, 172
Rosing, K., 63

Rothstein, A. H., 192
Rutile, 182

Saccorhiza, 187
Sakhalin, 57
Salt brine, 16
Sand and gravel, 23, 31
Sarawak, 51
Sardinia, 22
Saudi Arabia, 54-56
Sawhill, J. C., 55, 103
Scheelite, 15
Schlesinger, A., 45
Scotland, 13, 73, 108; Aberdeen, 83, 154; *Act of Union*, 73; Cromarty Firth, 81; Cruden Bay, 86; Department of Industry, 83; Drumbuie, 85-86; Firth of Clyde, 84; Firth of Forth, 84; Flotta petroleum terminal, 74; Hebrides, petroleum exploration, 65; Highlands and Islands Development Board, 86; Kishorn, 83; Lerwick, Shetlands, 83-84; Lyness, Orkneys, 83; Loch Kishorn, 84; Orkneys, 74, 83, 88; Peterhead, 83-84; Petroleum Industry Training Board, 83; St. Fergus, 86; Shetland County Council, 84; Shetlands, 65, 67, 83, 88; socioeconomic impacts of offshore petroleum, 81-84; spatial-environmental impacts of the oil industry, 84; Sullom Voe Association, 86; Sullom Voe oil terminal, 74, 86; Yell Sound, 138; *see also* United Kingdom
Scour, 158-59
Scripps Institution of Oceanography, 179, 200
Sea Gem, 156
Sea of Marmara, 54
Seabed exploration techniques, penetrometers, 114-15; seismic, sonar and magnetometer profiling, 113-14; trenching devices and methods, 137
Seasat, 159
Sedco/Phillips SS, 70
Sediment, data bank, 200; plumes, 220; sampling devices, 204
Sedimentary basins, 35
Selden, J., 243
Senegal, 54
Senkaku Islands (Tiao-yu T'ai), 47
Seychelles, 53
Sharjah, 55
Shyam, M., 249
Side-scan sonar, 204
Sierra Club, 255
Singapore, 251

Silicates, 170
Siliceous clays, 189; ooze, 179, 183, 209
Silicon, 182
Silver, 168, 239
Single point mooring systems, 139-40, 218
Skinner, B. J., 166-67
Smale-Adams, K. B., 192
Smith, H. D., 83
Societe LeNickel of New Caledonia, 229
Sodium, 182
Solent, aggregate dredging, 24
South Africa, 25; *see also* South-West Africa (Namibia)
South Atlantic, 177
South China Sea, 60; map, 46
South Korea, 50
South Vietnam, 48
Southeast Asia, 37, 98; archipelago petroleum frontiers, 51-52
Southeast Georgia Embayment sale, 96-97
South-West Africa (Namibia), 22
Soviet-Iranian Defense Agreement of 1921, 56
Soviet Union, 19, 27, 56, 230-31, 237, 252, 258; Andreyev Bank, 56; Black Sea, 54, 56; Caspian Sea, 41, 56, 58, map, 57; Golitsnaya Uplift, 56; Kara Sea, 56; Laptev Sea, 56; Sea of Azov, 56
Spain, 54
Spar I, 140
Spitzbergen, 257
Sport Fisheries Institute, 255
Spratlys (Nan-sha), 48
Sri Lanka, 20, 53
States vs. U.S. federal government on the continental shelf, 92-93
Statoil (Den Norske Stats Oljeselskap), 70, 79, 138; *see also* offshore petroleum producers and Norway
Stenlandet Island, Finland, 12
Storms, 155-58
Subduction and mineral formation, metals, 172; petroleum, 35-36
Submersible vehicles, 117-19
Subsea completion systems, 128-31; dry systems, 130
Sudan, 171
Sudanese Minerals, Ltd., 171
Sulphate ionizing bacteria, 186
Sulphur, 16
Sultanate of Oman, 55
Sumitomo Group, 194, 206
Summa Corp., 197
Sweden, 24, 153
Sweeney, R. J., 240
Switzerland, 199

Taiwan, 18, 31, 48, 51; Strait, 47, 49
Takashima coal mine, 13
Tankers, 133, 139-41, 144
Tasman Sea, 175
Tasmania, 15
Teng, H., 47
Texas, 92-93
Thailand, 19, 24, 51
Thames Estuary, 24
Tierra del Fuego, 58
Thermosbottle barges, 16
Tin, 15, 19
Tinsley, C. R., 193, 196, 198, 230
Timor Sea, 52
Titanium, 183, 229
Tizard Bank, 49
Todorokite, 180
Topographic surveying, 117
Translation zones, 165
Troodos Massif, 167, 172
Truman, H. S., 92, 244; Proclamation, 91-92, 244, 259
Tuamotu Archipelago, 177
Tungsten, 172
Tunisia, 22, 54
Turkey, 54
Two-hundred mile exclusive economic zone, 252

Union Carbide Exploration Corp., 197
Union Corp., Ltd., 19
United Kingdom, 13, 24, 58, 64-65, 80-81, 88, 194; Admiralty, 72; aggregate dredging, 24; Bacton, 86; British Gas Corp., 134; coal, 13; Conservative Party, 86; *Continental Shelf Act*, 72; Cornwall Peninsula, 10, 15, 19; Department of Energy, 67, 72, 74-75, 134, 157; Economic Development Committee, 72; environmental regulations, 153-54; Liverpool Bay, 24; National Economics Development Council, Offshore Operators Association, 73-75; Petroleum Revenue Tax, 74-75; petroleum use policies, 72-74; sixth-round licensing, 75; Teesside, 79, 86; tin, St. Ives Bay, 19; *see also* North Sea and Scotland
United Nations, 26, 171, 199; Conference on Trade and Development, 233, 246; Seabed Committee, 246; Third Law of the Sea Conference (UNCLOS III), 164, 197, 230, 240, 243-58
United States, Bureau of Land Management, 28-29, 94, 96-97, 107, 146; Bureau of Mines, 197, 229, 234, 236; Central Intelligence Agency, 45, 230; Coast Guard, 28, 149-50; Commission on Marine Sciences, Engineering and Resources, 254-55; Congress, 145, 254-58; Corps of Engineers, 28-29, 146; Council on Environmental Quality, 98; *Deep Seabed Hard Minerals Act*, 256; Department of Commerce, 28-29, 234; Department of the Interior, 4, 28-29, 93-98, 100-05, 111, 146; Department of State, 42, 257; Environmental Protection Agency, 28; Federal Hard Minerals Leasing Program, 25; *Federal Water Pollution Control Act*, 29; Geological Survey, 28, 92-93, 104-05, 146, 148, 197; Government Accounting Office, 98; Land and Water Resource Fund, 101; Library of Congress, 193; *Marine Protection, Research and Sanctuaries Act of 1972*, 28; *Marine Resources and Engineering and Development Act*, 254; National Academy of Sciences, 144; National Advisory Committee on the Oceans and Atmosphere, 255; National Council on Marine Resources and Engineering Development, 254; *National Environmental Policy Act*, 31; National Oceanic and Atmospheric Administration, 197, 219-20, 255-56; National Science Foundation, 197; OCS Advisory Board, 100; Ocean Mining Administration, 230, 255; Office of International Energy Affairs, 78; Office of Technology Assessment, 109, 146, 149; Office of Marine Minerals, 255; *Outer Continental Shelf Lands Act of 1953*, 92-93, 145; ... *of 1974*, 28; Overseas Private Investment Corp., 256; *Submerged Lands Act of 1953*, 92; *Water Quality Improvement Act of 1970*, 145
United States, eastern seaboard, 91, 104-11; diagrammatic cross-section of the Atlantic continental shelf, 106; import dependency, 2-7; mid-Atlantic coast, 96, 105-08, 146-50; OCS well log data bank, 93; offshore aggregate production, 214; offshore leasing programs, 93-102, 105; bidding systems and comparative statistics, 94-98; pipelines, 133, 136; potential offshore petroleum areas, map, 99
Uruguay, petroleum resources, 58

Vanadium, 182, 227
Varley, E., 74
Venezuela, petroleum resources, 58; Lake Maricaibo, 145, 246
Vietnam, 60
Virginia, 107, 136-37

Volcanism, 165

Wakefield, S. A., 104
West Germany, 49, 171, 194; Emden, 137; petroleum resources, 64, 67-68, 134, 153
Willums, J-O., 45
Woods Hole Oceanographic Institution, 25
Washington, 93, 98
Wilson, B., 256
World Bank, 87, 248
World Court (International Court of Justice), 43, 64

World Federalists, 255

Yamani, A. Z., 55
Yellow Sea, 49-50
Yugoslavia, 54

Zaire, cobalt, 233-36, 238; copper, 233; petroleum resources, 53
Zambia, 231; cobalt, 233; copper, 233-34, 238
Zinc, 10, 168, 170, 182, 188, 222, 234